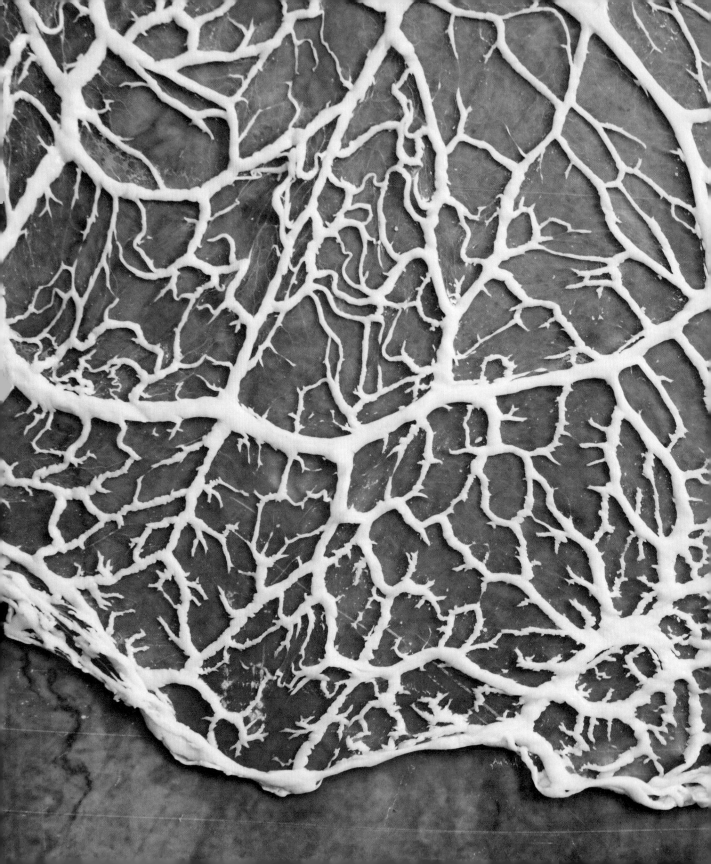

HOME
SAUSAGE
MAKING

HOME
SAUSAGE
MAKING

From Fresh and Cooked to Smoked,
Dried, and Cured
100 SPECIALTY RECIPES

Charles G. Reavis & Evelyn Battaglia,
with Mary Reilly

Photography by Keller+Keller

Storey Publishing

The mission of Storey Publishing is to serve our customers by publishing practical information that encourages personal independence in harmony with the environment.

Home Sausage Making was first published in 1981 and was revised in 1987 and 2003. All of the information in the previous editions has been revised and updated. This edition includes new text and 120 new recipes.

Edited by Deborah Balmuth and Corey Cusson
Art direction and book design by Alethea Morrison
Text production by Kristy L. MacWilliams
Indexed by Andrea Chesman

Cover photography by © Keller + Keller Photography
Interior photography by © Keller + Keller Photography
Additional interior photography by Adam Albright of Albright Photography, 86, 87; Amy Ager, 61; Ben Siegel, 217; Broken Arrow Ranch, 192; Chelsea Lane Photography, 60; © Christopher Hirsheimer, 130; Courtesy of David Samuels, 65; Courtesy of Fossil Farms, www.fossilfarms.com, 186, 187; Courtesy of Old Sturbridge Village, 92; Courtesy of Ryan Plett, 112; Creminelli Fine Meats, 42, 182, 183; Dominic Perri, 69 (right); Emily Julka, 126, 127; Greg DuPree Photography, 74; © Jupiterimages/Getty Images, 162; Marcelino Vilaubi, 15, 41, 68, 69 (left), 174, 175, 241, 247, 333; © Marilyn Angel Wynn/Getty Images, 4; © Matthew B. Moreland, Wayne Jacob's Smokehouse, LLC, 104, 105; © Neil Burton/iStockphoto.com, 181; © rogertrentham/iStockphoto.com, 180; Paul Bertolli, Founder Fra' Mani Handcrafted Foods, 164, 165; Nick Simonite, 216; Rachel Watson, 79; Reid Rolls, 208, 209; © Sohl/iStockphoto.com, 16; Southside Market & Barbeque, 134, 135; Straw Stick & Brick Delicatessen, 146; The Herbivorous Butcher, 255, 256
Food styling by Mary Reilly
Photo styling by Christina Lane
Illustrations by Alethea Morrison, 54, 116, 140, 178, 204, 226, 242, 264; Elayne Sears, 51, 97; Elena Bulay, 159

Storey books are available for special premium and promotional uses and for customized editions. For further information, please call 800-793-9396.

Storey Publishing
210 MASS MoCA Way
North Adams, MA 01247
storey.com

Printed in China by R.R. Donnelley
10 9 8 7 6 5 4 3 2 1

Library of Congress Cataloging-in-Publication Data
    Names: Reavis, Charles, 1948– author. | Battaglia, Evelyn, author
Title: Home sausage making : from fresh and cooked to smoked, dried, and cured : 100 specialty recipes / Charles G. Reavis and Evelyn Battaglia ; with additional recipes by Mary Reilly.
Description: 4th edition. | North Adams, MA : Storey Publishing, 2017. | Includes index.
Identifiers: LCCN 2017011747 (print) | LCCN 2017012798 (ebook) | ISBN 9781612128702 (ebook) | ISBN 9781612128696 (pbk. : alk. paper) | ISBN 9781612129853 (hardcover : alk. paper)

CONTENTS

PREFACE
TO THE 4TH EDITION

*L**ike the first person to eat an oyster,** history's first sausage maker is anonymous. But we owe this person a great debt, for sausage is up there with wine and cheese and bread as one of the world's greatest foods. It's more than just a way of preserving meat and using up the scraps — just consider all the myriad ways you could do that and *not* end up with sausage. It's also intrinsically satisfying in its shape and form and texture (or mouthfeel), which somehow cuts across all walks of life to please every type of eater — omnivore, pescatarian, poultry preferrer, and vegan alike — and manages to be an ideal vessel for flavor in all its forms. How else to explain the popularity of sausages that venture far from their origins, be they made with pork, beef, oysters, alligator, or tofu. Just think: Sausage has been made for centuries in a great many cultures and countries around the world, with great variation and ingenuity in taking advantage of local conditions and ingredients.

This book is well suited for sausage makers of all skill levels. If you have never made sausage before, this book will walk you through the steps to do so successfully and confidently. With even minimal equipment, anyone can make a simple fresh sausage: It's sort of like meatloaf stuffed into a casing (or, much more simply, formed into a patty). If you choose the traditional route, invite family and friends to pitch in; creating an assembly line of the various steps will make the process (from butchering to grinding to stuffing and cooking) that much easier and more enjoyable.

Even the more experienced sausage makers will find inspiration among the recipes for 100 different types of sausages — everything from classic fresh pork or poultry sausages to old-world cured and smoked specialties, and in this edition, a selection of sausages with up-to-date flavor profiles. Game sausage using elk, venison, boar, and rabbit (among other creatures) will please hunters — even those of the armchair variety. (Come to think of it, that first sausage maker might well have been a hunter faced with hundreds of pounds of woolly mammoth, yards of intestines, and the prospect of a long winter and an expansive family to feed.) There are also sausages made with seafood as well as utterly delicious ones that are vegetarian.

Once you get the hang of sausage making, there's no turning back. You'll find excuses to put your skills to excellent use anytime you can. Befriend a local farmer and you'll have a steady supply of quality pastured meats with which to hone your craft. Let your own ingenuity be your guide in improvising on the basic formula, knowing that you're in good company.

The years since this book was first published have seen an explosion of interest in ethnic foods and flavors. People have focused on sausage as both an old-fashioned favorite and a satisfying food with great flavor possibilities. In retail markets, sausage sales have boomed, and the number of home sausage makers has also risen, inspired by the resurgence of old-world, small-batch products that are popping up at farmers' markets and specialty grocers across the country. People want sausage that is both tasty and healthful, and that is made with meat that is responsibly raised. When you make your own sausage, you can ensure that your product also passes muster.

In this edition, contributions from sausage makers in the United States have resulted in added recipes for ramp and caraway pork sausage, venison and lamb cevapi, and Moroccan goat sausage, just to mention a few. We hope these recipes inspire you to try your own variations.

— **Evelyn Battaglia and Charles Reavis**

PART I

THE BASICS

CLOCKWISE FROM TOP LEFT: Lamb, Rosemary, and Pine Nut Sausage, page 121; Liverwurst, page 96; Smoked Cheddar Summer Sausage, page 172; Sicilian-Style Turkey Sausage, page 224; Southern Serrano and Pork Crepinettes, page 76; Salamette, page 109

SAUSAGE 101

Everything has an end except a sausage, which has two.

— Danish proverb

Sausage was born of necessity, a way of preserving meat in times of plenty to eat when life turned lean. The fact that it tasted good, made efficient use of a slaughtered animal, and could be seasoned and shaped according to the sausage maker's taste meant sausage was a real keeper in the larder, right next to cheese, wine, beer, dried lentils, and other staples.

The word *sausage* comes to us by way of the Middle English *sausage* and the Old French *saussiche*, all from the Latin word *salsus*, meaning "salted." The ancient Romans served highly seasoned sausages at every festive occasion. Not only did the salt, pepper, spices, and fat enhance the flavor, but they also helped to preserve the sausages, while the intestines into which the ingredients were stuffed helped to keep out microbes. Sausages that weren't consumed immediately were smoked or left hanging in the hot, dry Mediterranean winds.

Homer mentioned the Greeks' love for grilled sausage in the *Odyssey*. The legionnaires of Imperial Rome wouldn't march without their little bottles of *garum* (a fermented fish sauce) and long strings of dried or smoked sausages.

Sausage making really took off in Europe during the medieval period, when an energetic spice trade and returning Crusaders brought exotic seasonings and new cooking techniques to sleepy farms and villages. Medieval towns all across Europe — Bologna, Frankfurt, Vienna, and many others — gave their names to distinctive sausages that are still enjoyed today.

In North America, Native Americans dried and smoked venison and buffalo meat to make jerky, and they stuffed meat, suet, and berries into skins to make pemmican (see page 4).

Part of the original portable feast, sausage still delights our palates. It is a sturdy, nourishing comfort food, a link, so to speak, to the past. Today, homemade sausages are still popular among hunters, who like to make good use of the wild game they bring home. People who raise livestock also turn to sausage as a delicious way to make economical use of their animals at slaughtering time. But as our recipes will demonstrate, you can live in a tiny city apartment and shop in a supermarket and still make your own tasty, distinctive sausage. It's true: Sausage making is for everyone.

Types of Sausages

There are literally thousands of varieties of sausage in the world, but in the inimitable words of the United States Department of Agriculture (USDA): "Sausages are either ready to eat or not." Indeed, the USDA groups sausages into two types: uncooked, including fresh bulk sausage, patties, links, and some smoked sausages; and ready to eat, including dry, semidry, and/or cooked sausages.

Classifying sausages into categories is difficult because sausages are produced by so many different methods. Following is a simple classification of various types and how they should be stored and cooked.

Fresh sausages, such as breakfast and Italian sausages, must be kept refrigerated and used within a few days, or frozen for up to three months (they won't spoil after this time but their flavor will begin to wane). Fresh sausages must always be cooked before serving.

Cooked sausages are just that — fully cooked, as by poaching or smoking, during processing. Examples include frankfurters and bologna. Similar to fresh sausages, cooked sausages must be stored in the refrigerator or freezer; however, these may be eaten without any further heating or cooking, though many are intended to be heated again before serving.

Other ready-to-eat sausages, also called preserved or cured sausages, are treated with salt and other additives to impart different flavors and extend storage time. Some sausages, including pepperoni, are cured, or preserved, by drying, and some are smoked as well during processing. These sausages need no further cooking and are considered shelf stable, meaning they can be kept at a cool room temperature, such as hanging in your cold cellar, until slicing. After that, they will keep for a few weeks in the refrigerator or months in the freezer.

PEMMICAN

The Cree tribes of the North American woodlands called animal fat *pimyi,* from which we get the word *pemmican*. The Cree and other tribes dried strips of meat from buffalo, deer, rabbits, squirrels, and antelope, creating a product like jerky. This dried meat was nourishing and easy to carry, but it became moldy in wet weather. To solve the mold problem, the Cree made rawhide bags about the size of a pillowcase, filled them with pulverized bits of dried meat, dried berries, herbs, and nuts, and poured in hot melted bone marrow and fat. Then they sewed the bags shut and walked on them to compress the mixture and drive out air. The bone marrow and fat cooled and congealed around the meat, effectively sealing it. The Cree had, in effect, created a very large sausage.

It would be futile to try to catalog all of the varieties of sausage in the world. Some kinds are made only in a small region or even a single household. As you learn to make all manner of sausages, you'll no doubt be inspired to personalize the recipes to create your own unique variety. After all, sausage is a state of mind — and a delightful matter of taste.

A Few Ground Rules

Keep in mind that the texture of sausage is first determined by the amount of meat (or other base ingredients), fat, and liquid. What every maker strives for is an ideal meat-to-fat ratio that will produce the desired taste and, more important, mouthfeel. Too little fat and you'll end up with a tough product. Plus, fat carries flavor. But use too much fat and the result is, not surprisingly, a greasy link. What experienced sausage makers know is how to achieve that ratio no matter which part of the animal they are using — leaner pork tenderloins require more pork fat, while pork shoulder may not need any additional fat at all (that's why pork shoulder is a sausage maker's best friend).

The other factor that determines texture is how those elements are combined. Starting with clean equipment in good working order, with sharp blades (called "plates" or "disks") and dies, is a must; a dull blade will not only cause the meat to heat up during a slow grind but also turn the meat into mush.

The practice of stuffing may seem like much ado (especially for fresh sausage), but it is beyond doubt worth every ounce of energy and any measure of time spent in the endeavor. Besides lending sausage its signature shape, the casing helps to improve the texture of the finished product and meld the flavors. That sublimely satisfying *snap!* just doesn't happen when you shape the mixture into patties, no matter how well prepared.

Supplies and Sundries

The tools of the trade are few but critical to success. You'll need to know the different types of equipment, as well as how the steps will vary depending on your choice of equipment. You'll also need to absolutely commit to keeping your tools clean at all times, and not only to protect your investment.

What's more, you'll need to understand the various ingredients that go into making sausage. No surprise here: Always seek out the best-quality ingredients available to ensure that you get the best-tasting sausage. That's never more true than when shopping for meat, and we highly encourage you to shop locally to find the freshest meats possible. (See Meat Lover's Mea Culpa, page 13.)

Once you've procured all the essentials, you'll want to take a cue from the French, who have a term for culinary preparedness: *mise en place*. This means you should have everything gathered together and "in its place" before you start a recipe. By "everything" we mean all of the equipment and ingredients you will need. So where better to begin than with a description of "everything."

Equipment

Sausage making requires only one or two pieces of specialized equipment: a meat grinder and sausage stuffer, plus a smoker if you are going to take that step. You probably already have the rest of the gear in your kitchen. There is a bevy of online sources for new equipment, but yard sales, auctions, and flea markets are all good places to pick up secondhand versions — perhaps exactly what your grandparents would have used to make sausage back in the day.

Meat Grinders

If you are going to go to all the effort of making sausage, then you must start by grinding the meat yourself. Using preground meat from the supermarket will simply not yield the desired texture and flavor (and there's no way to guarantee what's actually in that shrink-wrapped package). There are many different options available; the one you choose largely depends on how often you plan to make sausage, and in what size batches.

Grinder attachment for heavy-duty mixer. For beginning or occasional sausage makers, we find that a grinder that attaches to a sturdy stand mixer is an excellent gateway tool — and a good alternative to purchasing a separate grinder, especially if you've already invested in the mixer. Check the instruction manual and warranty to make sure your mixer's motor is powerful enough to handle the heavy work of grinding meat. You'll want to buy extra plates (or disks) if you start to increase your output; dull ones will mar the final outcome. You can also get the blades and dies sharpened by a professional knife sharpener as necessary.

MEAT GRINDERS

Grinder attachment for heavy-duty mixer

Hand grinder

SAUSAGE STUFFERS

Push stuffer

Crank stuffer

Hand grinders. An old-fashioned cast-iron hand grinder is another inexpensive option and will last for generations. Many are still made in Eastern Europe, where sausage making is a long-standing tradition. A heavy-duty hand grinder will either clamp or bolt onto a countertop. Be sure that you can achieve a sturdy grip either way, for grinding meat does involve a good deal of elbow grease and torque.

Electric grinders. If you don't want to get "sausage elbow" grinding meat by hand, or if you intend to make a lot of sausage frequently, an electric grinder is a worthy and wise investment. Kitchen stores, restaurant-supply shops, and many online sources have grinders in a range of models and prices.

Food processors. If you own a heavy-duty food processor, it can do a splendid job of chopping meat. Models with a wide feed tube work best for sausage making. Check the instruction manual and warranty to see if the motor on your model is strong enough for chopping meat. (Models with smaller motors can overheat easily and burn out.) The major drawback to using a food processor is that it is all too easy to overprocess meat: A few seconds too many, and a pound of meat cubes will be reduced to the consistency of toothpaste.

Grinder plates (or disks). All grinders and grinding attachments come with a choice of grinding plates: ½ inch (extra coarse), ¾ inch (coarse), ¼ inch (medium), and ³⁄₁₆ inch (fine). To sharpen grinding plates, place a sheet of emery cloth (available at hardware stores) on a flat surface; rub the plates over the cloth, using circular motions. Or take the plates to a professional knife sharpener.

Sausage Stuffers

Unless you plan on forming your sausage base into patties, you will need a way to stuff your sausage mixture into casings. It's worth noting that some advanced grinders also come with sausage stuffers, but many people (including advanced makers) prefer one of the following over that option. And if your heavy-duty electric stand mixer has a grinding attachment available, the manufacturer probably also sells a sausage-stuffing attachment, though the range of casing sizes that mixers can accommodate is generally more limited. This might be fine for the beginning sausage maker who plans to make only a few batches a year.

FOR BEGINNERS ONLY: SAUSAGE FUNNELS

Also called hand stuffers, these are hand-held funnels that you use to push the meat into casings. The funnels range in size from ½ inch to 2 inches, and each relates to the size of the sausage casing. For example, a ½-inch funnel is appropriate for hot dogs, while a 1¼-inch model is good for Italian sausages. Sausage funnels are straight, not tapered, and are usually about 5 inches long. Most are plastic and inexpensive. (Don't be tempted to make do with an ordinary kitchen funnel — most are too tapered on the small end and not long enough to gather up the casings.) Hand-stuffing sausages takes a long time, and it's worth exploring one of the other options to make the task that much easier.

There are two basic types of stuffers, and the one you choose will depend on your skill level and how often you plan to use it:

Push stuffers. If you are just starting out and only plan to make smaller batches, you may want to try your hand with a horn stuffer, named for the shape of its shaft. Although it requires more muscle power, it is easier to control the pace in which you add the meat, for more even stuffing (if you go too slow with an electric one, the casing could be too loose, and vice versa).

This manual stuffer has traditionally been made of chrome- or tin-plated cast iron, though stainless steel models are now also available. Capacities range from 3 to 5 pounds of sausage. To work it, you must push down on a handle to force out the meat, which requires a fair amount of muscle power. Some experienced sausage makers still prefer to use this type of manual stuffer when making emulsified sausages or those with cheese added to the mixture, as the heat of the auger in electric stuffers can cause the mixture to warm up during stuffing.

Crank stuffers. If you find yourself making a lot of sausage, or if you want to save yourself from the effort of using a push stuffer, it may be worth investing in a crank model. These offer a gear ratio to make it easier to crank out the meat. There are two main types of crank stuffers, based on the orientation of the crank to the work surface, and each comes in manual or electric models and ranges in capacity from 5 to 15 or 20 pounds.

Horizontal stuffers have a large pistonlike cylinder for holding the meat that is mounted horizontally, and the meat is extruded when you turn the crank at the other end of the cylinder or turn on the electric motor. The manual version must be put near the edge of a table or counter to allow enough clearance to turn the crank, making it less than ideal for some folks.

Vertical stuffers are hands down the most favored stuffers by beginners and professionals alike (especially the latter, who prefer the larger-capacity, commercial-grade electric models). With these models, the cylinder that holds the sausage mixture is mounted vertically, and it extrudes the meat from the stuffing tube at the bottom of the unit.

Other Essentials

Almost as important as the grinder and stuffer are the tools you will use in preparing and measuring the ingredients and also in testing for doneness in cooked sausages.

Scales. A reliable scale for weighing meats and other ingredients is essential. For small batches of sausage, a scale often termed a 1 by 11 (1-ounce increments, up to 11 pounds) should suffice. For the more ambitious, a 2 by 22 (the standard meat scale, 2-ounce increments, up to 22 pounds) may be more desirable. You can find inexpensive scales (even digital ones) at kitchen shops and online. Be sure to buy one that measures in both metric and imperial units (grams and ounces). You will find many other uses for the scale around the house once you have it.

If you plan to make cured sausages, the USDA recommends that you also have a smaller scale that can weigh curing ingredients (primarily sodium nitrite) to the nearest one-tenth of a gram. (For more information about curing agents in sausage, see page 19.)

Measuring cups and spoons. A good set of measuring spoons, from ¼ teaspoon up to 1 tablespoon, will come in handy with every recipe. A glass measuring cup for liquids and a set of dry measuring cups, ¼ cup to 1 cup, will also be useful.

Knives. A good knife may be the single most important tool you use in making sausage because so much of the job involves cutting, boning, and trimming the meat, poultry, and fish that will be ground into your sausage. No doubt you already have a prized knife that you keep sharp with a stone or other sharpening device. If you have both a **chef's knife** and a **boning knife**, you are already well equipped.

If you don't already own at least one excellent knife (by this we mean one made from high-carbon stainless steel, which can be sharpened to a razor's edge), now is the time to make an investment. Wash it by hand only and store it in a knife block or separated from other utensils. The high-quality knife will last a knifetime . . . make that a lifetime.

Sausage pricker. Air pockets in filled sausage casings can cause spoilage. Prick any air bubbles with one of these tools, prized by many expert sausage makers.

Butcher's twine. Cotton butcher's twine or kitchen string is used to make some types of sausage links. It can be found wherever professional culinary equipment is sold and at sausage-supply companies; it can also be ordered from your butcher. If necessary, you can use other 100 percent cotton string, but it tends to be stiff and difficult to work with. Do not use plastic string or a polyester-cotton blend, as it could melt.

Instant-read thermometer. It's essential that you test sausage for the proper degree of doneness. This applies to when making hot-smoked sausage as well as cooking fresh sausages. An instant-read thermometer quickly measures the temperature of a food in 15 to 20 seconds. It is not designed to remain in the food while it is cooking, but should be used near the end of the estimated cooking time to check for final temperatures. To prevent overcooking, check the temperature before you expect the food to be done.

For accurate temperature measurement, insert the probe of the thermometer the full length of the sensing area (usually 1½ to 2 inches) into the food. If you are measuring the temperature of a thin piece of food, such as a sausage link, insert the probe through the side of the food rather than from above, so that the entire sensing area is positioned through the center of the food. Some models can be calibrated; check the manufacturer's instructions.

Oven thermometer. If you have a smoker, ideally it will have a built-in thermometer; if not, set an oven thermometer on one of the racks.

RECOMMENDED: STEAMER BASKETS

Steamer baskets, found at kitchen-supply stores and many department stores, adjust to fit most 2-quart and larger saucepans and are a handy gadget in the kitchen. Food cooked in a steamer over simmering water retains more nutrients and does not get waterlogged. When steaming, always cover the pan tightly during cooking, and use no more than 1 inch of water (for the proper distance from heat source to steamer basket).

Steamer baskets are available in stainless steel or bamboo, the type used in Asian cooking. If you don't have one, you can make do with a colander, so long as you can fit a cover over it and the pot tightly (or cover with aluminum foil instead).

Ingredients

The chief reason to make your own sausages? You get to have ultimate say in what goes into them. You may not guess it by reading the list of ingredients in a package of commercial sausage, but you can make a delicious homemade link without inordinate amounts of fat and sodium (though both are essential components) and without any preservatives. This applies to sausages made from pork, beef, lamb, game, and seafood, and even meat-free sausages, which typically contain a host of unrecognizable ingredients.

You also control the quality of the meats used as well as where they come from — important for those committed to eating only meat that has been raised humanely and sustainably (see Meat-Eater's Mea Culpa on page 13).

Striking the right balance of salt and seasonings is also important, and rarely does anyone, even seasoned sausage makers (pardon the pun), get it right the first — or fifth or umpteenth — time. But you'll gain valuable insight along the way and even less-than-perfect batches will be delicious all the same, and still a better option than many commercial varieties.

Meat

While pork may be the most familiar choice for sausage, there are many equally appropriate and authentic options. Beef, lamb, poultry, and veal are also traditional, depending on where you look. Indeed, many Americans have been exposed to a great many varieties during travels across the globe — or in the neighborhood joint where the owner makes his or her own native specialties. Then there's the proliferation of sausages that tap into current culinary trends — everything from new flavor profiles that mimic a favorite meal-in-one (bacon blue-cheeseburger

sausage?) to vegetarian options that incorporate beans or protein sources beyond tofu. The possibilities for sausage are truly vast and easily achieved with a bit of culinary imagination and practice-makes-perfect experience.

When you make your own sausages, be they traditional varieties such as kielbasa, chorizo, or andouille, or more novel creations (including the ones you'll find in these pages), here's the rub: You get to handpick each of the ingredients — and, in particular, the cuts of meat that will yield the desired taste and texture (among other concerns). Just always remember that sausage is based on a fairly consistent lean-to-fat ratio, so choosing the cut of meat will determine what else needs to be included in the mix. And know that when making meat-based sausages, the quality of the meat you start out with will determine the quality of the end result; find a local butcher you can trust, or go straight to the source and buy meats directly from the farmer, at farmers' markets or farm stands.

Pork. The pork butt (or shoulder) is the best cut for making sausages, as it is about 75 percent lean. Boston butts have a slightly thicker fat cover than other types of pork butts; the picnic ham, or the forequarter, has a fat ratio and texture similar to the Boston butt. A rib or loin roast substitutes admirably for the butt when the price is right, but you will want to grind some extra fat into the mix so your sausage isn't too dry; ask your butcher for some pork trimmings or pork belly, or use bacon or pork fat, available at most supermarkets.

Beef, lamb, veal, and goat. Beef chuck and rump are the most economical cuts and contain about 20 percent fat, making them good choices for sausages. Blade-cut beef chuck is what you will find in many of the recipes in this book. The first cut (or flat cut) of brisket is about 85 percent lean, while the

Pork and Apple Sausage,
page 71

second cut (or point end) of brisket is about 70 percent lean, as are short ribs.

Lamb is typically about 20 percent fat (except for the chops, which are higher in fat). Veal and goat are leaner than lamb and beef and have a more delicate flavor; they are therefore most often used in combination with fattier cuts of beef or pork, or the trimmings from the calf, if available.

Game. Venison, elk, bison, moose, boar, and game birds from doves to ducks all make excellent sausage. Because wild animals move around more than domestic ones, their meat tends to be dark, lean, and muscular. The quality and food-worthiness of the meat depends on the health of the animals and the skill of the butcher. Game sausages can be cooked and eaten fresh, or they may be cured and/or smoked. Making sausage is a great way to use up the carcass.

Poultry. Chicken, turkey, duck, and goose are all common in sausage making. When using chicken or turkey, which are leaner than other birds, you'll need to make sure to watch the lean-to-fat ratio, by either adding pork or pork fat or using skin-on chicken thighs or turkey legs, which have the proper fat ratio (emphasis on the skin). Duck and goose have higher fat ratios, but they are still often used in combination with pork to boost the flavor and fat.

Seafood. Lobster, scallops, shrimp, and oysters make excellent sausage, thanks to their relatively high amount of fat, as do salmon and other fatty fish such as trout, mackerel, and tuna. You can even use smoked salmon or trout to lend extra flavor to sausage, without having to actually smoke the links. Firm white-fleshed fish — including halibut, cod, haddock, and grouper — are also good options; because they are leaner than other types of seafood, they need additional fat, usually heavy cream or eggs (or both). The relatively mild taste of the seafood serves as a canvas for regional flavorings, from Thai and Mexican to Italian and New England.

Vegetarian staples. All the staples of a vegetarian diet — beans and legumes, whole grains, fresh vegetables, nuts, soy, and dairy products — can be used to make delectable meat-free sausages. It's easy to make these into patties, but you can also wrap the sausage base in an edible casing like leeks or cabbage leaves, or in plastic wrap or parchment paper, before cooking to shape the sausages. There are now vegetarian casings that you can use to make traditional links, too.

Kasekrainer, page 153

MEAT-EATER'S MEA CULPA

For many home (and professional, for that matter) sausage makers, the commitment to buying only meat that has been raised — and slaughtered — humanely is fundamental. Buying responsibly raised meat, they say, is not only to honor the animals that gave up their lives for our consumption but also to support the small, usually family-owned farms that willingly choose to forgo commodity (read: more profitable) farming.

What *humanely* means exactly can depend on whom you ask. The Animal Welfare Audit, designed by animal activist Temple Grandin and used by the USDA, has become the industry standard. Yet some farmers are going beyond that and adopting more rigorous standards. Consequently, some independent farms are volunteering to be audited by the Animal Welfare Association, which established the more stringent 5-Step Animal Welfare Rating System, among other certifications (see Food Labels, Demystified, right).

What can the home sausage maker do? Whenever possible, buy directly from the source, whether that means visiting a nearby farm that sells its meat to the public or shopping at the local farmers' market. Butcher shops are another excellent avenue; many proudly proclaim their policy on sourcing animals on their websites, but if not, you can ask the proprietors where the meat comes from (and perhaps persuade them to find meat that is responsibly raised). If you are looking for wild game or exotic meats, consider ordering online (see Resources, page 356). There are also some larger national stores that stock only humane meats.

Food Labels, Demystified

When shopping at a supermarket, where you have to rely on the label on the package, the choices can seem promising, but you have to learn to read between the lines. Here's a primer on common food labels and what they really mean:

AMERICAN HUMANE CERTIFIED. This voluntary program, the oldest animal welfare certification in the United States, was established by the American Humane Association in 1958. The program is based on the Five Freedoms of Animal Welfare: Animals must be healthy, comfortable, well-nourished, safe, able to express normal behavior, and free from unpleasant states such as fear, anxiety, and distress. Producers are audited by independent third parties to ensure compliance. It's important to note that the certification allows animals to be kept in cages large enough to allow natural behaviors like nesting and perching, rather than being pastured or free-ranged. Antibiotics are allowed to treat sick animals in conformance with Food and Drug Administration (FDA) guidelines.

ANIMAL WELFARE APPROVED. This certification is granted by the Animal Welfare Institute (AHC), a nonprofit organization, to independent family-owned farms. It goes farther than AHC by mandating that animals are raised fully outdoors on pasture or range and can express natural behaviors (according to standards devised by a board of veterinarians and farmers). The animals must also be slaughtered in a humane manner. Sick animals who have been treated with antibiotics can still be certified as AWA so long as the drug has had sufficient time to leave the body before slaughter.

CERTIFIED HUMANE. According to Humane Farm Animal Care, the parent nonprofit organization that administers the program, 76.8 million of the roughly 10 billion farm animals raised for meat in 2012 were raised under Certified Humane standards. That means that the animals were allowed enough space to exhibit natural behaviors and were never housed in cages, crates, or stalls, though it doesn't mean they were pastured or that poultry were allowed to even go outside. Certified Humane is the only other program (besides AWA) that regulates slaughtering practices. Meat, dairy, and eggs can bear the Certified Humane label.

CERTIFIED ORGANIC. For meat and poultry to be labeled as organic, it must meet the USDA's standards: The animals cannot be treated with hormones or antibiotics; they must be fed only organically grown feed with no animal by-products; and they must have year-round access to the outdoors (ruminants such as pigs, cows, and sheep must be pastured).

GRASS-FED. Grass-fed animals are allowed to eat grass (or hay) rather than being forced to eat a grain diet, as at many factory farms. The benefits of being grass-fed are many: The animals are much less likely to develop digestive diseases — and require antibiotic treatment — as when on a grain diet; many grass-fed animals are pastured and allowed to roam free for the grazing season (though this is not required); and the practice is much more sustainable in terms of environmental impact. Some products will be called grass-fed even though the animals were fed grains during the last weeks of their lives (in a practice called

grain finishing), so look for "100% grass-fed" on the package. This label is for beef, lamb, and goat products only; pigs and poultry require some grain in their diets.

PASTURE-RAISED. Pasture-raised animals are primarily raised in a pasture where they can roam freely and eat grasses and other plants, as they have evolved to do, with less risk of digestive disease and parasites. Animal Welfare Approved products must be made with pasture-raised animals, as must those that are certified organic (except for eggs and poultry).

CAGE-FREE AND FREE-RANGE. These labels only apply to poultry. So long as hens are not raised in cages, they are considered to be cage-free, even if the animals were never allowed access to the outdoors. Some cage-free eggs will bear the American Humane Certified (AHC) label; other products may not have been certified at all. The USDA regulates whether poultry can be designated as free-range or not. The label means only that the animals have been "allowed access to the outside," but there is no requirement concerning how long that access must be for each day, nor for the size or quality of the outside range, so this should not be confused with the pasture-raised designation.

NATURAL. All fresh meat qualifies as natural according to the USDA definition: no artificial ingredients or added color and only minimally processed. But the label does not indicate anything at all about how the animal was raised or whether it was treated with antibiotics or growth hormones. It does not mean that the animal was raised organically, nor does it mean that it is healthier than other animals not bearing the "all-natural" label. In essence, it is meaningless.

HORMONE-FREE/NO ADDED HORMONES. The USDA prohibits chickens, turkeys, and hogs to be treated with any hormones, but not other animals, which is how it's possible for many factory farms to give growth hormones and other hormones to cattle and dairy cows to boost production. Indeed, in 2011, as much as 80 percent of all feedlot cattle were injected with hormones. (It's interesting to note that the European Union has banned the use of hormones in cattle since 1988, due to studies linking their use with a higher risk of infection in animals and an increased risk of cancer in humans.) There is no designated hormone-free certification, but products bearing the organic or grass-fed labels, and those with humane certifications, will be derived from hormone-free animals.

FISH-BUYING GUIDELINES

As with beef, pork, poultry, and other meats, your best course is to buy fish and seafood as close to the source as possible, but this can be a bit harder to do if you aren't near rivers, lakes, or the coast. No matter where you live, you can look for fish at farmers' markets, trusted fishmongers, and specialty grocers. Some supermarket chains set high standards for their fish purveyors (both wild and farmed). Don't be put off by fish that's been frozen; most fish and all shrimp must be flash frozen on the boat to preserve it before it gets to market. So long as it has been frozen properly and kept properly frozen, it will be as good as fresh upon thawing (preferably overnight in the refrigerator).

The concern over mercury levels in wild fish has led to a rapid increase in fish farming, or aquaculture, in the United States and across the world. There are also concerns over the sustainability of certain species of wild fish due to overfishing. On the other hand, aquaculture has been responsible for polluting local

waterways and negatively impacting the surrounding environment, so deciding which way to go can be tricky. Here's a brief overview of the criteria you should consider when choosing which type of fish to buy.

Farmed Fish

According to the National Oceanic and Atmospheric Administration (NOAA), which is responsible for overseeing aquaculture in the United States, approximately 90 percent of fish consumed here is imported, and over half of all fish consumed in the United States is farmed. Concerns over the rapid rise of fish farming mostly deal with the negative impacts on surrounding marine ecosystems, the spread of disease and parasites to the local environment, and the resulting decline of wild species of fish in nearby waters. But the aquaculture industry is vastly improving its practices and there are more and more eco-friendly options. A good place to start is at NOAA's FishWatch.gov or the Environmental Defense Fund's website.

Wild-Caught/Wild Fish

There is an important difference between these two seemingly similar labels: A fish that is "wild-caught" can have been harvested from or have lived for some time in a fish farm before being released or returned to the wild and eventually caught. "Wild fish," on the other hand, must have spent the entirety of its life in the wild. While many people claim to only buy wild fish, this is not always the most ecologically sound decision. According to the Monterey Bay Aquarium's website, 90 percent of the world's fisheries are fully exploited or overexploited or have collapsed. As of this writing, Atlantic populations of halibut are at an all-time low. The choice to only buy wild also comes with some consequences. The key is to choose fish that are sustainable, whether in the wild or through aquaculture.

Sustainable

The types of fish considered sustainable — those that are harvested in a way that can be maintained over the long term without jeopardizing the ocean's resources — fluctuate regularly. Sustainable wild fisheries target species that are in abundance, including smaller fish that can quickly reproduce and replenish their supply. Aquaculture operations can also be sustainable if they establish practices that limit environmental impacts from pollution, disease, and damage to coastal ecosystems that support wild species, and avoid using wild-caught fish as feed, which puts additional strain on wild fish stocks.

Check the Monterey Bay Aquarium's website for the current list of the most sustainable seafood choices at any given time. While not used in the United States as often as across Europe, the Marine Stewardship Council (MSC) label on seafood indicates that the fish meets their rigorous standards for sustainability.

Herbs, Spices, Salt, and Sugar

Herbs and spices are a small percentage of sausage by weight or volume, but they can decide the character and piquancy of the recipe. The variations in the flavorings make the difference between bockwurst and bratwurst, knockwurst and kielbasa.

Herbs. The best of all possible worlds is one in which you grow your own herbs, even if on your windowsill. If this is not practical, buy fresh herbs at a supermarket, or, better yet, at a farmers' market, where you will also find a larger variety. Some sausage recipes call for dried herbs, and you can dry your own fresh herbs for this purpose. If buying dried herbs, do so from a market with a high turnover to ensure freshness, store them in a cool, dark spot away from sunlight, and replenish them whenever they lose their fragrance — usually after about six months.

GIY (GRIND IT YOURSELF)

Freshly ground spices will always, without exception, have a more pronounced flavor than those that are preground. You can buy whole spices in bulk much more cheaply, too, and in small quantities so they don't lose their taste before you deplete your supply. Many experts prefer to grind spices the old-fashioned way: with a mortar and pestle. Choose a ceramic or marble mortar and pestle, because wooden sets absorb food odors easily. If you prefer a faster, hands-free method, use an electric coffee grinder, reserved just for spices.

Spices. It's a good idea to buy spices in small quantities from a reliable supplier (see Resources on page 356 for mail-order spice companies) and store them in a cool, dark place for no more than six months. While it may be handy, never store your spices in a hot spot like over the stovetop; they'll lose flavor faster. Whenever possible, buy whole spices and grind them as needed (see previous page). Nutmeg in particular is worth buying whole and then grating with a rasp grater (or special nutmeg grater) as you go.

Salt and pepper. Salt is essential for adding flavor to sausage. Use kosher salt or sea salt (fine or coarse) in fresh sausage, rather than salt with any additives — table salt included (it has iodine added). As for pepper, buy whole peppercorns and grind them yourself. Freshly ground pepper has a sharper flavor than preground pepper.

Sugar. Often added to sausage mixes to balance the sharp taste of salt and other flavorings, sugar contributes its own qualities to the finished product. Plain granulated sugar will do the job, but when you experiment with recipes, you might try brown sugar, honey, maple syrup, or molasses for a subtle change of flavor. In the form of dextrose, sugar is an ingredient in premixed cures.

Cures

If you are planning to make cured sausages, such as salami and soppressata, you will need to use premixed curing salts (aka pink salts) to prevent botulism and retard spoilage. Be sure to use these only as directed in a specific recipe or on the label of the salts.

Insta Cure #1 (or Prague Powder #1) is a common all-purpose cure containing 6.25 percent sodium nitrite; the recommended quantity is a level teaspoon per 5 pounds of meat. This is used for cured

sausages (and bacon and hams) that will eventually be cooked, such as cold-smoked sausages.

Insta Cure #2 (or Prague Powder #2) contains 6.25 percent sodium nitrite and 1 percent sodium nitrate mixed with a salt carrier; this is used for curing meats that do not require cooking or refrigeration, such as air-dried sausages. Nitrate itself has no efficacy against spoilage until it breaks down to nitrite, so it essentially kicks in once the sodium nitrite has run its course.

Saltpeter, or potassium nitrate, is no longer allowed to be used in commercial sausages here in the United States, though it is still used in Europe. It should also be avoided by the home sausage maker, as it is not as reliable as sodium nitrite.

SAFETY GUIDELINES FOR USE OF CURES

- Never exceed the amount of cure called for in the recipe (or the amount advised by the supplier).

- Be sure to measure carefully: A teaspoonful is always measured level, not heaping.

- Always keep nitrates and nitrites out of the hands (and mouths) of children and away from pets.

- Premixed cures replace the saltpeter and salt in older recipes; do not exceed 1 teaspoon cure for every 5 pounds of meat, even if the recipe calls for more.

Extenders and Binders

Sausage extenders and binders include bread crumbs, rice and other grains, and nonfat dry milk powder. There is a common misconception that they are added to sausage to "extend" the meat and keep the price down, like watering down the soup. This isn't always the case. These ingredients are added to some sausages to make them more moist and juicy or to give them a particular texture.

What homemade sausage *doesn't* include are additives, colorants, extra water, monosodium glutamate (MSG), odd pieces of meat tissue, stabilizers, and other ingredients that go into many commercial sausage. No more mystery meat or hunks of gristle in your sausage when you make your own.

Casings

Unless you are making patties, sausage has to be stuffed into something. If you think of sausage as the world's first "convenience food," and with edible packaging to boot, you can understand why the intestinal tract of a pig, cow, or sheep makes a handy holder for chopped meat and seasonings. But there are alternatives to natural casings, especially when you desire to make fully organic sausages (natural casings are only available from commodity pork producers), seafood sausages, or vegetarian sausages.

Natural casings. Before you wrinkle your nose about intestines, rest assured that they are scrupulously cleaned before use. Once flushed out and packed in salt to keep them fresh, the innards — we'll call them "natural casings" at this stage — are usually sealed in airtight bags and kept frozen or refrigerated. Sheep and hog casings are digestible and are permeable to moisture and smoke.

Buy natural casings based on the diameter of the sausages you want to make: Sheep casings are the smallest of the natural casings, ranging from lamb (¾ inch; 20 mm) to adult (just over 1 inch; 26 mm in diameter). Sheep casings, the most delicate of the natural casings, are often used for hot dogs.

Hog casings come in several sizes for home sausage making. The smallest casings are about 1¼ inches (32 to 35 mm), used for bratwurst and Italian sausage. The intermediate size, approximately 1½ inches (35 to 38 mm), is used for knockwurst. Larger-diameter casings are 2 inches (42 mm) or more. The recipes in this book generally specify a type of casing; however, you have the option to make a smaller- or larger-diameter sausage than suggested. Hog casings are considered to be the "all-purpose" option.

The most common beef casings are called beef bungs, beef rounds, and beef middles. Beef casings are larger, 2½ to 4 inches (6.4 to 10 cm) in diameter. Bungs and middles are generally used for bologna, veal sausage, and cooked salami, and the rounds for mettwurst or ring bologna. Because beef casings tend to be tough, they are usually peeled away and discarded before the sausage is eaten.

Edible collagen casings. Also called synthetic casings, collagen casings are made from edible protein derived from animal connective tissues that is formed mechanically into casings. They are made in small sizes ideal for breakfast links, and they can be used for fresh, smoked, and dried sausages. Take care not to overfill them, as they do not stretch the way natural casings do. Do not rinse collagen casings or get them wet, as they are much easier to work with when they are dry.

Artificial casings. Fibrous casings are popular among commercial sausage makers because they are uniform in size and easy to use. Cellulose and plastic casings are relative newcomers to the artificial casing field. These must be removed before eating.

Muslin casings, sometimes used for large salamis and summer sausage, can be purchased or homemade, using simple sewing skills. (For instructions, see the box on page 97.)

Vegetarian casings. There is also a plant-based casing that can be used for vegetarian and vegan sausages (or kosher or halal sausages). Unlike cellulose casings, these are completely edible. Vegetable casings are made from polysaccharides (as are cellulose casings) and are sold pleated onto a tube; one tube is usually sufficient to stuff about 10 pounds of sausage. You can store them for a few years as long as they are kept in an airtight container.

Vegetarian casings are a little different to work with than natural casings: Do not presoak them, as they will disintegrate. Handle with care when stuffing, as they are not as elastic; a small stuffing horn works best. When stuffing sausages, be sure to fill them well, but not so much that they are compacted tightly — leave a little slack to allow for twisting. They do not hold a twist well, so for best results, twist and then tie butcher twine between each link.

CASING SOURCE

If you are in a pinch, ask the butcher where you buy your meat if they can sell you some casings. Most will have them in abundance and are happy to share a few feet — along with some tips for making sausage. Your local supermarket can also be a good place to get a small batch of casings; just ask the butcher in the meat department to sell you some. Find out if the casings have already been soaked; if so, you will need to use them within a day or two. Or you can put the casings in a container, cover with salt, and refrigerate until needed. Do this, too, with any soaked casings left over from a batch of sausage, wringing them out well first.

You can skip the recipe step that calls for letting the stuffed sausages air-dry in the refrigerator before storing or cooking, as these vegetarian casings are already fairly dry. Cook the sausages in a dry pan over medium-low heat, or over indirect heat on a hot grill, until the casing is crisp and the sausage is cooked through. Poaching can cause the casing to dissolve, so this cooking method should be avoided. You can, however, smoke or dry-cure sausages with these casings.

Vegetable casings can be purchased from sausage-supply houses, meatpacking companies, butcher shops, and ethnic groceries. See our Resources list on page 356 for names and addresses of some suppliers.

Nitrates and Nitrites: For Better or Wurst

If you plan to make dried or semidried sausages, such as pepperoni or others that the USDA would categorize as ready to eat, you will want to read on. Sodium nitrate and sodium nitrite are used to make many of these sausages (the USDA mandates that none of these ingredients may be used in *fresh sausage*). Of all the additives in food, few have attained the notoriety of the nitrates and nitrites, which in high doses and under certain conditions are known to be carcinogenic.

What isn't often known is that nitrites occur naturally, in fairly high amounts, in many vegetables that are harvested from the ground, notably celery. That's why you'll find "celery powder" or "celery juice" in the ingredients on a package of "nitrate-free" hot dogs (meaning they aren't actually free of nitrates after all). Indeed, according to the Centers for Disease Control and Prevention (CDC), we derive a whopping 80 percent of our nitrate consumption from plant sources and a mere 6 percent from cured meats (the rest come mostly from water).

When this book was first published in 1981, there was great confusion over the role of these chemicals in making cured sausages. Some people advocated eliminating them altogether; others declared them necessary because they prevent spoilage (they are particularly effective in stopping the formation of botulism in processed meats) and give cured meat its characteristic rosy hue. They also contribute to the flavor of the finished product.

Today, nitrates are used only in sausages that undergo a slow cure. During the curing process, nitrate breaks down to form nitrite. Commercial sausage makers must meet strict regulations for the amounts of nitrate and nitrite they add to cured meats. The FDA limits the amount of sodium nitrite that can be used in the curing of meat and meat products (including poultry and wild game) "to not more than 200 parts per million in the finished meat product, and the amount of sodium nitrate to not more than 500 parts per million in the finished meat product." This means that nitrites are not allowed to exceed ¼ ounce, or 7 grams, per 100 pounds of meat; nitrates must not exceed 2¾ ounces, or 77 grams, per 100 pounds of meat. Nitrates are banned from most kinds of bacon and from any food intended for babies and toddlers (the label must say to "keep out of reach of children"). In meat-processing plants, nitrates and nitrites used as curing agents are strictly controlled and regulated.

Nitrites and nitrates are still the most controversial cured-sausage ingredients, and the most tightly regulated at the commercial level. They should always be handled with great caution and are tinted pink so they do not get confused with salt (hence they are commonly referred to as pink salts). They are regarded as safe as long as they are used at the prescribed levels set by the USDA, which has ruled that in curing meats, the known benefits of nitrates and nitrites outweigh the potential risks.

It is extremely difficult to meet these requirements when you make cured sausage on a small scale at home. You know, for example, that sodium nitrite may not exceed ¼ ounce (7 grams) per 100 pounds of meat. For a batch of sausage using 5 pounds of meat, this would mean weighing out 0.35 gram of pure sodium nitrite and figuring out how to distribute it evenly throughout your meat. For this reason, we recommend that you use only premixed cures (see page 19) when making cured sausages.

In some recipes for cured sausage, ascorbic acid (vitamin C) is added to help with color retention. It does not prevent spoilage and does not substitute for curing salts. Buy pure crystalline ascorbic acid at a pharmacy (vitamin C tablets contain binders and other ingredients that you would not want in your sausage).

In the end, there's just no getting around using nitrite and nitrate if you are making (or buying) dry-cured sausages. So if you are concerned about limiting the amount of these additives in your diet, you may decide to stick with the fresh sausages, of which there are many delicious choices.

Safety Guidelines

Whether you buy your sausage meat from the supermarket or farmer, raise and butcher your own livestock, bag your own game, or catch your own fish or shellfish, there is one cardinal rule that must be followed faithfully: All meat — any source of protein — must be kept optimally fresh.

Ground-up meat has a proportionately greater surface area than the same weight of meat before it is ground. The more surface area, the larger the breeding ground for bacteria. Bacteria thrive at temperatures between 40 and 140°F (4 and 60°C). This means that your fresh ingredients must be kept refrigerated at all times — before, during, and after grinding and stuffing — so that bacteria will not have a chance to reproduce and taint the meat.

When you make sausage at home, you take on the responsibility of providing food that is both safe and delicious. Following are the basic rules:

Ingredients. Starting at the place of purchase (be it supermarket, butcher shop, or farmers' market), be sure that raw meat and poultry are packaged securely and kept separate from any foods that will be eaten without further cooking. To avoid cross contamination, take extra care that meat juices do not drip onto other food or onto countertops or utensils. (If they do, wash food in soap and hot water and rinse thoroughly or discard; disinfect countertops and utensils.)

Refrigerate anything that can spoil. Keep it cold right up to the moment you start preparing the sausage. Never let meat warm to more than 40°F (4°C). It's best to keep it as close to the freezing point as possible to bring bacterial action to a standstill. (Chilling the meat also makes it easier to grind.)

Contrary to age-old wisdom, do not rinse raw meat or poultry before using, as this increases the risk of juices splashing into the sink or onto countertops where other foods may come into contact with them.

Contact surfaces. Scrub with hot water and detergent all surfaces that will be in contact with the meat. Be particularly careful of your cutting board. Disinfect wooden boards and countertops with a solution of 1 tablespoon chlorine bleach in 1 gallon of water. (It's also advisable to designate separate cutting boards for raw meat, raw poultry, and raw seafood.) Rinse everything thoroughly and allow them to air-dry. Be sure the room is cool — under 70°F (21°C) — when you are making sausage; have a large bowl of ice water standing by to chill the meat and fat as it is extruded from the stuffer (into another bowl) as an extra precaution.

Utensils. Pour boiling water over all utensils and your grinder — anything that will come into contact with the meat. Allow everything to cool completely before proceeding so that the residual heat does not warm up the meat and encourage the growth of bacteria. Most experts also advise chilling the grinder and stuffer in the refrigerator to prevent this from happening.

You. Remove rings and other jewelry that might come in contact with the meat and wash your hands carefully, scrubbing under your fingernails with a nail brush. Wash your hands again if you are called away from your sausage making to answer the phone, put the cat out, or perform any other activity.

Only now are you ready to begin making sausage.

ESSENTIAL TECHNIQUES

Laws are like sausages — it is better not to see them being made.

— Otto von Bismarck

With all due respect to Herr von Bismarck, we disagree! Quite the contrary: Seeing sausage being made gives you a deeper appreciation of the end result.

Once you have mastered the basic steps in sausage making, you will be able to follow any recipe in this book with ease. In this chapter, we break that down into ten simple parts, with helpful how-to photos. You'll also learn to make fresh sausage; the best ways to cook and store your homemade sausage; and how to dry, cure, or smoke your sausage.

Most of the recipes in the book call for stuffing the sausage into casings — an easy enough process to master. And while many fresh sausage recipes can be left in bulk form or shaped into patties, if preferred, they will be that much more satisfying when in the universally recognized — and beloved — link form.

HOW TO MAKE FRESH SAUSAGE
AN OVERVIEW IN 10 BASIC STEPS

The method for making fresh sausage is simple and straightforward, as demonstrated by the following at-a-glance instruction. We've broken down the process into ten steps to help you visualize what's going to happen before you ever get started. Seeing, as they say, is believing — here, in the possibility of preparing your own sausage from scratch, and with ease.

(For detailed instructions, see pages 29 to 31.)

1. Prepare the natural casing by rinsing, flushing, and soaking. (Do not rinse collagen or vegetarian casings.)

2. Make the sausage mixture by cutting the meat (including poultry or seafood) and fat into 1-inch cubes; for vegetarian sausages, prepare all ingredients. Measure all seasonings and mix together the protein, fat, and seasonings.

3. Grind the meat mixture (once or twice) using your preferred method. (With a hand grinder, grind the sausage ingredients together twice, adding the seasonings after the first grinding.)

4. Fry a small amount of the mixture and taste to see if you want to adjust any seasonings. (Note: At this point, you could wrap the sausage mixture well and use as a bulk sausage.)

5. Gather the sausage casing over the stuffer tube (or nozzle) and pack the ground mixture into the canister of the stuffer until it reaches the opening of the tube. Tie the end of the casing into a knot and begin feeding small amounts of meat mixture through the tube, maintaining an even thickness and filling the entire length of casing and coiling the sausage as you go.

6. Prick any air bubbles with a sausage pricker. (Note: If you are making coiled sausages, you can skip steps 7 and 9.)

7. Twist off the sausage links, beginning at the tied end. Grasp the desired length of sausage and give it about five twists to form a link. When the entire casing is done, tie off the other end.

8. Refrigerate the sausages overnight, or preferably for 24 hours, to meld the flavors and firm the texture; leave uncovered to allow surface area to dry, for the desired snap.

9. Cut the sausages to separate the links using kitchen shears.

10. Cook the sausages thoroughly. Enjoy!

The Details of Making Fresh Sausage

The basic method described here can be used for most of the sausages in this book (except for vegetarian sausages, which vary in their methods and many of which are simply formed into patties). Refer to the preceding chapter for detailed information about the equipment and ingredients, including casings. Remember to keep the meat cold, cold, cold at all times — before, during, and after making the sausage.

To help you plan, it takes about 45 minutes to soak and flush the casing; during this time you can prepare the sausage mixture. Filling the casing will take another 30 to 45 minutes. You can easily make, cook, and eat your sausage on the same day, but it will have a much improved taste and texture if allowed to rest in the refrigerator for a day before cooking, especially if stuffed into casing.

Preparing the Casings

These instructions are for traditional natural casings packed in salt. If you are using a casing packed in brine or another type of casing, follow the supplier's directions for preparing it.

NOTE: *Never rinse collagen or vegetarian casings before using.*

1. Snip off about 4 feet of casing and rinse it under cool running water to remove any salt. Place in a bowl of cool water to cover and let it soak for at least an hour, or preferably overnight.

2. Rinse the casing again under cool running water: Holding one end of the casing open under the faucet, turn on the cold water gently, then more forcefully, to flush out any salt in the casing and pinpoint any tears or breaks. (Should you find a tear or break, simply cut out that section.)

3. Soak the casing again. This time, add 1 tablespoon white vinegar for each cup of cool water in the bowl. The vinegar softens the casing and makes it more transparent, resulting in a better-looking finished sausage. Leave the casing in the vinegar-water solution until you are ready to stuff it, then rinse well and drain.

Grinding the Meat

A clean grind is the key to sausage success: You want to extrude the meat in uniform strands, each separate from the other. This allows the meat to better combine with the seasonings for optimal taste and texture. Some recipes call for first grinding the meat and fat before adding seasonings and other ingredients; others call for grinding the meat first, adding seasonings, and grinding again. Here, we use the most standard process of mixing the meat with the seasonings before grinding.

1. Measure out the seasonings and any other ingredients that will be added to the meat, leaving the meat and fat in the refrigerator until ready to begin.

2. Cut the meat and any fat into 1-inch cubes, or small enough to fit easily in the opening of your grinder without pushing. Uniform-size pieces will grind at the same rate for the cleanest grind. Place in a mixing bowl.

3. Add seasonings and other ingredients to the meat and toss to combine. Spread the mixture evenly on a rimmed baking sheet and freeze, uncovered, for about 30 minutes, or until the meat is firm but not solid. (You can also opt to refrigerate the meat mixture in an airtight container for several hours or overnight.)

4. Meanwhile, chill the grinder plates in the refrigerator (not the freezer) until cold to the touch.

5. Fit the grinder with the appropriate plate as directed in the recipe; when the meat will be ground just once, you will usually use the

medium plate, but that depends on the desired texture of the final product.

6. Remove the meat from the freezer and push it through the grinder one piece at a time. You will develop a feel for when to add the next piece, so don't rush the process or else the meat will begin to heat up.

7. Allow the ground meat to fall into a bowl that's set in a larger bowl of ice water to keep it properly cold until you finish grinding all the meat.

8. If you are grinding the mixture a second time, return the meat and clean grinding parts to the freezer for another 30 minutes and repeat, using the plate called for in the recipe.

TEMPERATURE MATTERS

Professional sausage makers carefully monitor the temperature of the meat before they begin grinding (see Temperatures Related to Safe Handling and Consumption of Meat, page 33), and you should follow their lead in your own sausage making. The process of grinding necessarily involves heating up the meat, so you want to start with meat that is properly chilled to prevent the meat from becoming mushy while grinding. The same goes for the grinding plates. If at any point the strands start to look mushy or "smeared," put the meat back in the freezer (also freeze the ground portion), and clean and chill the grinder parts, before proceeding.

STOP SPOILAGE BEFORE IT STRIKES

Besides literally tasting rotten, spoiled food poses serious health risks to anyone unlucky enough to eat it. Spoilage is caused by the action of microorganisms on food. These organisms include molds, yeasts, and bacteria, the latter of which are the main problem for the home sausage maker. When some bacteria are allowed to reproduce in an uncontrolled environment, they can cause illness and even death.

The five bacteria responsible for most food-borne illness are *Salmonella*, *Clostridium perfringens*, *Staphylococcus aureus*, *Campylobacter*, and *Clostridium botulinum*. All are found throughout our environment and in most food, but the trouble arises when they are allowed to multiply freely.

SALMONELLA bacteria are the most common source of food poisoning in humans. These bacteria can survive in frozen and dried foods, but do not reproduce at temperatures below 40°F (4°C) or above 140°F (60°C). They are destroyed if food is held above 140°F (60°C) for 10 minutes.

CLOSTRIDIUM PERFRINGENS (commonly referred to as the "food service germ") can strike if food is held between 70°F (20°C) and 140°F (60°C) for an extended period of time.

STAPHYLOCOCCUS AUREUS, like *Closteridium perfringens*, is inactive at temperatures below 40°F (4°C) and above 140°F (60°C). Staph germs that are allowed to multiply form a toxin that cannot be boiled, baked, or otherwise cooked away.

CAMPYLOBACTER grows best at 108°F (42°C) to 113°F (45°C) and is the most common bacterial cause of diarrheal illness in the United States. Most cases of illness from this bug are associated with handling raw poultry or eating raw or undercooked poultry meat. Even one drop of juice from raw chicken meat can infect a person. Freezing reduces the number of bacteria on raw meat, and they are killed by thorough cooking to 165°F (74°C).

CLOSTRIDIUM BOTULINUM organisms, though rare, are the strongest villains in the arsenal. They love room temperature and moisture, and they are anaerobic, meaning that they thrive and produce toxin in the absence of air. Under certain conditions, the bacteria produce spores; when the spores reproduce, they give off a powerful, deadly toxin. Two cups of the toxin cold kill every human being in the city of Chicago. The toxin itself can be killed by 10 to 20 minutes of boiling, but the spores require 6 *hours* of boiling to stop them from reproducing. One of the main reasons nitrates and nitrites are used in making cured sausages is that those agents are effective against *Clostridium botulinum*.

TEMPERATURES RELATED TO SAFE HANDLING AND CONSUMPTION OF MEAT

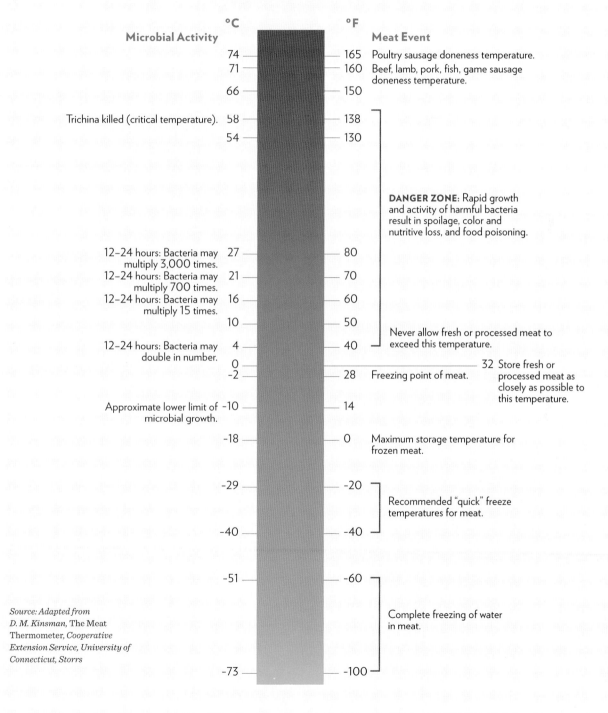

°C **°F**

Microbial Activity   **Meat Event**

74 — 165   Poultry sausage doneness temperature.
71 — 160   Beef, lamb, pork, fish, game sausage
66 — 150   doneness temperature.

Trichina killed (critical temperature). 58 — 138
54 — 130

DANGER ZONE: Rapid growth and activity of harmful bacteria result in spoilage, color and nutritive loss, and food poisoning.

12–24 hours: Bacteria may 27 — 80
multiply 3,000 times.
12–24 hours: Bacteria may 21 — 70
multiply 700 times.
12–24 hours: Bacteria may 16 — 60
multiply 15 times.
10 — 50

Never allow fresh or processed meat to exceed this temperature.

12–24 hours: Bacteria may 4 — 40
double in number.
0
-2 — 28   Freezing point of meat.   32 Store fresh or processed meat as closely as possible to this temperature.

Approximate lower limit of -10 — 14
microbial growth.

-18 — 0   Maximum storage temperature for frozen meat.

-29 — -20   Recommended "quick" freeze temperatures for meat.

-40 — -40

-51 — -60   Complete freezing of water in meat.

Source: Adapted from D. M. Kinsman, The Meat Thermometer, Cooperative Extension Service, University of Connecticut, Storrs

-73 — -100

Make the Farce (or Bind)

Many experts suggest using your hands to knead the mixture into a paste; the theory is that the warmth from your hands heats up the meat just enough to allow it to better absorb the flavors. If you do this, be sure your hands are thoroughly clean; wearing food-safe gloves will prevent the heat from transferring. Some sausage makers prefer to use a stand mixer with the paddle attachment at this stage, especially when making smoother-textured sausages.

1. Knead the mixture until well combined and a paste forms (this is called a farce or bind). You want it to be moist but not so wet that it won't hold together. Keep a bowl of cold water nearby and rinse your fingers as needed to keep the fat from sticking while you work.

2. At this point, you can quickly fry up a small amount of the sausage mixture to determine if you've achieved the right taste and texture. If the patty is too dry and crumbly, you may see the fat leaching out into the pan; this means you will need to continue to mix the meat until it is better emulsified. Also taste the cooked patty for seasonings and adjust as desired.

BULK SAUSAGE

If you plan to make your sausage in bulk form, stop here! You can put it in an airtight storage container and refrigerate for up to 3 days or freeze for up to 3 months; thaw overnight in the refrigerator before using. Or better yet, shape it into logs (like icebox cookie dough), about 3 inches in diameter, wrap well in plastic wrap, and refrigerate until firm, or freeze for longer (and thaw). Then slice off disks as needed in the desired thickness to make patties.

Stuff the Sausage

The following steps apply to any of the sausage stuffers discussed on pages 7 to 8, with some modifications for each; it's always a good idea to check the manufacturer's instructions before using any specialized equipment. If using a motorized stuffer, it will be much easier with two people: one to feed the stuffer and the other to handle the casing.

1. Coat the stuffing tube with olive oil or nonstick cooking spray, then gather the entire length of the casing over the tube until the end of the casing is even with the opening of the stuffer tube.

2. Begin cranking the stuffer (or, if using a hand stuffer or mixing attachment, push the ground meat mixture through the funnel or feed spout with a meat pusher) until the mixture reaches the lip of the tube opening. Pull about 2 inches of casing off the end of the tube.

3. Holding the end of the casing with one hand, slowly crank the stuffer until the end is filled, then squeeze the end of the casing to release air and tie into a knot.

4. Feeding small amounts of meat through the stuffer at a time, continue stuffing the entire casing, arranging the sausage in a coil on the clean work surface (we used a baking sheet). Pack the casing firmly but not to the bursting point, maintaining an even thickness throughout the length of the casing. When all the meat has been used, slide any leftover casing off the stuffer.

5. Inspect the length of the filled casings, looking for any air bubbles or pockets. Prick the bubbles with a sausage pricker. Air pockets can fill with fat during cooking; in dried sausage, they can allow mold to grow. If you notice any tears in the casing, cut that part out and discard, adding the sausage mixture back into the stuffer (or frying it off as a patty).

Twist into Links

When twisting sausage into links, make sure the casing isn't filled too tight, or it will burst; some sausage makers like to alternate the direction of the twisting after each link.

1. Beginning at the first end of the stuffed casing, grasp the desired length of sausage and gently twist it five or six times in one direction to form a link.

2. Continue twisting off links until you reach the other end of the casing.

3. Tie off the open end in a tight knot, if desired, and cut away any excess. Note that some sausage makers prefer to just twist off the end and cut as the other links, to avoid anyone having to eat that knot down the road.

Let Sausage Air-Dry Briefly

Fresh sausages benefit from a short aging period, which allows the flavors to meld and the casings to become glued to the meat mixture inside, for the desirable dry-out and snap.

1. Arrange the links in a single layer on a wire rack set on a rimmed baking sheet and refrigerate for at least a few hours, or preferably overnight (or as directed in the recipe). Turn the sausages frequently to allow them to dry evenly.

2. When you are ready to separate the links, use kitchen scissors (or a very sharp knife) to cut them apart. The casing will fit the sausage mixture like a glove and the meat will not squeeze out.

Cooking and Storing Sausage

When you've invested time and effort to make delicious sausage, you'll want to be sure to cook it and store it properly.

Cooking to the Proper Temperatures

A cooked fresh sausage link will be firm to the touch and hot all the way through. But always remember that you can't tell if meat is really, truly cooked to the proper temperature just by looking at it. Instead, fresh sausage should be cooked slowly and thoroughly until its internal temperature, as measured with an instant-read thermometer, reaches the following temperatures recommended by the USDA and the FDA:

- **Beef, lamb, pork, fish, or game sausage: 160°F (71°C)**
- **Poultry sausage: 165°F (74°C)**

One good rule is to cook sausage made from pork or beef for about 20 minutes for a 1-inch diameter link, 15 minutes for poultry, and 10 minutes for seafood and vegetarian sausages. For fatter sausages, cook for another 5 to 10 minutes, then measure the internal temperature to be sure the sausage is cooked through. Because wild game is usually quite lean, sausage made from it can be on the dry side. Treat it as you would any fresh sausage, but do not overcook.

Avoid testing sausage for doneness by cutting into it, as this will allow all the juices to escape. For the same reason, we suggest handling links with tongs rather than spearing them with a fork.

Cooking Sausage Patties

Sausage patties may be panfried over medium-low heat, grilled, or broiled as you would hamburgers.

Their cooking time will vary considerably depending on the type of meat or vegetarian ingredients they are made of, as well as their thickness. They should be hot all the way through, their juices should run clear, and meat sausages should no longer be pink in the center. Turn them once during the cooking time.

Cooking Bulk Sausage

When using bulk sausage in soup, stew, casseroles, and other such dishes, you'll need to cook it first: Sauté with a little oil or butter in a skillet, breaking it up with a wooden spoon or the back of a fork and cooking until the meat is no longer pink. Shake the pan frequently for even browning.

Storing Homemade Sausage

Fresh sausage will always taste its best within two or three days of being made, and should be kept refrigerated until cooking. You can also freeze sausage for longer storage. Wrap links and patties individually in plastic wrap and then freeze together in plastic freezer bags. Bulk sausage can also be frozen, wrapped well and tucked in a heavy-duty freezer bag. Frozen fresh sausage is best used within about three months; it won't be spoiled after this time, but it will likely begin to decline in taste and quality. Cooked sausages may be refrigerated for up to two weeks and frozen for up to three months; let cool before storing.

Always label and date sausage when freezing for future reference, and be sure to note whether it's fresh or cooked. Thaw frozen sausage overnight in the refrigerator before cooking or heating.

A perfectly cured sausage can theoretically be hung dry for weeks without spoiling especially if you live where it is cool and dry. Once you cut into a sausage, however, it must be refrigerated, wrapped well, and consumed within three weeks.

COOKING METHODS

Panfrying

When in doubt, panfry your sausages. It's fairly fool-proof so long as you cook low and slow, and it works with just about any variety, except those that are specifically designed for poaching (as directed in the recipe). This is by far the best choice for sausages encased in vegetarian casings. Panfrying is also an excellent way to reheat sausages that have been poached or smoked, since it will heat them through gently and slowly.

THE METHOD: Heat a heavy skillet (preferably cast iron) over medium heat, then swirl in oil to coat thinly. Add the sausages, leaving room between each, and cook until browned on all sides and cooked through, turning with tongs, 10 to 15 minutes. Check for doneness with an instant-read thermometer.

NOTE: Some people like to finish by pouring in a bit of water at the end, once the sausages are browned, to allow them to finish cooking all the way through, but this is not essential so long as you cook them slowly over moderate heat. Others add liquid at the end to deglaze the pan, creating a delicious pan sauce to serve along with the sausage; wine, vermouth, broth, or even water can be used, depending on the flavor profile of the sausage.

ALTERNATIVE METHOD (BEST FOR THICK FRESH SAUSAGES): This is sort of a cross between steaming and panfrying, in that you start by placing the sausages in a skillet with enough cold water to come about one-quarter of the way up the sides of the links. Bring to a simmer, cover, and cook until done throughout and the water evaporates, 10 to 15 minutes. Then remove the lid and cook the sausages, turning with tongs, until crisp and golden brown on all sides, 3 to 5 minutes longer.

Poaching

Gently simmering sausage in liquid ensures that it stays moist and juicy, and it allows you to cook sausages all the way through without having to worry about overcooking the outside. The downside is that you won't get the textural contrast of panfrying or grilling, unless of course you poach first to cook through, then panfry or grill to brown and crisp the outside — something many avid sausage makers do (see note below).

Poaching is best for emulsified fresh sausages such as bratwurst, weisswurst, knockwurst, and frankfurters and hot dogs. It's not so good for coarse sausages, such as Italian sausages, as the fat will melt and render out during cooking, leaving the sausages too dry, or those that are crumbly by nature, such as Fresh Chorizo (see page 67) or merguez.

THE METHOD: Bring enough water (or beer, wine, stock, or a combination) to cover the sausages to a boil in a large pot, then add the sausages. Once the liquid returns to a simmer, reduce the heat and cook at a bare simmer until cooked through, 10 to 20 minutes for 1-inch (2.5 cm) diameter sausages. Allow more or less time for larger or smaller sausages. Check for doneness with an instant-read thermometer. Let the sausages stand for 5 to 10 minutes before serving. Or let cool completely and refrigerate for later.

NOTE: This is a great way to cook sausages ahead of time so all that's needed is to rewarm them (say, when throwing a backyard barbecue for many guests). It's also a particularly good option for more delicate sausages that you plan to grill, since simmering them first in liquid will keep the casings from bursting on the hot grates.

Braising

Use this method for thicker, coarser sausages that can hold up to long, slow cooking: Italian sausages, kielbasa, bratwurst, and bangers.

THE METHOD: Heat enough oil to coat the bottom of a Dutch oven or heavy pot over medium-high heat. Add the sausages and cook until browned on all sides, about 8 minutes. Add aromatics, such as onion and celery, and cook until softened, about 8 minutes. Add enough braising liquid to cover the sausages, such as beer, cider, or wine mixed with water, or use a flavorful broth, and bring to a boil. Cover and reduce the heat; simmer until the sausages are cooked through, 10 to 20 minutes longer. Check for doneness with an instant-read thermometer. Remove the sausages from the pot; boil the cooking liquid until reduced and thickened, about 15 minutes. Return the sausages to the pot and heat through. Serve the sausages with the sauce.

Roasting

Cooking sausages in the oven is a relatively hands-off way of achieving the same textural contrast of panfrying, without the spatter or need for frequent turning. It's also just as universally appropriate for practically any type of sausage (again, except those where a particular cooking method is essential to the recipe).

THE METHOD: Preheat the oven to 425°F (220°C). Line a rimmed baking sheet with a wire rack and arrange the sausages evenly on the rack. Roast for 10 to 20 minutes, or until evenly browned, turning halfway through. Check for doneness with an instant-read thermometer.

Grilling

Most types of fresh sausages are excellent when cooked on the grill, but you may want to poach more delicate ones first, to keep them from bursting. You can also grill smoked and even dry-cured sausages.

THE METHOD: If using a charcoal grill (our preference), start by heating the coals until medium-hot, then move about one-third of them to the side to create a section with indirect heat, often called a cool zone. (You can also achieve this with a gas grill, usually by moving the food onto the raised grates, farther from the heat source.)

Take time to allow the coals to reach the right temperature; too hot and the meat in the sausages will swell, the casings split, and the fat will drip off, resulting in dry, shrunken sausages. This can also cause flare-ups, which will cause the casings to char too quickly. (If you notice that happening, move the sausages to the cool zone.)

The coals are medium-hot when you can hold your palm about 5 inches above the surface of the grates for 5 to 7 seconds only. Once the grates are hot, wipe them down with a little oil to prevent the sausages from sticking. You may also want to rub the links with oil for the same reason.

Start the sausages over direct heat until browned, turning with grilling tongs. Then move to indirect heat to finish cooking through, if necessary. Use an instant-read thermometer to determine doneness; never cut into sausage or you'll lose the flavorful juices.

ALTERNATIVE METHOD: Here's a trick that many experienced grill masters use to avoid the concern about flare-ups when cooking a bunch of sausages at once; it also lets you cook the links all the way through before getting too charred, while allowing them to take on some of the smoky quality of being cooked over hot coals. Place the sausages in an aluminum baking pan and add some water, beer, or other liquids to come about halfway up the links, and cook on a covered grill, turning as needed.

Broiling

Because the broiler works only with super-high temperatures, it's generally suggested that you precook fresh sausages by poaching them. This step is not necessary for smoked sausages that are already cooked to the proper internal temperature.

THE METHOD: Preheat the broiler to high. Place the sausages on a broiler pan or a wire rack set on a rimmed baking sheet and broil, about 5 inches from the heat source, for about 5 minutes, until crisped and heated through, turning with tongs so they brown evenly. Always watch carefully when broiling foods.

Dry-Cured Sausages

Dry-cured meats were born of necessity: The idea that one could butcher an animal in the fall and still eat the meat the following spring must have been revolutionary. From these early experiments in meat preservation — and dry curing is the oldest method of preserving meats known to man — came all the wonderful types of traditional, regional, unique cured sausages we enjoy today.

Since the last edition of this book was published, there has been a full-scale renaissance in dry-cured sausage making among the new breed of purveyors who are devoted to preserving old-world ways. Restaurant chefs and artisanal butchers and smaller scale producers alike have all embraced this what's-old-is-new-again philosophy. Many have studied at the hands of skilled masters, both here and abroad, and are using those centuries-old methods to create products that were at risk of becoming extinct. This resurgence in handcrafted charcuterie has awakened our taste buds to the possibilities that this technique can yield. That's good news for all of us: Not every sausage today is extruded from a huge machine and packaged in plastic. Some of the best things in life are still made by hand.

Dried sausages are admittedly a challenge for even the most experienced sausage makers. They should not be attempted on a whim but rather because you desire to experience firsthand an incredible journey. The road starts with raw meat and minimal seasonings and ends, without any cooking or heat, but with your utmost attention and care, in an utterly delectable slice of life.

How It's Done

Dry-curing sausage is just another form of preserving — in this case, meats. To do it, you are basically mixing salt and curing salt with the sausage mixture before stuffing into casings, then hanging in a cold, damp spot with some air movement for a few weeks or months. This preservation relies on four things for flavor *and* food safety: salt, curing agent, humidity, and temperature. Well, and one more critical factor: you! Only with your careful attention and control will you be able to pull off this culinary feat.

Salt is used to draw moisture from the meat (a process called osmosis) to the surface, and then the moisture on the surface is pulled into the air and out of the drying room or chamber. The reasoning behind this technique is simple in theory but tricky in practice: Pathogens require the presence of water to survive, so if you remove the water you'll also remove the risk of contamination. That's why dried sausages are safe to eat even though they've never been cooked. (It's also why dried sausages are so singularly appealing: That elemental taste of the meat is allowed to shine through without being tamped down by cooking.)

Here's where the humidity of the drying chamber — typically around 80 percent in the initial phase, then 70 to 75 percent from then on out — is important. There must be enough moisture in the air to allow the water in the center of the sausage to be "pumped" out of the sausage. It may seem counterintuitive, but you need to keep the surface of the sausage moist enough to be permeable to the water inside, so that it can escape. If the sausage is too dry, the water inside the casing will be trapped there,

DRY SAUSAGE SAFETY

Dry sausages — such as pepperoni and summer sausage — that are preserved but never cooked have had a good safety record for hundreds, if not thousands, of years. But during the 1900s, some children and adults became ill from the *Escherichia coli* (*E. coli*) bacteria after eating dry-cured salamis from a sausage plant. Somehow, the bug survived all of the processing, even in a carefully controlled commercial atmosphere.

The USDA's Food Safety and Inspection Service (FSIS) had not, as of this writing, changed recommendations for consumer handling of dried sausage products, but the agency has required commercial producers to use new protocols in making dried, uncooked sausages.

What's a home sausage maker to do? The first step is to make sure to use the curing salt in the amount prescribed by the manufacturer. The next step is to always cook or smoke sausages to an internal temperature of 160°F (71°C) — or 165°F (74°C) for poultry sausage — at some point before consuming them to ensure that all potentially harmful bacteria are killed.

If you do choose to make dry sausages that are not cooked, you should follow the USDA's recommendation that people at risk (the elderly, very young children, pregnant women, and those with weakened immune systems) avoid eating them. And note: Home-dried pork sausages will be free from trichinae if you've prefrozen the meat (see page 44). But you can't be certain that other potentially harmful microrganisms or bacteria have been killed.

TRICHINOSIS UPDATE

Trichinosis (or trichinellosis) is a food-borne illness caused by a parasitic roundworm, *Trichinella spiralis* (or trichinae), that can be found in some pork and wild game meat, including wild boar, deer, and bear. The symptoms, which typically appear within a few days of eating contaminated meat, include nausea, diarrhea, vomiting, fever, fatigue, joint and muscle pain, difficulty breathing, and (in severe cases) heart problems and even death.

The incidence of trichinosis has sharply declined in the United States due to improved hog-feeding practices and increased public awareness. The Centers for Disease Control and Prevention (CDC) reported only 84 confirmed cases between 2008 and 2012 (and none fatal); of those, only 22 cases could be attributed to pork. Wild game in fact is more likely to be the culprit. Regardless of the source, most of the cases were contracted after eating raw or undercooked pork or wild game.

Preventing it is easy if you heed the guidelines set forth by the USDA and the CDC:

1. Always cook meats to the recommended internal temperature as measured with a food thermometer inserted in the thickest part of cuts or the center of ground-meat products such as sausages and patties. Include any rest time, during which the temperature will remain constant or continue to rise and thereby contribute to destroying pathogens.

> **FOR WHOLE CUTS OF PORK AND BEEF:** Cook until at least 145°F (63°C) in the thickest part of the meat and allow the meat to rest for 3 minutes.

> **FOR GROUND MEAT (INCLUDING WILD GAME):** Cook to at least 160°F (71°C).

> **FOR ALL WILD GAME (WHOLE CUTS AND GROUND):** Cook to at least 160°F (71°C).

> **FOR ALL POULTRY (WHOLE CUTS AND GROUND):** Cook to at least 165°F (74°C) and allow to rest for 3 minutes.

2. Wash your hands with warm water and soap after handling raw meat; even better, wear food-safe gloves and discard after each use.

3. Clean meat grinders thoroughly each time you grind meat.

4. Never, ever taste raw pork, poultry, or wild game or sample uncooked sausages made with these meats. Pork that will eventually be consumed raw, as in a dried uncooked sausage, should be prefrozen to make it safe to eat. This is especially true for pastured, free-range pork. The CDC calls for prefreezing pork that's less than 6 inches (15 cm) thick at 5°F (−15°C) for 20 to 30 days to kill the *Trichinella* larva. To shorten the freezing time, you can lower the freezing temperature: −10°F (−23°C) for 12 days and −20°F (−30°C) for 6 days. If you do this, an accurate freezer thermometer is a must!

When you are ready to make sausage, partially thaw the meat, then cut and grind it as the recipe directs. (It's easier to cut and grind very cold, firm meat.) According to the CDC, some worms found in wild game are freeze-resistant, meaning you may want to stick with sausages that will be cooked (including smoked varieties) before eating when using venison, wild boar, or other game meats.

resulting in a "dried" sausage that is still moist — and prone to spoilage — in the middle. Unfortunately, you won't find out until you slice into it, and by then it will be too late and you will have to scrap that batch and start again. (Or you can use a pH monitor.) That's why developing a feel for when the sausage is too dry on the outside is essential; only then will you be prompted to regulate the humidity in the drying chamber to promote more even drying of the sausage. (See pages 166 to 168 for tips from an expert on how to master the art of dry curing.)

Curing agents are another critical component of dry-cured sausages. They help prevent spoilage, give sausage an appealing rosy hue, and contribute flavor of their own. They are also intended to be used in the amounts prescribed on the label — not a milligram more or less. See pages 21 to 22 for a discussion about the reasons why.

Last but not least is the temperature of the drying chamber in combating pathogens. It is imperative that you keep the meat in a cool environment throughout the drying process to impede the proliferation of microorganisms that cause rancidity, which cannot multiply below 58°F (14°C). You will need to monitor the temperature of the room or chamber at all times, along with the humidity.

Another often overlooked element in drying sausage is the use of airflow to help regulate the drying conditions and promote more even drying. For instance, if the room is too hot, you can use a fan (or a cool breeze through an open window) to bring the temperature down quickly or to help keep the moisture in the air moving, which will also help promote evaporation.

Time is yet another factor, and how long the salami needs to age depends on several factors. In general, the slimmer the link, the shorter the drying time, but other considerations include the type of

SEMIDRY SAUSAGES

Summer sausages are typically referred to as semidry, but they are not prepared in the same manner as truly dry-cured sausages. Instead, they are usually smoked to fully cook through and, at the same time, to partially dry them. These sausages are typically semisoft and keep well because of the way they are processed.

casing, the coarseness of the sausage meat, and of course the drying environment. Be patient; you can't rush the process, but you can make sure the time is well spent by paying careful attention every step of the way. Promote the right conditions to allow science to do its magic, and you, too, will experience firsthand the joy of participating in this age-old culinary tradition. That's worth all the wait in the world.

Where to Dry

If you already have all the equipment necessary to make fresh sausage, you only need a few more tools to make dried — most especially a cold, damp place where you can hang the sausage to dry for a few weeks. While you can find some sausage-drying chambers small enough for home use on the market, they are a hefty investment (and the reviews are mixed on how well they work).

If you live in a cold enough climate, you might be able to use your unheated cellar or garage to dry sausage, assuming it has the optimum environment, with a temperature of about 60°F (15°C) and 70 to 80 percent humidity. If you go this route, you'll need to keep a close eye on the temperature with a thermometer if there's no thermostat already installed. You'll also need to be able to control the airflow, so consider cordoning off a corner of the room if it's a particularly large space.

Otherwise, most people rely on an extra refrigerator for a drying chamber. (Look for one at tag sales or secondhand shops, or find an inexpensive model on sale.) You can remove all but the top and middle racks from which to hang the sausages. You'll need to plug the refrigerator into a separate temperature-controlled outlet (this outlet is then plugged into your regular outlet). What this enables you to do is set the thermostat to a specified temperature; when the inside of the refrigerator rises above that temperature, the thermostat comes on to cool it down again.

Whether you use your cellar or a refrigerator, you will also need a few other basic items that you might already have on hand: an oscillating fan with variable speeds for controlling the flow of air, and a small dehumidifier and humidifier. (Yes, you need both! They keep each other in check when controlling the moisture levels in the air.)

How to Smoke Sausage

Smoked sausages take extra effort and time — and special equipment or your own ingenuity — to make, but you'll be rewarded with that incomparable taste and characteristic color. They can either be cooked or uncooked, depending on the method used. Some sausages rely on both methods to achieve the desired outcome: cold smoking for intensifying the flavor, then hot smoking for cooking through.

Hot Smoking

This method involves cooking the sausage at temperatures above 175°F (80°C) — hot enough to prevent spoilage. The combination of smoke and heat sets the color of the sausage and also promotes that desirable "snap" (the casings better adhere to the sausage). Just like other types of cooked meats, smoked sausage is perishable and should be kept in the refrigerator (or freezer).

Cold Smoking

This method calls for smoking meat at temperatures below 175°F (80°C), meaning the sausage is still essentially uncooked (unless of course you started with precooked or dry-cured sausage). The benefit of cold smoking over hot smoking is that you end up with a more pronounced smoky taste, since uncooked sausage can absorb flavors better than cooked. For this reason, some sausages are first cold smoked for taste, then hot smoked for texture — and to cook through.

Note: You can hot-smoke most any of the fresh sausages in this book, but those that are more delicately seasoned can be overwhelmed by the smoky flavor. Keep in mind that you must always add a curing salt to the meat mixture to prevent spoilage because the sausages will linger too long in the "danger zone," where harmful bacteria thrive. Follow this formula set forth by the USDA for food safety: Add 1 teaspoon (6 g) Insta Cure #1 (sodium nitrite) to the seasoning mixture for every 5 pounds (2.25 kg) meat and fat.

How It's Done

Smoking is a technique, the basics of which can be taught. It is also a craft that you can only master through much practice until you develop the feel that will help you create a consistently good product. You will also need to pay careful attention during the smoking process to prevent spoilage. The following guidelines are based on the recommendations set forth by the USDA.

NON-SMOKING SECTION

For some people, using a smoker (or charcoal grill) is just not a viable option. Never fear: There are still ways to impart smokiness to your sausages (and other meats) without so much as striking a match.

LIQUID SMOKE: Liquid smoke is made by distilling the steam that's created during the smoking process into a super-concentrated liquid that you can add to the sausage mix after grinding. Despite all the studies, these products have been determined safe as of this writing. They have even been found to have some of the same antimicrobial properties as the real deal (though they are not the same as preservatives).

SMOKED MEATS: Try incorporating some cooked bacon, pancetta, tasso, or other smoked meats into your sausage mixture for a subtle smokiness.

DARK BEER: In recipes that call for liquid, replace some of it with stouts or porters; or seek out craft beers that have a bacony, smoky flavor.

SMOKED SEASONINGS: Use pimentón (Spanish smoked paprika) or dried chipotle chiles (these are smoked jalapeños), which are available whole and ground. You could also add the adobo sauce from canned chipotles in place of some of the liquid in a recipe for a sausage.

MOLASSES: Maple syrup is commonly used for adding sweetness and flavor to breakfast (and other) sausages, and molasses can function in much the same way. It has a smoky sweetness that works in some flavor profiles, especially those with warming spices like cardamom and ginger.

SMOKED SALTS: Because these are so intensely flavored, you will need to be careful using smoked salts other than as a finishing salt; try replacing just some of the salt in the seasoning mix, or just sprinkle over the sausages (or finished dishes) before serving.

Guidelines for Smoking Sausage

1. The first, critical step is to allow the sausages to air-dry until the surface is tacky (almost sticky to the touch) before smoking, as only then will the smoke cling to the surface. Spread the links on a rimmed baking sheet fitted with a wire cooling rack and place in the refrigerator; this will take at least a couple hours, but overnight is even better. Avid sausage makers often empty out an old refrigerator just for this purpose, so they can hang the sausages to dry before smoking (or to cure them, as described on pages 166 to 168).

2. When dry, hang the sausages in the smoker or place them on the racks, making sure they do not touch each other or the smoker wall. The entire surface of the meat must be exposed to ensure an even color.

3. For your fire, burn hardwood, such as hickory, oak, apple, cherry, pear, peach, beech, chestnut, pecan, or maple. Mesquite, a hardwood shrub from the Southwest, is popular for its unique aroma. Dry corncobs may also be used. Many sausage makers like to use damp sawdust because it produces a good smudge — that is, a fire that produces a lot of smoke with relatively little heat and is easy to control. You can also buy hardwood briquettes that are specially formulated so they don't flare up. Some suppliers also sell chips from old oak wine barrels; wine-saturated oak gives the sausage a rich, unique taste. Never burn softwoods, such as pine, cedar, spruce, hemlock, fir, or cypress. They create oily, sooty smoke that turns the sausage dark and bitter.

4. Heat the smoker to 225 to 275°F (107 to 135°C) and monitor it throughout the smoking process. Remember: The goal of smoking is to cook the sausage low and slow; if the smoker is too hot, it can cause the fat in the sausage to melt and drip off, resulting in dry, shriveled sausages.

5. Cook meat and seafood sausages to a minimum internal temperature of 160 to 165°F (71 to 74°C), or 165°F (74°C) for poultry sausage, using a food thermometer (see next page). The cooking time will depend on the type of sausage, the distance of the food from the heat source, the temperature of the coals, and even the weather. Plan on anywhere from 4 to 8 hours to smoke sausages made with meat, game, and poultry, and less for seafood and vegetarian sausages.

6. Once the sausage has reached the safe temperature and is the desired color, remove it from the smoker. If desired, immediately immerse the links in an ice-water bath to stop the cooking and prevent shrinkage. Swish them around until cool to the touch, then remove and lay in a single layer to dry.

   Be sure to allow the sausages to air-dry again thoroughly before storing.

7. Some makers leave the sausages out at room temperature to "bloom" and take on an even deeper, richer brown color. Others like to refrigerate the sausages overnight, uncovered, in a single layer on a wire rack on a rimmed baking sheet to allow the flavors to continue to meld.

8. Smoked sausages will keep in the refrigerator, wrapped well, for 3 to 4 days. They can also be frozen in freezer bags for up to 3 months, or longer if sealed in vacuum packages; thaw overnight in the refrigerator before using.

THERMOMETERS RULE

To ensure that meat and poultry are smoked safely, you'll need to monitor the internal temperature of the smoker *and* the food being smoked. The air temperature in the smoker (or grill) should always remain between 225 and 275°F (107 to 135°C) throughout the cooking process. Many smokers have built-in thermometers; use an oven thermometer in those that don't (and in your converted charcoal grill).

Use a food thermometer to determine the temperature of sausage made with meat or poultry. Oven-safe (probe) thermometers can be inserted into the meat and remain there during smoking; we prefer to use one with an alarm that will let you know when the desired temperature has been reached without having to open the chamber (and let all that flavorful smoke and the moist steam escape). You also don't want to risk losing all the juices by continually piercing the links with an instant-read thermometer. Probe thermometers are easy to find and inexpensive — and a sausage smoker's best friend.

Finding the Right Smoker

Smokers come in a wide range of styles and prices, so which one you choose will depend on (a) how often you plan to use it and (b) whether you are partial to the ease of gas or electric models or prefer the more traditional (and some would say authentic) results of charcoal. If you are a "weekend smoker" who will be smoking a relatively small amount of sausage, you can even convert a simple covered charcoal grill into a makeshift smoker. Otherwise it's worth investing in a unit that will help you achieve consistent results without a lot of hassle and that will take much of the guesswork out of the process.

Covered Charcoal Grill

Converting a kettle grill into a smoker is easy to do, and you may be familiar with the process if you've ever barbecued a whole chicken or larger cuts of meat. Basically, it requires setting up dual temperature zones for direct and indirect heat.

Use a chimney starter to light the briquettes. Once they are covered with gray ash, drop them onto the lower heat grate and push them to one side. Then put an aluminum pan on the other side and fill with water. The steam from the water will create a moist environment, which will keep the sausage from drying out during cooking; it will also help regulate the heat for more even cooking over an extended period of time.

You may also want to add hardwood chips for added flavor; you'll need to experiment to see how much and which type works for different sausages (better to start with a little and work your way up to adding more). Place the sausages on the grill over the water pan. Close the lid and keep the air vents open.

To hot-smoke sausage, add about 10 new briquettes every hour, checking the temperature frequently to maintain the heat between 225 and 275°F (107 to 135°C) throughout the process. Follow a specific recipe for times and temperatures.

Note: For longer, slower cooking, such as when smoking larger links or cuts of meat, place unlit briquettes in a single line around the edge of the grill to form a "snake," leaving space between the head and tail, then add another two layers of briquettes above those. Place the aluminum pan with water in the center. Light about eight briquettes in a chimney starter and, when they are ready, pile the lit coals at the head of the snake; the other coals will slowly burn down the line for hours. If necessary, add more unlit coals to the tail end.

Charcoal Smokers

For those who prefer the distinctive taste that comes from smoldering charcoal, there are a couple of options: **Vertical smokers** (aka barrel or drum smokers) are relatively small, easy to use, and affordable, making them a safe, convenient choice for novices. They have two bowls: one on the bottom for holding the charcoal and wood, and one above for holding water (or other liquid) to help keep the food moist during smoking and also help regulate the temperature.

For more money, look for **cabinet-style smokers**, which function similar to the barrel models but have a couple of advantages: They are roomier inside and have removable racks, meaning they can accommodate larger sausages and cuts of meat and can smoke a greater variety of foods at the same time. These also typically have more vents than vertical smokers, making heat regulation easier.

Gas Smokers

Once spurned by barbecue aficionados for not being the real deal, gas smokers are now even being used in some commercial establishments. They've always been embraced by home smokers for offering hands-off cooking and for being reasonably priced and reliable. To create the smoke, you add chips to a pan that's situated over the burner or to a special wood or chip box. A potential downside is that low and slow cooking can run up the cost of propane and you risk running out of gas before you're done, so you might want to pick up an extra propane tank as a backup.

Wood Smokers

Experienced sausage smokers tout the superior flavor that these smokers (also known as stick burners) produce, but they are not for the faint of heart. When wood is burned at lower temperatures, as when smoking, it can emit a dirty smoke that will cause food to taste bitter, so you have to monitor the heat fairly frequently over the course of smoking to prevent this from happening. The larger submarine models have horizontal barrel-shaped chambers; so-called offsets have a box that's attached to one side for burning wood (or charcoal). Buy only high-quality (read: expensive) models, as the cheaper versions aren't rugged enough for regular use.

Electric Smokers

These set-and-forget models are great for beginners since you don't have to maintain a fire or worry about regulating the temperature (so long as you follow the manufacturer's directions). Premoistened wood chips create the smoke. There are a variety of models to choose from, including portable smokers and those that rival commercial ones in their capacity. You should weigh your options based on how much you plan to smoke at any given time. Electric smokers are now much more affordable than in the past and many have built-in racks for suspending the sausage, making them an efficient all-in-one option for aspiring sausage smokers.

ANATOMY OF A SMOKEHOUSE

A temporary smokehouse can be made with a barrel or drum connected by a stovepipe to a fire pit. Bury a 10- to 12-foot length of 6-inch stovepipe underneath about 6 inches of dirt, orienting it so it slopes down slighty and connects to an opening in a fire pit, also embedded in the ground. Cut both ends off the barrel and insert the bottom end a bit farther into the barrel to serve as a baffle for the flow of the smoke. Rest the barrel on the top of the upper end of the stovepipe. Control the heat in the pit by covering it with sheet metal and mounding dirt around the edges to cut off the draft. You want a lot of smoke and not much flame. Fasten cleats to the top of the barrel lid and rest it on pieces of a broomstick (or other wooden dowels) that support the hanging sausage. Hang an oven thermometer from a hole bored in the lid or from one of the broomsticks. Drape a piece of clean muslin or burlap over the top to protect the meat from insects. Inside the barrel, set a water pan on two bricks.

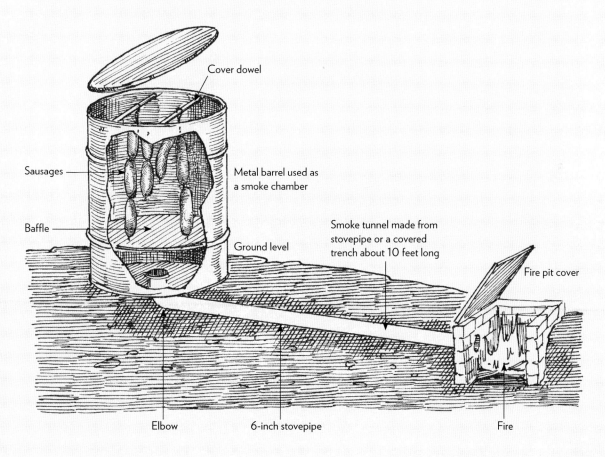

Cover dowel

Sausages

Metal barrel used as a smoke chamber

Baffle

Ground level

Smoke tunnel made from stovepipe or a covered trench about 10 feet long

Fire pit cover

Elbow

6-inch stovepipe

Fire

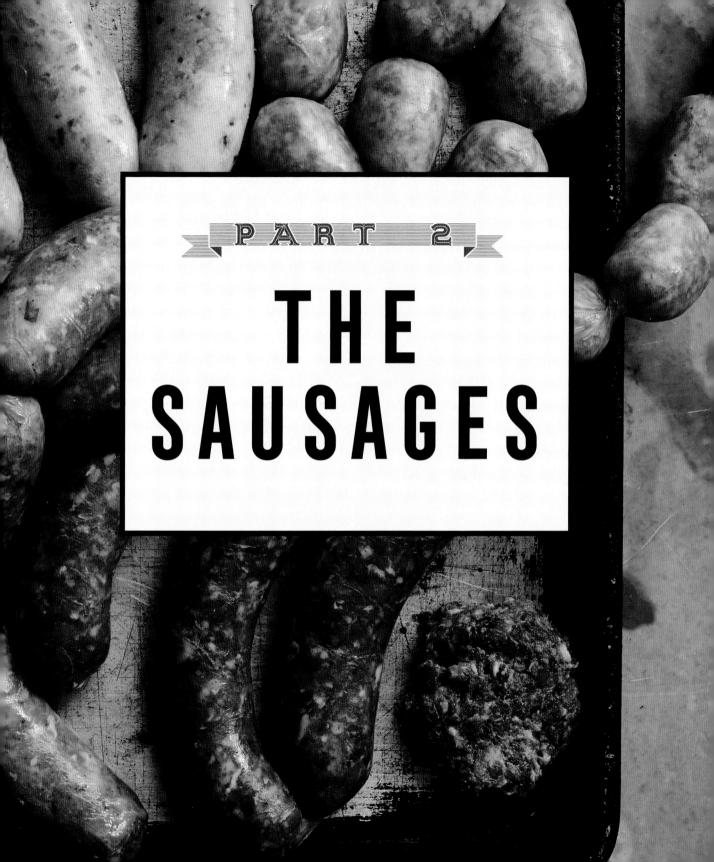

PART 2
THE SAUSAGES

leg

loin

Boston butt

head

belly

picnic

Luganega, page 66

PORK SAUSAGES

The autumn pig slaughter was until recently the most important gastronomic event on the European peasant calendar.

— Elisabeth Luard, *The Old World Kitchen*

Pork still reigns as the king of sausage meat for one simple, unarguable reason: The pig is the perfect meat animal, able to pack on the most pounds relatively easily (and cheaply, for the farmers). Pigs also provide the widest array of cuts — hams, chops, ribs, and tenderloins, to name the most common. You can also use the meat to make bacon, sausage, pâtés, and other charcuterie, plus there's the belly, trotters, skin, fatback, caul fat, and intestines, which double as natural casings for sausage. Nearly every last shred of the animal can be used, including the blood (for boudin noir) and the bristles (for paintbrushes). Everything, as they say, but the squeal.

Over the centuries, France, Germany, Italy, and Spain developed the largest repertoires of pork sausages, from andouille to bratwurst to soppressata. Butchering was always done in the late fall to take advantage of cool weather, and a good store of sausages in the larder made the difference between hunger and satisfaction for the coming winter.

Because pork is a mild-tasting meat, you can season and spice it to create a number of different identities. Pork fat, creamy and smooth, helps to bind together the ingredients well and creates a rich, juicy sausage. It's as elemental as the meat itself.

Your homemade pork sausage will be just as satisfying to you as it has been to generations of home sausage makers around the world. Review the instructions for sausage making before getting started.

COUNTRY-STYLE BREAKFAST SAUSAGE

MAKES 3 POUNDS

Despite their name, these savory sausages are very versatile and are just as welcome at brunch, lunch, or dinner as they are on the breakfast table. Make sure to use the freshest dried herbs for the best flavor, or substitute fresh for dried using the standard formula: 1 tablespoon fresh for every 1 teaspoon dried.

1. Prepare the casing (see page 29).

2. Cut the meat and fat into 1-inch cubes. Freeze the cubes for about 30 minutes to firm them up before grinding.

3. Grind the meat and fat together through the fine disk of a meat grinder.

4. In a large bowl, combine the meat mixture, salt, sage, white pepper, sugar, thyme, marjoram, cloves, and crushed red pepper. Mix well, using your hands. Freeze the mixture for 30 minutes.

5. Grind the seasoned mixture through the fine disk of the meat grinder.

6. Stuff the mixture into the prepared casing, prick air pockets, and twist off into 3-inch lengths. (Or, shape into 16 patties.)

7. Place the links on a baking sheet fitted with a wire rack and refrigerate, uncovered, for a few hours or preferably overnight.

8. Cut the links apart. Refrigerate, wrapped well in plastic, for 2 to 3 days, or freeze for up to 3 months; thaw overnight in the refrigerator before using.

9. Cook as desired (see methods on pages 38 to 40) to an internal temperature of 160°F (71°C).

Ingredients

- 4 feet small hog or sheep casing
- 2½ pounds boneless lean pork butt or shoulder
- ½ pound pork fat
- 1 tablespoon kosher salt
- 1½ teaspoons dried sage
- ¾ teaspoon freshly ground white or black pepper
- ¾ teaspoon brown sugar
- ½ teaspoon dried thyme
- ¼ teaspoon dried marjoram
- ⅛ teaspoon ground cloves
- ⅛ teaspoon crushed red pepper

- 4 feet small hog or sheep casing
- 3 pounds boneless lean pork butt or shoulder
- 1 small onion, chopped (about ½ cup)
- 1 tablespoon kosher salt
- 2 teaspoons dried sage
- 1 teaspoon dry mustard
- ¾ teaspoon freshly ground black pepper
- ¼ cup pure maple syrup
- 2 tablespoons milk

FRESH

MAPLE BREAKFAST SAUSAGE

Consider these links the perfect accompaniment to pancakes or French toast, since they already have a touch of sweetness from maple syrup. This is an ideal sausage for making in bulk (see page 34). *See photo, page 58.*

1. Prepare the casing (see page 29).

2. Cut the meat into 1-inch cubes. Freeze the cubes for about 30 minutes to firm them up before grinding.

3. Grind the meat through the fine disk of a meat grinder. Add the onion and mix well, using your hands. Freeze for 30 minutes.

4. Grind the meat mixture through the fine disk of the meat grinder. Add the salt, sage, dry mustard, and pepper. Mix well, using your hands. Stir in the maple syrup and milk, and toss lightly to blend. Do not overwork the meat mixture.

5. Stuff the mixture into the prepared casing, prick air pockets, and twist off into 3-inch lengths. (Or, shape into 16 patties.)

6. Place the links on a baking sheet fitted with a wire rack and refrigerate, uncovered, for a few hours or preferably overnight.

7. Cut the links apart. Refrigerate, wrapped well in plastic, for 2 to 3 days, or freeze for up to 3 months; thaw overnight in the refrigerator before using.

8. Cook as desired (see methods on pages 38 to 40) to an internal temperature of 160°F (71°C).

EASY SLICING

You can shape bulk fresh sausage into logs (like cookie dough), wrap them well in plastic wrap, and refrigerate until firm. Then slice off disks in the desired thickness before frying (see page 34).

BREAKFAST SAUSAGES

Vegetarian Sausage and
Kale Hash, page 306

Country Chicken Sausage, page 206

Smoked Salmon
Sausage, page 232

Maple Breakfast
Sausage, page 57

Merguez, page 132

Shakshuka with
Sausage, page 272

Vegetarian Breakfast
Sausage, page 245

Meet the MAKERS

⚘

Jamie and Amy Ager

Hickory Nut Gap Farm

Fairview, North Carolina

FARMING FOR THE FUTURE

In a serene setting about 20 miles southeast of Asheville, North Carolina, in the small town of Fairview (population 2,678) sits Hickory Nut Gap Farm, which held its centennial celebration in 2016. "The farm has been in my family for 100 years," explains Jamie Ager, who, with his wife, Amy, rejuvenated the 90-acre property in 2000, as recent college graduates (and sweethearts). "My great-grandparents moved here from Chicago in 1916 and started with an apple orchard, so we're definitely a family farm, for better or for worse, as I always say."

Despite having grown up on the land, Jamie says he "wasn't encouraged to come back to the farm because financially it's hard to make a living." That all changed when he went to the local college to study sustainable agriculture and worked on the school's farm, where they put sustainable policies into practice. "That opened the idea for me to be able to raise and sell meat that was good for the environment, that was humane, and that was healthier. I thought I could do this at home and try to make a living on the farm. I was right out of college and young and idealistic, and now I'm 38 and have three boys and am just as idealistic as I was then."

Such idealism has evidently paid off. Not only have the Agers built a successful farm, where they raise about 500 pastured pigs (and also poultry) and 60 grass-fed steers each year, but they also have achieved their other goal: to have a place where people could get to know the farm they get their food from and where the community could come together, share the farm experience, and just have fun. The farm hosts Friday night barn dances, tours of the property, kitchen classes in butchery and sausage making, and seasonal activities including corn mazes and pumpkin patches in the fall and U-pick blueberries and blackberries in the summer.

As for the farming operations: "We started with grass-fed beef and happy pigs. That was the primary focus," says Jamie. "We would sell retail cuts of beef and pork at our little dairy parlor, a holdover from the farm's dairy operation, one day a week, then we started selling them at farmers' markets." Some 15 years later, they have a full-fledged restaurant and butcher shop where all the processing is done on site, rather than at a processing facility. "One of the frustrations we had was being so close to Asheville, where there was this cultural interest in the farm-to-table movement, and people would drive down for the day to see the animals but there was no place nearby for them to actually eat the meat," recalls Jamie. Now their steady stream of visitors can enjoy house-made fresh sausages and house-cured pastrami, hams, and bacon — all made from meat that's been raised right there on the farm. "It's a total experience when you see where your meat is coming from that you simply can't get by walking into your typical grocery store. It just adds to the story," says Bryan Bermingham, who oversees the butchering for the farm.

Besides, for Jamie, being a farmer is an opportunity to educate consumers about more than just where their meat comes from. "I see it as an opportunity to change the conversation around food production and the economics of farming. In the past, people would buy meat solely based on price, and that drove farming to only be profitable if making a commodity product, but I like being on the front edge of the discussion that's changing that relationship between farmer and consumer."

Here's to another 100 years of "serving as an example of healthy land steward-ship" — and to passing the torch to the fifth generation of Agers, who are growing up surrounded by rolling hills, pastured pigs and poultry, grazing cows, and a strong sense of community.

TRADE SECRETS

"You get out what you put in, so to speak. So when I teach the sausage-making course, I encourage people to get out-side their comfort zone. If you just go and buy already ground pork, you may not be happy with the results. I always rec-ommend that you buy a large cut, which lets you make those added-value products in your own kitchen. You save money, too; for $2.99 a pound you can buy a whole shoulder instead of spending $6.99 a pound for a Boston butt. You'll have a lot of work ahead of you, but you'll be putting a lot more love into it when you've broken it down yourself. Then you'll have all the trimmings and other cuts that can be used in making sausage."

— Bryan Bermingham,
Head Butcher

CIDER AND SAGE BREAKFAST SAUSAGE

MAKES 5 POUNDS

This sausage is from Hickory Nut Gap Farm (see page 60), which has a certified organic apple orchard that produces a lot of cider, the remains of which are put to excellent use in making this seasonal specialty. (In the summer, they make blueberry-maple sausage with the surplus of berries from their U-pick operation.) According to Bryan Bermingham, head butcher, "The sausage is great for breakfast, an hors d'oeuvres platter with apple butter, and also dinner with fried skillet potatoes. It's just a good all-purpose sausage." They usually make the sausage in bulk and cook it as patties, but you can also stuff it into small hog or lamb casings; you'll need about 6 feet, or 8 ounces.

1½ cups apple cider

5 pounds boneless pork butt or shoulder, about 70% lean

3 tablespoons kosher salt

1 tablespoon rubbed dried sage

2 teaspoons freshly ground white pepper

1½ teaspoons granulated garlic

1. In a 1-quart saucepan, bring the cider to a boil. Let it reduce until you have about ⅓ cup. Remove from the heat, let cool, and chill.

2. Cut the meat into 1-inch cubes. Freeze the cubes for 30 minutes to firm them up before grinding.

3. Combine the pork cubes, reduced cider, salt, sage, pepper, and garlic. Mix well, using your hands. Grind the meat mixture through the medium disk of a meat grinder. Freeze for 30 minutes.

4. Grind the meat mixture through the medium disk of the meat grinder.

5. Form the mixture into ½-inch-thick, 3-inch patties (or leave in bulk form; see Note below).

6. Refrigerate, wrapped well in plastic, for 2 to 3 days, or freeze for up to 3 months; thaw overnight in the refrigerator before using.

7. Cook as desired (see methods on pages 38 to 40) to an internal temperature of 160°F (71°C).

NOTE: *This is an excellent sausage to leave in bulk form, shaping it into a log the same diameter as the patties you will slice it into (about 3 inches). For ease, divide it into smaller portions so you can thaw only what you need at one time. Wrap well in plastic wrap, and store in a sealable plastic bag.*

COTECHINO

MAKES 3 POUNDS

2–3 feet small hog casing

2½ pounds boneless lean fresh ham

½ pound pork skin with fat, cooked until lightly browned

2 tablespoons grated Parmesan cheese

2 teaspoons freshly ground black pepper

1½ teaspoons kosher salt

1 teaspoon ground cinnamon

1 teaspoon freshly grated nutmeg

½ teaspoon cayenne pepper

½ teaspoon ground cloves

If you're fortunate enough to spend New Year's Eve in Italy, you'll likely be offered that country's good luck dish: lentils with cotechino — the sausage courtesy of the city of Modena, in the Reggio-Emilia region of northern Italy. Pork skin is essential to giving the sausage its characteristic flavor and creamy texture, as well as its name (*cotiche* is the word for "pig skin"). A fresh ham (or fresh hind pork leg) is the cut of choice for making it. Cotechino is also part of bollito misto, another Italian tradition. *See photo, page 215.*

1. Prepare the casing (see page 29).

2. Cut the meat and cooked skin into 1-inch cubes. Freeze the cubes for about 30 minutes to firm them up before grinding.

3. Grind the meat and skin through the coarse disk of a meat grinder.

4. In a large bowl, combine the meat mixture, Parmesan, black pepper, salt, cinnamon, nutmeg, cayenne, and cloves. Mix well, using your hands. Freeze for 30 minutes.

5. Grind the seasoned mixture through the fine disk of the meat grinder.

6. Stuff the mixture into the prepared casing, prick air pockets, and twist off into 8- to 9-inch lengths.

7. Place the sausage on a baking sheet fitted with a wire rack and refrigerate, uncovered, for 1 to 2 days.

8. Cut the links apart. Refrigerate, wrapped well in plastic, for 2 to 3 days, or freeze for up to 3 months; thaw overnight in the refrigerator before using.

9. Poach the sausages as described on page 39 to an internal temperature of 160°F (71°C).

HOT OR SWEET ITALIAN SAUSAGE

MAKES 3 POUNDS

If there's any sausage that's universally adored, it would have to be the classic Italian link. It's redolent with fennel seed and made in both sweet and hot varieties. Either one works well — and interchangeably — in any number of dishes, and using a variety of cooking methods (but panfrying is probably our favorite). When making hot Italian sausage, use more or less crushed red pepper to suit your heat preference (and omit entirely for sweet sausage). This is a good recipe to leave all or part of the sausage as bulk sausage, for sautéing and adding to tomato sauce or a ravioli filling.

- 3 feet medium hog casing
- 2½ pounds boneless lean pork butt or shoulder
- ½ pound pork fat
- 1 tablespoon kosher salt
- 2 teaspoons ground coriander
- 2 teaspoons freshly ground black pepper
- 1 tablespoon crushed red pepper (omit for sweet sausage)
- 2 garlic cloves, minced

1. Prepare the casing (see page 29).

2. Cut the meat and fat into 1-inch cubes. Freeze the cubes for about 30 minutes to firm them up before grinding.

3. Grind the meat and fat together through the coarse disk of a meat grinder.

4. In a large bowl, combine the meat mixture, salt, coriander, black pepper, crushed red pepper (if using), and garlic. Mix well, using your hands.

5. Stuff the mixture into the prepared casing, prick air pockets, and twist off into 3- or 5-inch lengths.

6. Place the sausage on a baking sheet fitted with a wire rack and refrigerate, uncovered, for 1 day.

7. Cut the links apart. Refrigerate, wrapped well in plastic, for 2 to 3 days, or freeze for up to 3 months; thaw overnight in the refrigerator before using.

8. Cook as desired (see methods on pages 38 to 40) to an internal temperature of 160°F (71°C).

TRADE SECRETS

"The key is to use the proper temperature and quality ingredients, and that's true whether you're making 20,000 pounds or 2 pounds. Now people have the opportunity to go to the butcher shop and buy quality pork and beef or other meats, and they can go to the farmers' market and buy fresh herbs and other ingredients for improved flavor. But none of it will work if you're not watching the temperature."

— David Samuels

TAKING ON A LEGEND

For many New Yorkers, Esposito's is a familiar and trusted name in sausage. The company got its start as Giovanni Esposito and Son's (or make that Sons, since there were three), a butcher shop located in what is known as Hell's Kitchen that still stands today. As the demand for their sausages grew, thanks largely to being touted by local chefs and others in the food industry whose opinions mattered, so too did their supply, and they eventually spun off the wholesale business that was run by Armand Esposito (one of the sons). When he passed away, his only offspring — a daughter — was living in California and unable to carry on the family business, and that's where David Samuels, a distant cousin of Armand's son-in-law, enters the picture. "When I first walked into the place, back in 2002, it looked like they were taking these beautiful steaks and making sausage out of them, which in a way they were because they were using whole muscles and none of the scraps and trimmings that were standard at the time. It was just a great opportunity to be part of such a wonderful business."

As such, his main ambition was to simply "do no harm." "We had three goals. The first was to improve efficiency and produce more volume to meet the increasing demand, without changing any of the recipes and especially without affecting the texture, which is such an important quality of sausage," he explains. The second goal was to become a little more modern and to expand the product line, starting with poultry sausages, which were starting to become popular, and since then adding other classic varieties including fresh bratwurst and kielbasa, plus chorizo and Irish bangers as part of their Sausage of the Month Club. "Lastly we had a really great reputation in New York and to some degree outside, but we wanted to give people the opportunity to get it around the country. So now you can find Esposito's sausages in hotels in most major cities, thanks to distributors." But David says they also establish direct relationships with smaller operations, such as family-run inns and bed-and-breakfasts. They also sell directly to customers through their website.

"We produce an average of 60,000 pounds of sausage each week, and it's made fresh and sold fresh the same day, never frozen. That's what a large commercial place can do in a few hours. And we are still using the same recipes and following the same methods that have been used to make Esposito sausages since the beginning, over 80 years ago, only we make a lot more of it."

David Samuels
Esposito's Finest
Quality Sausage
New York, New York

LUGANEGA

This ancient sausage was named by the Romans, who first came across it in Basilicata, in southern Italy, which was then known as Lucania (the sausage has as many spellings as Italy has dialects). Today, the sausage is made throughout Italy, where it is formed into one long coil and sold by the meter (*salsiccia de metro*). Skinned and crumbled, it is considered the quintessential sausage for risotto and is also wonderful incorporated into pasta sauces. *See photo, page 54.*

- 4 feet medium hog casing
- 3½ pounds boneless lean pork butt or shoulder
- ½ pound pork fat
- 1½ teaspoons kosher salt
- 1 teaspoon grated lemon zest
- 1 teaspoon grated orange zest
- 1 teaspoon freshly ground black pepper
- ½ teaspoon ground coriander
- ½ teaspoon freshly grated nutmeg
- 1 garlic clove, minced
- ½ cup dry vermouth
- 1 cup grated Parmesan cheese

1. Prepare the casing (see page 29).

2. Cut the meat and fat into 1-inch cubes. Freeze the cubes for about 30 minutes to firm them up before grinding.

3. Grind the meat and fat together through the fine disk of a meat grinder.

4. In a large bowl, combine the salt, lemon and orange zests, pepper, coriander, nutmeg, garlic, and vermouth. Add the meat mixture and Parmesan; mix well, using your hands.

5. Stuff the mixture into the prepared casing, coiling the sausage into one long spiral. Prick air pockets.

6. Place the sausage on a baking sheet fitted with a wire rack and refrigerate, uncovered, for 1 day.

7. Refrigerate, wrapped well in plastic, for 2 to 3 days, or freeze for up to 3 months; thaw overnight in the refrigerator before using.

8. Cook as desired (see methods on pages 38 to 40) to an internal temperature of 160°F (71°C).

FRESH CHORIZO

5 pounds boneless pork butt or shoulder, about 70% lean

¼ cup kosher salt

¼ cup granulated garlic

1 tablespoon dried oregano, preferably Mexican oregano

1 teaspoon freshly ground black pepper

¼ teaspoon ground cinnamon

¼ teaspoon freshly toasted and ground whole cloves

1 ounce dried ancho and/or pasilla chiles (use one type or a combination)

4 feet medium hog casing

¼ cup red wine vinegar

You can find chorizos both fresh and dried. This fresh version, adapted from one prepared at Sutter Meats (see page 68), is the crumbly (fresh) Mexican kind. Owner Terry Ragasa drew on his Mexican and Filipino heritage to create his signature sausage, with its vibrant blend of pork, spices, and chiles. *See photo, page 248.*

1. Cut the pork into 2-inch cubes. In a large bowl, combine the pork, salt, garlic, oregano, pepper, cinnamon, and cloves. Cover and refrigerate overnight.

2. Meanwhile, make the chile paste by soaking the chiles in hot water. After they soften, remove the stems and as many seeds as desired (the more seeds that remain, the spicier the sausage). Purée the chiles in a blender, adding as much water as needed to make a smooth paste. Portion out ¼ cup of paste and set aside any extra for another use (see Note). Refrigerate until ready to use.

3. Prepare the casing (see page 29).

4. Grind the meat mixture through the medium disk of a meat grinder into a chilled bowl. Mix in the chile paste and vinegar until well blended. Freeze for 30 minutes.

5. Stuff the mixture into the prepared casing, prick air pockets, and twist off into 6-inch lengths.

6. Place the sausage on a baking sheet fitted with a wire rack and refrigerate, uncovered, for 1 day.

7. Cut the links apart. Refrigerate, wrapped well in plastic, for 2 to 3 days, or freeze for up to 2 months; thaw overnight in the refrigerator before using.

8. Cook as desired (see methods on pages 38 to 40) to an internal temperature of 160°F (71°C).

NOTE: *The chile paste is very easy to make and is easily doubled or tripled. This recipe calls for ¼ cup of paste. Any leftover chile paste can be used as a marinade or frozen in an airtight container for future use, for up to 3 months.*

Meet the MAKERS

⚓

Terry and Susan Ragasa

Sutter Meats

Northampton, Massachusetts

LINK TO THE COMMUNITY

When Terry and Susan Ragasa, the husband-and-wife team behind Sutter Meats, were looking for a spot to open their butcher shop in 2013, they looked in cities large and small all around the country before settling on Northampton, Massachusetts. In the heart of what is known as the Pioneer Valley, which abuts Berkshire County, this vibrant college town offered a chance for the Ragasas to be a hub of the community. "This is a highly educated place where people are concerned with sustainable farming, which is essential to our business," says Terry. "We also wanted to be closer to the farming community and to support them by sourcing all of our meats from them. Local in New York City is 150 miles, but local here is 15 to 20 miles." That means they can visit their suppliers on a regular basis and also reach out to new farmers, including Sage Farm, owned by Tyler Sage, who was the head butcher at the shop and who provides about eight heritage hogs each month. (According to Terry, these Hereford-Berkshire crosses have the best fat for the best sausage mouthfeel: "It's like a memory foam pillow that springs back when pushed.")

Indeed, the Pioneer Valley is agriculturally rich, with many small, family-run farms and a strong sense of connection through the steadfast vision and support of CISA (Community Involved in Sustaining Agriculture), a nonprofit group that has supported local farmers for over 20 years and that Terry said was an important element in their decision to plant roots there.

Terry has years of experience in butchery, including an apprenticeship at Fleisher's Grass Fed and Organic Meats in the Hudson Valley of New York, where Susan also later worked after making a career switch. It was there that they both gained an appreciation for the importance of sustainable sourcing in the quality of the meats. Terry then went on to open one of the first whole-animal butcheries in Brooklyn — and that model still drives him today. "We only buy whole animals from local farms who share our standards and practices. We also strive to use the entire animal, out of respect for the care the farmers took in raising it."

So besides selling a much greater variety than the usual cuts of meat, Sutter Meats also fabricates its own fresh sausages (200 pounds a week on average) and uses the rest to make bone broth, dog food, and other products that they keep in their freezer and sell in their shop. Terry says they usually receive five to eight animals a week, and you can find out which farms supplied the different meats by visiting the website. That's all part of "narrowing the gap between the eater and their food source," an essential tenet of the shop.

You can also visit their website each day to see what sausage is available, usually 10 to 12 varieties but always their popular chorizo (see recipe, page 67). They also make a green chorizo with roasted tomatillos, or essentially a salsa verde, mixed in; a goat sausage with Pacific Rim flavors; and "rockin' Moroccan lamb skewers," basically casing-free sausages formed around a stick. They'll use the fattier parts of beef to make their filler-free bologna that's seasoned with mustard and pepper as well as capicola and other whole-muscle meats.

The shop itself is a nod to old-fashioned butcher shops (and a reflection of their commitment to following old-world traditions) with its crisp white-and-black beadboard walls and hand-painted signage. Because they sell a great many obscure cuts of meat not available elsewhere, they organize the meat on display by cooking method. And if you stop by on any workday, you can find Susan and Terry in the shop with their team of skilled butchers, passing their two-year-old daughter back and forth seamlessly (to her great delight) while they scurry around the tight quarters. They joke that she'll probably grow up to be vegan, but if she does she'll be missing out on some of the best sausage around.

TRADE SECRETS

"Don't be afraid of fat and salt. Both ingredients are your friends when it comes to flavorful and well-textured sausage. And even though you're grinding the meat, don't skimp on quality. In the end, the flavor of the pork is the star. Your seasonings are the supporting characters."

— Terry Ragasa

VIETNAMESE PORK AND LEMONGRASS SAUSAGE

MAKES 5 POUNDS

Inspired by the vibrant flavors and fragrances of Vietnamese cooking, these sausages are every bit as delicious as they sound. Tuck them into a crusty French demi-baguette, pile on cilantro and pickled daikon and carrots, and you have the makings of the Far East's contribution to the sandwich world: banh mi. *See photo, page 85.*

1. Prepare the casing (see page 29).

2. Cut the meat and fat into 2-inch cubes. Freeze the cubes for about 30 minutes to firm them up before grinding.

3. Grind the meat and fat together through the medium disk of a meat grinder.

4. In a large bowl, combine the ground meat, scallions, ginger, shallots, lemongrass, sugar, salt, and pepper. Mix well, using your hands. Freeze for 30 minutes.

5. Stuff the mixture into the prepared casing, prick air pockets, and twist off into 3- or 6-inch lengths.

6. Place the sausage on a baking sheet fitted with a wire rack and refrigerate, uncovered, for 1 day.

7. Cut the links apart. Refrigerate, wrapped well in plastic, for 2 to 3 days, or freeze for up to 2 months; thaw overnight in the refrigerator before using.

8. Cook as desired (see methods on pages 38 to 40) to an internal temperature of 160°F (71°C).

- 4 feet medium hog casing
- 3½ pounds boneless lean pork butt or shoulder
- 1½ pounds pork fat
- ¼ cup minced scallions
- ¼ cup minced peeled fresh ginger
- ¼ cup minced shallots
- 1 tablespoon finely minced lemongrass (see box)
- 2 tablespoons brown sugar
- 1 tablespoon kosher salt
- 2½ teaspoons freshly ground white pepper

LEMONGRASS

A staple of Vietnamese, Thai, and Indian cooking, lemongrass adds a citrusy note to marinades, curry pastes, soups, stir-fries, and more. You can find lemongrass at Asian grocers, natural food stores, and some supermarkets. Avoid any stalks that are browning or appear dry. To use lemongrass in cooking, trim both ends of the stalk, then peel away the tough outer leaves until you see the pale, soft flesh. The bottom, thicker part is for eating, while the top, more fibrous part can be used to infuse sauces and broths. Thinly slice the bottom 4 inches and mince as you would shallots or garlic. To use the tops, make a few slits, then bend a few times to bruise them and release the aromatic oils.

2½ feet medium hog casing

2¾ pounds boneless lean pork butt or shoulder

¼ pound pork fat

1 cup apple cider

1 tablespoon canola oil

2 small leeks, cleaned and chopped (white parts only)

1 Granny Smith apple, peeled, cored, and chopped

1 tablespoon kosher salt

½ teaspoon freshly ground black pepper

½ teaspoon grated lemon zest

2 tablespoons chopped fresh flat-leaf parsley

2 tablespoons chopped fresh rosemary

<< FRESH >>

PORK AND APPLE SAUSAGE

There's pork chops and applesauce, and then there's this sausage, which is built on the famous pairing. Sautéed apple and leeks add depth of flavor, as does apple cider that's been reduced to a syrup. A Granny Smith apple works best, keeping the sweetness factor in check (aided by the lemon zest and fresh herbs). Roast some potatoes, wilt some spinach or kale, and dinner is served. *See photo, page 73.*

1. Prepare the casing (see page 29).

2. Cut the meat and fat into 1-inch cubes. Freeze the cubes for about 30 minutes to firm them up before grinding.

3. While the meat chills, simmer the cider in a saucepan, uncovered, until it reduces to ¼ cup of syrupy liquid, about 20 minutes. Remove from the heat; set aside.

4. Meanwhile, heat the oil in a medium skillet over medium heat. Add the leeks and apple; sauté until the apple is golden, 3 to 5 minutes.

5. Grind the meat and fat through the fine disk of a meat grinder.

6. In a large bowl, combine the meat and fat with the salt, pepper, lemon zest, parsley, and rosemary. Mix well, using your hands. Freeze for 30 minutes.

7. Grind the seasoned mixture through the fine disk of the meat grinder. Add the leeks and apple and the cooled reduced cider. Mix well, using your hands.

8. Stuff the mixture into the prepared casing, prick air pockets, and twist off into 4-inch lengths.

9. Place the sausage on a baking sheet fitted with a wire rack and refrigerate, uncovered, for a few hours or preferably overnight.

10. Cut the links apart. Refrigerate, wrapped well in plastic, for 2 to 3 days, or freeze for up to 3 months; thaw overnight in the refrigerator before using.

11. Cook as desired (see methods on pages 38 to 40) to an internal temperature of 160°F (71°C).

PERFECT FOR PICNICS

Spanish Tortilla, page 278

Panzanella
with Salami,
page 306

Genoa Salami,
page 161

All-Beef Summer
Sausage, page 137

Pork and Apple
Sausage, page 71

Sausage and Beer
Hand Pies, page
319

Linguiça, page 107

Kevin Ouzts
The Spotted Trotter
Atlanta, Georgia

DEFYING THE ODDS

Proof positive that size doesn't matter: In a space that's less than 2,000 square feet, Kevin Ouzts and his team of butchers at The Spotted Trotter manage to produce anywhere from 1,800 to 2,000 pounds of sausage a week, including whole-muscle items like coppa and prosciutto, all starting with whole pigs they butcher on site. "I've had the same crew for the past four years, and it's mind-numbing to see what these people manage to do in this space. That they've always been able to get this amount of product out there is pretty outstanding. The USDA comes in and sees what we do and tells us we're one of the cleanest spots they've ever seen."

The tight quarters aren't the only obstacle facing Kevin and his dedicated team — they also contend with the particularly steamy climate in Atlanta. No surprise, then, that The Spotted Trotter is the first charcuterie to open its doors in Georgia. "There have been people who made sausages, but no one had ever thought of producing salami here. It's so hot and sticky, and we made a lot of mistakes along the way. If it rained for 10 days straight, we had to figure out how to deal with that. Creating a safe and controlled environment in a more humid and hot climate is something we had to overcome, and once we figured it out we've had to work hard to maintain that."

Another considerable hurdle was overcoming the hesitancy of the South to adopt new food trends, Kevin says. "It's one thing to have an audience in the downtown area, but outside of that you start discovering how many people don't even know what salami and pâtés and terrines are, which I just took to mean that we have an opportunity to bring something new to the table." Plus he says Southerners have been reluctant to embrace their culinary heritage until recently. "Now we try and honor the rice and grains and other foods that were brought here during the antebellum era."

In that spirit, he imbues sausages with flavors that are indicative of the South — smoking the meats and sausages over pecan wood, and making dried chile powder out of local peppers. "We had some farmers grow Espelette for us here, and now we supply that to other restaurants. We use lots of Espelette and garlic in our most popular fresh sausage, a rabbit and pork boudin that's spiked with brandy and Madeira." The signature salami is the toasted black pepper and sorghum, which turned out to have another redeeming quality. "What I discovered over a lot of trial and error is that sorghum, a cereal grain that's rich in flora and fauna, allows the salami to reach the proper pH level much faster than with the chemically derived dextrose that is often used in making salami."

The Spotted Trotter turns out an impressive array of classic fresh, dry-cured, and smoked Italian, French, and other old-world offerings, as well as pâtés, terrines, jerky, scrapple, lard, and much more. As if that weren't enough, they also offer one or two crepinettes a week. "I learned how to make crepinettes from Taylor Boetticher at The Fatted Calf in Napa Valley. You can take any kind of forcemeat, wrap it up in these beautiful packages, and roast them over hot coals or fry them up in a skillet. People go crazy for them here and we sell out of them quickly. The caul fat has a really unique texture to it and the flavor is fantastic." They have a rotating lineup, including a couple that feature local crops: One crepinette has local salt-roasted peaches and the other bright coral-colored mayhaw berries that are cooked down to a jam. (See their recipe for another version on page 76.)

So how did Kevin ever decide to defy all these odds and open a charcuterie in Atlanta? It was while he was doing a six-month stint at The French Laundry in Napa, making up for lost time after switching careers. While there, he would drive by "this interesting charcuterie called The Fatted Calf" on his drive to and from work. Courage gathered, he walked in one day and said he wanted to learn the craft, and they gladly accepted. Soon enough Kevin was calling his wife with his plan: "'I know what we're going to do, we're going to open a charcuterie in Atlanta!' And she kind of giggled and said, 'What the hell is that?'"

As Kevin tells it, the idea of borrowing money back in 2008 to open a restaurant just wasn't happening. But he planned to follow Taylor's lead and start small, selling at the local farmers' markets — two of which had just opened in Atlanta — and he was convinced it was worth a go. When he got back to Atlanta, he bought a grinder and a stuffer, rented a 10-square-foot space out of a small café in Decatur, and began making about 1,000 pounds of sausage each week. "It went off like wildfire."

Eventually, thanks to some small business loans, and despite it being at the height of the economic downturn, he was able to expand to the space they are in today. He started selling to his many contacts in local restaurants, and eventually to wholesale clients in 17 states. "The idea that I can come to work every day and do something no one has ever done before in the state of Georgia, and for The Spotted Trotter to still be doing what it's doing to this day, is pretty exciting for me and it keeps me going." So much so that he's opened a second location, where he can proudly supply the growing demand.

TRADE SECRETS

"At the risk of sounding totally cliché: Have fun. Whether you're doing it with your kid or brother or best friends, make the whole process a journey — going to the market to pick out the ingredients, visiting the farm or the butcher to buy your proteins, getting the casings. It's not something you can do in a couple quick hours; you have to think it through, so make it enjoyable. There are so many ways to go about it, but at the end of the day it should be fun. I've got a group of guys who regularly come in and buy a whole animal, and it's a social thing. It's great to see the camaraderie and how they treat it as a learning experience. If they make a mistake, they're always really cool about it, because of what they've learned. That's the right approach to take."

— Kevin Ouzts

SOUTHERN SERRANO AND PORK CREPINETTES

Essentially sausage patties that are wrapped in a layer of lacy caul fat, crepinettes — the name means "caul fat" in French — are a wonderful way to ease into sausage making for the uninitiated, since you don't have to contend with stuffing and yet the webbed caul fat presents a more polished parcel than a regular patty would. Crepinettes were historically coated with melted butter and then dredged in bread crumbs before sautéing, but Kevin Ouzts, who contributed this recipe based on one of his offerings at The Spotted Trotter (see page 74), prefers to cook them as is. "The caul fat has a really unique texture to it and the flavor is fantastic." In this version, caramelized onions add depth of flavor and wonderful sweetness to a mixture of coarsely ground pork, copious fresh herbs, and two kinds of ground dried chile pepper — Aleppo and serrano, found at many supermarkets or from specialty stores.

- 8 pounds boneless lean pork shoulder
- 3 yellow onions
- 2 tablespoons vegetable oil
- 4 garlic cloves, minced
- ¼ cup plus 1 tablespoon kosher salt
- 2 bunches flat-leaf parsley, stems discarded, leaves chopped
- 1 bunch chives, minced
- ¾ cup sorghum syrup
- 3 tablespoons dry sherry
- 2 tablespoons sherry vinegar
- Finely grated zest of 1 orange
- 2 tablespoons dried savory
- 2 tablespoons ground Aleppo pepper
- 1 tablespoon dried ground serrano pepper
- 1 pound caul fat

1. Cut the pork into 1-inch cubes and freeze for about 30 minutes to firm them up before grinding.

2. Meanwhile, cut the onions in half through the root end and then cut into thin slices. Heat the oil in a large skillet over medium heat and cook the onions over low heat, stirring occasionally, until nicely caramelized, about 40 minutes. Let cool.

3. While the onions are cooking, smash the garlic cloves on a cutting board or with a mortar and pestle, then sprinkle with some of the salt and mash to a paste. You should have about 2 tablespoons paste.

4. Grind the pork through the coarse disk of a meat grinder into a large bowl. Add the onions, remaining salt, parsley, chives, garlic paste, sorghum syrup, sherry, sherry vinegar, orange zest, dried savory, and ground peppers. Mix well until all the ingredients are incorporated.

5. Using a ¼-cup measuring cup, form the mixture into pucklike patties.

6. Wrap each patty with one layer of caul fat, trimming the fat as needed. Place on a baking sheet fitted with a wire rack and refrigerate, uncovered, overnight.

7. Panfry or grill as described on pages 38 to 40.

PORK AND PISTACHIO SAUSAGE

MAKES 4 POUNDS

5 feet medium hog casing

3½ pounds boneless lean pork butt or shoulder

½ pound pork fat

1 tablespoon freshly ground black pepper

1 tablespoon kosher salt

1 teaspoon freshly grated nutmeg

1 tablespoon chopped fresh rosemary or 1 teaspoon dried

½ cup chopped, toasted unsalted pistachios

Pistachios (and other nuts) are a classic addition to sausages, adding textural contrast as well as taste. Toasting them first deepens their flavor. Fresh rosemary is worth using in this recipe, rather than dried. Fry or grill the links to serve.

1. Prepare the casing (see page 29).

2. Cut the meat and fat into 1-inch cubes. Freeze the cubes for about 30 minutes to firm them up before grinding.

3. Grind the meat and fat together through the fine disk of a meat grinder.

4. In a large bowl, combine the meat mixture, pepper, salt, nutmeg, and rosemary. Mix well, using your hands. Freeze for 30 minutes.

5. Grind the seasoned mixture through the fine disk of the meat grinder. Add the pistachios and mix well, using your hands.

6. Stuff the mixture into the prepared casing, prick air pockets, and twist off into 4-inch lengths.

7. Place the sausage on a baking sheet fitted with a wire rack and refrigerate, uncovered, for a few hours or preferably overnight.

8. Cut the links apart. Refrigerate, wrapped well in plastic, for 2 to 3 days, or freeze for up to 3 months; thaw overnight in the refrigerator before using.

9. Cook as desired (see methods on pages 38 to 40) to an internal temperature of 160°F (71°C).

TOASTING NUTS

To toast nuts, spread in a single layer on a rimmed baking sheet and heat in a 350°F (175°C) oven, tossing occasionally, for 10 to 15 minutes, until they are slightly darkened and fragrant. You can also toast nuts (especially smaller ones like pine nuts) in a dry skillet over medium heat; stir frequently to prevent burning.

Jonah Shaw

*Catskill Food
Company*

Delhi, New York

RETURNING TO THE ROOST

Having grown up in the Catskills of upstate New York, where family farms and home-grown food are the norm, Jonah Shaw has been immersed in the food world for his entire life. When he was four years old, his mom founded the first whole-foods co-op in the area, in Delhi (aptly called Good Cheap Food), a place that steadily became a trusted source for locals and travelers looking for local products — and, starting about five years ago, some of Shaw's handcrafted sausages. After graduating from college and a brief stint as a graphic designer in New York City, Shaw fell back on his love of cooking and gained experience running several restaurants in Philadelphia before feeling the pull to move back to his hometown, where he opened the Quarter Moon Café. "The store and the restaurant share the same building, so they are connected."

Meanwhile, while opening a restaurant for a local brewery, he took an immersion course on charcuterie at The Culinary Institute of America, just down the road, to brush up on his sausage-making skills. "I was fairly new to sausage making. I had done some basic stuff, having worked in restaurants, but I'd never had any real instruction before." He developed a couple of beer-based sausages for the brewery and then started selling them in the co-op, "and from there people started asking about making them available in other places, and that's when I looked into selling them to other restaurants and also into wholesale opportunities." Today, some five years later, he sells 20 different sausages on the company's website (see Resources, page 355).

Shaw says the garlic-beer and the maple-date breakfast sausages are probably the most popular, followed by the Moroccan spice. He offers two cold-smoked varieties: andouille and kielbasa. "We use old warming cabinets, basically insulated metal boxes that you see in dining halls, to start the wood chips, and then the meats bask in the smoke. The finished product is still raw. I would need more heating elements in the smoker to make it hot enough to cook the product fully through, and I haven't invested in that yet."

He explains his process for developing new sausages this way: "I look at what flavor combinations work well and do a mash-up of the whole thing." Take, for instance, the tzatziki sausage, which folds the components of the Greek dip right into the pork mixture; same for the Chinese dumpling sausage, with the typical dipping sauce ingredients included as well. He also experiments periodically with more seasonal ingredients, like for the ramp and caraway sausage (see recipe, page 80), where the pigs used in the sausage foraged on the same supply of ramps. And sometimes you have happy accidents: "We were making an andouille sausage and the cayenne pepper turned out to be completely flat, with no heat at all. So instead of tossing it out, we called it smoked summer sausage and it was a huge hit. Unfortunately, we kept getting requests for it, but we had no idea how to make it again. This is why you must document everything you do so you can replicate it."

Shaw mainly uses pork and beef that can be sourced from local farms. "We have used some poultry and lamb, but working with local sources that becomes a bit cost prohibitive, because the raw ingredients are so expensive." He says that he prefers to make sausage that is leaner than most, or "about 80 percent lean, as opposed to a lot of commercial sausage, which is 65 to 70 percent lean. As long as you don't overcook it, the sausage stays plenty moist and you don't get the same shrinkage that you do when there's more fat. Plus, it's a healthy alternative to traditional sausage."

Besides ordering online, or driving to Delhi, you can find a selection of his sausages (and smoked bacon) at other shops throughout upstate New York and in New York City. "There are distributors that want to pick up our product and push it. I'm ready to go. I just need to have someone out there willing to buy it." Calling all takers!

TRADE SECRETS

"Try to experiment. Don't be afraid to try to your own combinations, and don't be beholden to a recipe. Keep certain ratios in mind like proper salt content, but other than that, try things that you think are going to taste good and fiddle around with that until it does."

— Jonah Shaw

Chinese Dumpling Sausage

Tzatziki and Sausage

RAMP AND CARAWAY PORK SAUSAGE

In the spring, when ramp season is in full swing, Jonah Shaw finds all sorts of ways to showcase the allium's garlicky flavors, including this sausage, which he makes for his company, Catskill Food Company (see page 78). "I use pigs who foraged on the very same field where the ramps were harvested." Since ramps are only available for a short time, this recipe is a good way to use up a surplus, as are pickled ramps (see opposite) for serving alongside. Ramps should be prepped similarly to leeks (which is what they are, only in the wild): First swish them around in a few changes of cold water to loosen all the grit, then pat dry, trim off woody ends, and slice or dice as desired (or as called for in a recipe). Ramps can also be roasted, sautéed, or used as most any other alliums.

1. Prepare the casing (see page 29).

2. In a small dry skillet, toast the caraway seeds until fragrant. Let cool.

3. Mix the caraway, salt, dry mustard, allspice, and pepper together. Stir in the vinegar and wine to make a wet paste. Mix the paste together with the pork, using your hands. Freeze for 30 minutes.

4. Grind the pork mixture through the medium disk of a meat grinder. Freeze for 30 minutes.

5. Grind a second time through the medium disk. Mix in the ramps and cheese. Freeze for 30 minutes.

6. Stuff the mixture into the prepared casing, prick air pockets, and twist off into 3-inch lengths.

7. Place the sausage on a baking sheet fitted with a wire rack and refrigerate, uncovered, for a few hours or preferably overnight.

8. Cut the links apart. Refrigerate, wrapped well in plastic, for 2 to 3 days, or freeze for up to 2 months; thaw overnight in the refrigerator before using.

9. Cook as desired (see methods on pages 38 to 40) to an internal temperature of 160°F (71°C).

Ingredients

- 4 feet medium hog casing
- ¼ cup caraway seeds
- 2½ tablespoons kosher salt
- 2 teaspoons dry mustard
- 1½ teaspoons ground allspice
- 1½ teaspoons freshly ground black pepper
- 1½ tablespoons apple cider vinegar
- ¼ cup white wine
- 5 pounds boneless pork butt or shoulder, about 70% lean, cut into 2-inch cubes
- ¾ pound ramps, trimmed and finely chopped (see headnote)
- 5 ounces Gruyère cheese, grated

PICKLED RAMPS

With such a brief season, it would be a shame to waste one single tendril of this flavorful allium. Hence, pickling:

1. In a pot, mix 1 cup water with 1 cup vinegar (apple cider, white wine, or plain white vinegar), ½ cup sugar, 1 tablespoon kosher salt, 1 teaspoon each coriander seeds, mustard seeds, and fennel seeds, ½ teaspoon whole black peppercorns, and 1 bay leaf (fresh or dried). Bring to a boil.

2. Add 2 pounds ramps, trimmed (save the dark green parts for another use, like making pesto), and return to a boil. Reduce the heat and simmer until the ramps are tender, about 5 minutes. Remove from the heat and let cool completely in the liquid.

3. If not serving immediately, refrigerate the ramps in their liquid, in a tightly sealed container, for about 1 month. By the way, this makes an excellent addition to any charcuterie board.

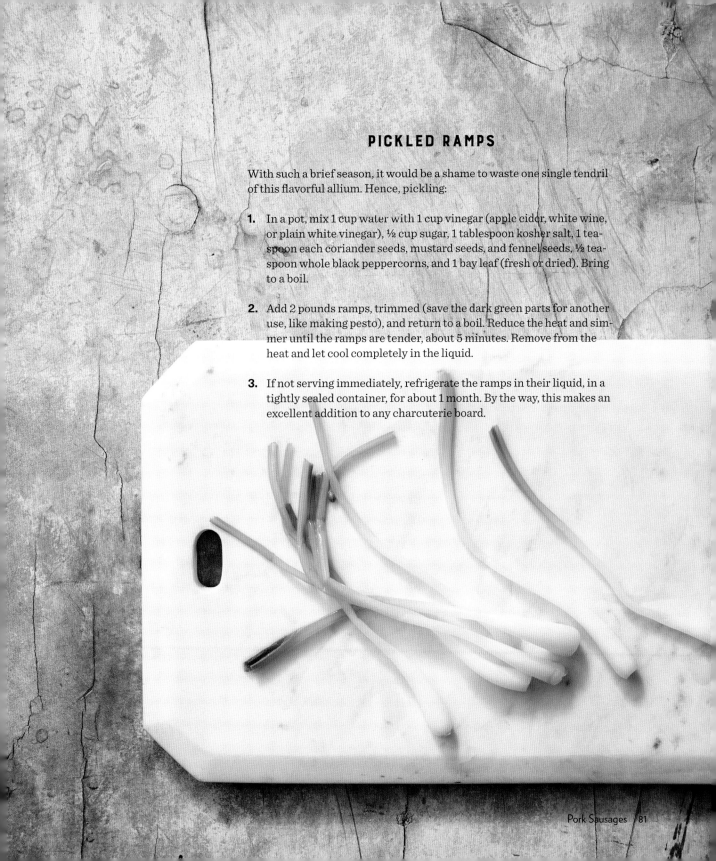

CURRY SAUSAGE
WITH CILANTRO AND LIME

MAKES 5 POUNDS

The vibrant flavors of red curry paste, which can be found in specialty food shops and many supermarkets, were the inspiration for this sausage. It's highly seasoned but not too spicy; adjust the amount of chile peppers (using more or less) to suit your taste.

1. Prepare the casing (see page 29).

2. Cut the meat into 2-inch cubes and the fat into 1-inch cubes. Freeze the cubes for about 30 minutes to firm them up before grinding.

3. Grind the meat and fat together through the medium disk of a meat grinder.

4. In a large bowl, mix the curry paste, cilantro, garlic, fish sauce, chiles, and lime zest and juice together and combine well with the ground meat mixture. Freeze for 30 minutes.

5. Stuff the mixture into the prepared casing, prick air pockets, and twist off into 3-inch lengths.

6. Place the sausage on a baking sheet fitted with a wire rack and refrigerate, uncovered, for at least a few hours or preferably overnight.

7. Cut the links apart. Refrigerate, wrapped well in plastic, for 2 to 3 days, or freeze for up to 2 months; thaw overnight in the refrigerator before using.

8. Cook as desired (see methods on pages 38 to 40) to an internal temperature of 160°F (71°C).

- 4 feet medium hog casing
- 3½ pounds boneless lean pork butt or shoulder
- 1½ pounds pork fat
- ¼ cup red curry paste
- ½ cup chopped fresh cilantro
- 10 garlic cloves, minced (about ¼ cup)
- ½ cup Asian fish sauce
- 2 teaspoons minced Thai or bird's eye chiles
- Zest and juice of 3 limes

SAI KROK ISAN (THAI SOUR SAUSAGE)

MAKES 4 POUNDS

4 feet medium hog casing

3½ pounds boneless lean pork butt or shoulder

½ pound pork fat

4 cups cooked white rice

½ cup minced garlic

1 tablespoon kosher salt

2 teaspoons brown sugar

18 grams (about 2 tablespoons) Bactoferm LHP, dissolved in ¼ cup water

Insta Cure #1 (use supplier's recommended quantity for 4 pounds of meat; see Notes)

½ teaspoon freshly ground white pepper

Although there are many versions of this traditional sausage, they all share one common characteristic: a distinctly funky, sour flavor. Most recipes, including this one (inspired by one that Bob del Grasso posted on his blog, *A Hunger Artist*), achieve this with a fermentation culture called Bactoferm LHP. When combined with curing salt, this culture encourages the proper results. *See photo, page 84.*

1. Prepare the casing (see page 29).

2. Cut the meat and fat into 1-inch cubes. Freeze for 30 minutes.

3. Grind the meat and fat through the medium disk of a meat grinder.

4. Using a food processor or food mill, grind the cooked rice to make a rough paste.

5. In a large bowl, combine the ground meat mixture, rice, garlic, salt, sugar, Bactoferm, curing salt, and pepper. Mix well, using your hands.

6. Stuff the mixture into the prepared casing, prick air pockets, and twist off into 2-inch balls or 3-inch lengths.

7. Hang the sausage in a cool area for 2 days to allow the sausages to ferment. Alternatively, place the sausage on a baking sheet fitted with a wire rack and refrigerate, uncovered, for 2 days, turning frequently.

8. If making 2-inch sausage balls, do not cut the small sausages apart. If making 3-inch links, cut the links apart. Refrigerate, wrapped well in plastic, for 2 to 3 days, or freeze for up to 3 months; thaw overnight in the refrigerator before using.

9. Cook as desired (see methods on pages 38 to 40) to an internal temperature of 160°F (71°C).

NOTES: *We recommend that you use a commercial premixed cure in any recipe for cured sausage. Premixed cures replace the saltpeter in older recipes.*

If not using Bactoferm LHP, omit it and the curing salt and add 2 tablespoons lime juice instead. Skip the fermentation step. Place the sausage on a baking sheet fitted with a wire rack and refrigerate, uncovered, for 1 day, turning frequently to allow for even drying. Refrigerate and use within 3 days, or freeze for up to 2 months. Cook as directed above.

ASIAN FLAVORS

Korean Lettuce
Wraps, page 314

Sai Krok Isan,
page 83

Lap Cheong, page 102

Ginger-Scallion
Seafood Sausage,
page 233

Thai Chicken
Sausage, page 219

Vietnamese Pork and
Lemongrass Sausage,
page 70

Meet the MAKERS

⚜

Herb and Kathy Eckhouse

La Quercia

Norwalk, Iowa

THE MEAT OF THE MATTER

The usual questions anyone asks when first hearing the name La Quercia is (a) how to pronounce it (easy: *la KWAIR-cha*), and then (b) what does it mean ("the oak" in Italian), followed by (c) why did the owners choose it for their company (read on to find out!)?

The owners, by the way, are Herb and Kathy Eckhouse. Together they have established what many consider to be the finest *prosciuttificio* in the United States that heeds religiously to the tenets and techniques of Parma, Italy, where the couple lived for over three years in the late 1980s, when Herb was relocated there for work — and where the oak is a symbol of the province (clue number one), and where acorns (from the oak trees) foraged by the pigs have been touted for giving the regional prosciutto its particularly prized flavor (clue number two). "When we moved back from Italy to Iowa, we were inspired by the beauty and bounty to try to create something delicious to show our appreciation," says Herb. Serendipity struck, as well: The oak tree is the state tree of Iowa (clue number three). Thus explains the name: "It unites Iowa, Parma, and prosciutto and is a symbol of persistence, integrity, and beauty — all values which guide us."

Another question the Eckhouses often get asked is, Why prosciutto? "We had barely eaten prosciutto before we lived in Parma, since it was not possible to import it at that time to the United States. But, while in Italy, we ate salumi and prosciutto several times a week, and the idea occurred to us when we moved back to Iowa that the opportunity looked good for producing our own." Plus, they both had extensive experience in agriculture — Kathy as a ranch hand and then a researcher, Herb in a variety of roles over the span of 30 years, including working cattle and developing commodity crops and vegetables. Ever the curious tinkerer (and perfectionist), Herb had also been teaching himself to make prosciutto the old-fashioned way at home for over 5 years before they even planted the seed for their business. "I love making prosciutto," he says. "It's like witnessing a miracle."

Besides creating food that can offer a memorable eating experience (their Speck Americano won the cherished Good Food Award in 2016, and is beloved by many of the leading chefs on the scene), the couple is committed to being part of what Herb describes as a "responsible food system," which he says is "making choices where they make things better." That includes purchasing only meat that is free from antibiotics, and never from farms where the animals are confined, which contribute to the air and water pollution in Iowa (an especially big problem there, according to Herb). But it goes beyond the meat itself; they also make responsible choices in their building materials and design, down to their refrigerant choice, and most important in the choices regarding their employees. "We offer a competitive vacation plan, sick days, health insurance, gain-sharing pay, and maternity leave. It's not a sustainable system if your employees can't live decent lives. We all get focused on the animals, but the people are important, too!"

When asked how they manage to maintain such a consistently excellent quality across all products, Herb explains: "It takes relentless and obsessive attention to detail, which is not always easy to live with." Not that anyone is complaining. "We've had a great reception from those at the top of the culinary pyramid, plus we hear from others all the time that they decided to mark the special occasions in their lives by eating our meats." And that, according to Herb, has been the most gratifying — and unexpected — part of the La Quercia journey.

- Start with making fresh sausage according to a recipe.

- Take careful notes.

- Change recipes according to your tastes.

- If you make cured meats, learn something about meat science and microbiology so you know what you're doing.

- Use cured meat recipes as a guide — you have to adapt them to your equipment and facilities.

- Meat curing is a very old craft and a robust process, so if you are careful, you can make something edible and delicious.

— Herb Eckhouse

'NDUJA AMERICANA

This recipe was provided by Herb Eckhouse of La Quercia (see page 86). Who better to describe this 'nduja than Herb himself: "The 'spirit of 'nduja' is making something good to eat from so-called poor cuts: those parts of the pig typically left over and considered not very desirable. In Calabria, when they were occupied by the French, they took these poor cuts and combined them with Calabrian peppers to make 'nduja. This is the spirit of our 'nduja, too, except that in our case we use the ends and miscuts of cured meat that's used for slicing, plus the trimmed fat from the prosciutto, all things that were looking for a good use. Our plant manager had the idea of making them into a form of 'nduja. After testing a handful of recipes, we settled on one we liked and then continued to work on it. Now we use three types of New Mexico chiles, not Calabrian, to make our 'nduja Americana." *See photo, page 151.*

- 1¼ pounds meat from lean cured meats, such as prosciutto or speck, preferably from La Quercia
- ¾ pound cured fat, such as trim from prosciutto or speck, or lardo
- ¾–1½ cups (2 to 4 ounces) crushed dried chiles

1. Cut the cured meat and fat into 1-inch cubes. Freeze for about 30 minutes to firm them up before grinding.

2. Grind the meat and fat through the medium disk of a meat grinder.

3. Mix in the chiles well (use more or less depending on your taste), using your hands.

4. Pack the 'nduja into an airtight container to protect from oxidation and refrigerate for 2 to 3 months. Bring to room temperature before serving.

DRIED CHILES

Chiles can vary greatly in their spiciness, and dried chiles will always be hotter than fresh and have a richer, more concentrated flavor. Mild chiles include ancho, California, cascabel, and poblano. Hotter ones include cayenne, pequin, tepin, and chile de Arbol. New Mexico and quajillo chiles land somewhere in between.

BOEREWORS

MAKES 5 POUNDS

- 4 feet pork casing
- 4 pounds boneless lean pork butt or shoulder
- 1 pound pork fat
- 1 tablespoon kosher salt
- 1 tablespoon sugar
- 1 tablespoon freshly ground black pepper
- 3 tablespoons coriander seeds, toasted and ground
- ½ teaspoon ground cloves
- ½ teaspoon freshly grated nutmeg
- ½ cup red wine vinegar or apple cider vinegar

A specialty of South Africa, boerewors (*boer* means "farmer" and *wors* is "sausage" in Afrikaans), or "boeries" as it is affectionately known, is a long, coil-shaped sausage that is the pride of the locals. While you can find the sausage throughout the year, *braaiing* boerewors — basically having a big cookout where you grill the coils of sausage — is especially popular during national holidays, and there's an annual braai celebration in September. Feel free to break with tradition and form individual links or smaller spirals if desired, but the large spiral will make quite a spectacle at your own backyard cookout, too.

1. Prepare the casing (see page 29).

2. Cut the meat and fat into 1-inch cubes.

3. In a large bowl, combine the meat and fat with the salt, sugar, pepper, coriander, cloves, nutmeg, and vinegar. Mix well, using your hands. Freeze for 30 minutes.

4. Grind the meat mixture through the medium disk of a meat grinder. Freeze for 30 minutes.

5. Stuff the mixture into the prepared casing, coiling the sausage into one long spiral. Prick air pockets.

6. Refrigerate, wrapped well in plastic, for up to 3 days, or freeze for up to 2 months; thaw overnight in the refrigerator before using.

7. Grill as directed on page 40 to an internal temperature of 160°F (71°C).

MORTADELLA

Studded with pork fat and pistachios, mortadella is a classic part of the Italian family of salumi. The recipe originated in Bologna, Italy, and provided a foundation for American "baloney," losing all trace of connection to its city of origin. Thin slices of mortadella are fantastic in sandwiches, of course, but for a real treat, cut it into 1-inch cubes and panfry it until crisp. Enjoy with spicy mustard, or sprinkle into salads — a little goes a long way toward adding richness.

Mortadella is traditionally stuffed in a beef middle (or bung). However, these casings can be expensive and hard to find outside of mail-order sources. In this recipe, the mortadella is poached in plastic-wrap casings, yielding three large round mortadella balls. If you have access to beef middles, use one large one instead of the plastic wrap. *See photo, page 150.*

1. Cut the meat into 1-inch cubes. Freeze for 30 minutes.

2. Freeze ¾ pound of the pork fat.

3. Bring a small pot of water to a boil, and blanch the remaining ¼ pound pork fat for 1 minute. Drain and set aside to cool.

4. Toss the meat with the vermouth, garlic, salt, and curing salt. Grind the meat mixture through the medium disk of a meat grinder into a metal bowl set into another bowl filled with ice (called an ice bath). Set aside in the refrigerator (reserve the ice bath).

5. Grind the frozen pork fat through the medium disk of the meat grinder into a metal bowl set in the ice bath. Set aside in the refrigerator.

6. Place the ground meat mixture, crushed ice, white pepper, paprika, nutmeg, bay leaf, cloves, and mace in a food processor. Process for 2 minutes. Check the temperature; it should be about 40°F (4°C). Process a little more if it's cooler than that.

7. Add the ground fat to the food processor. Blend until the mixture reaches 45°F (7°C), about 3 minutes.

- 1 pound boneless lean pork butt or shoulder
- 1 pound pork fat, diced
- 2 tablespoons dry vermouth or white wine
- 1 garlic clove, minced
- 1½ tablespoons kosher salt
- Insta Cure #1 (use supplier's recommended quantity for 2 pounds of meat; see Note)
- 10 ounces (about 2 cups) crushed ice
- 1 teaspoon freshly ground white pepper
- ½ teaspoon paprika
- ½ teaspoon freshly grated nutmeg
- ¼ teaspoon ground dried bay leaf
- Pinch of ground cloves
- Pinch of ground mace
- ½ cup nonfat dry milk powder
- ¾ cup pistachios
- 2 tablespoons coarsely crushed black peppercorns

8. Add the dry milk to the food processor. Blend until the mixture reaches 58°F (14°C), about 3 minutes. The mixture should look like fluffy pink mashed potatoes.

9. Scrape the mixture into a large bowl set into another ice bath. Fold in the blanched pork fat, pistachios, and black pepper.

10. Lay a 12-inch-long sheet of plastic wrap on your work surface and lay a second sheet crosswise across it, forming an *X*. Scoop one-third of the mortadella mixture into the center of the *X* and gather the plastic wrap up into a pouch. Using butcher's twine, cinch the pouch tightly and tie it shut. Set aside in the refrigerator. Repeat the process two more times with the remaining mortadella mixture. Store in the refrigerator while you prepare the poaching water.

11. Bring a large pot of water to a simmer. Drop the pouches into the water, and reduce the heat until the water is barely simmering; do not let it boil. Poach until the mortadella reaches 160°F (71°C) in the center, 15 to 20 minutes. Remove from the water and cool completely in an ice bath before storing in the refrigerator.

NOTE: *We recommend that you use a commercial premixed cure in any recipe for cured sausage. Premixed cures replace the saltpeter in older recipes.*

PRESERVATION SOCIETY
OLD STURBRIDGE VILLAGE

The Old Sturbridge Village (OSV), in central Massachusetts, is a beacon for anyone with an interest in how people lived in 1830s New England, drawing more than 250,000 visitors each year. This living museum is more than just a snapshot of that time and place — it's a viable working farm where there are heritage breeds of pigs, sheep, chicken, oxen, and cows being raised, harvested, and processed just as they were then (and in the way that many small farmers are committed to following today).

According to Debra Friedman, senior vice president, "In the past decade, there's been much more interest in historic foods and preservation. For chefs, it's a way for them to create something unique, which

is kind of ironic because they're just adopting what's been done for thousands of years." There has also been a resurgence of home canning and fermenting and other types of preservation methods that she says were appropriate in the early nineteenth century. "Things like cheese making, bread making, and sausage making are experiencing a huge comeback. I find it amazing that all this interest in local foods and the prevalence of CSAs (community supported agriculture) has spurred an interest in the preservation of foods."

Visitors to OSV have an opportunity to see the same tools, ingredients, and recipes that a New England farm wife of the 1830s would have used, and that includes preparing sausages in November and December, when the farm's pigs are slaughtered (humanely, and in the morning before the village is open). As Debra explains, "Here in New England you butcher in the fall, once it's cooler and the flies are gone; you also don't want to have to feed the animals during the winter. Historically, you would start with beef in late October, then go to mutton and lastly your pigs, most of which would be used to feed the family for the entire year." Basically, each pig would supply about 250 pounds of pork, including all the by-products, per person per year.

"For New Englanders, the basement was built for food preservation of all different types. The basement is not to keep food cold, but to keep it warm, because for six months out of the year it could be below freezing outside," she explains. Most pork was left to brine in large barrels in the cellar for about six weeks before it was smoked and then able to be preserved indefinitely. "When you go to Spain and Italy and see whole hams hanging, that's what you would find in New England as well."

The British had been making sausages for centuries before they brought that tradition over to New England. Blood pudding is the first recorded sausage, and practically every culture has its own version (morcilla in Spain, kiszka in Poland, sundae in North and South Korea . . .), most likely because it makes use of every part of the animal (except the meat, which goes into other sausages). "It is also the first type of sausage that's made after that animal is slaughtered, when the animal is bled before butchering. To make the sausage, they would add some of the pig fat to the blood along with a thickener — cornmeal for a New Englander — along with ample spices, and often some egg and milk. This mixture is then stuffed into the freshly scraped and cleaned intestines from the same pig. After that the sausages are simmered long enough to solidify and then hung to dry out, where they would keep until you boiled them again before using," explains Debra.

Today, many artisanal butchers and chefs are reviving these traditional preparations for the same reason: to leave no scrap behind. You can find handcrafted head cheese and blood pudding and other old-world sausages alongside more familiar offerings. What's old is definitely new again, and we can thank our forebears for that. Here's to history repeating itself, one delicious link at a time.

Old-Fashioned Recipes

"Three tea-spoons of powdered sage, one and a half of salt and one of pepper, to a pound of meat, is good seasoning for sausages."

The Frugal Housewife, Dedicated to Those Who Are Not Ashamed of Economy
Lydia Maria Francis Child, Boston, MA, 1830

"Take the piece of pork designed for sausages, and chop it up, and if it is too fat, add a little lean beef; season with sage or summer savory, salt, and pepper; then fry a small piece, to see if it is seasoned right.

"If you prefer not to stuff them into skins, you may take pieces of cotton cloth, eight or nine inches wide, and two or three feet long, and sew the sides together, and one end; then wet it, stuff your meat in as solid as you can, and hang them up in a cool dry place. It will keep as well, or better than in the skins: when used, peal [*sic*] the cloth down no farther than you slice off."

The New England Economical Housekeeper, and Family Receipt Book
Esther Allen Howland, Worcester, MA, 1844

"One pound of sausages, cut in pieces, with four pounds of potatoes, and a few onions, if they are liked, with about a table-spoonful of flour mixed in a pint of water and added to the dish, will make a sufficient dinner for five or six persons. The potatoes must be cut in slices and stewed with the sausages till tender."

The Good Housekeeper
Sarah Josepha Buell Hale, Boston, MA, 1839

BLOOD SAUSAGE (BLACK PUDDING)

MAKES ABOUT 4 POUNDS (FIVE
12-OUNCE SAUSAGES)

Based in Portsmouth, New Hampshire, Black Trumpet Restaurant and Bar is a farm-to-table restaurant in the truest sense. Chef Evan Mallett has made it his mission to use his region's meat and produce to bring the world's flavors to his door. His Black Pudding is an amalgam of styles he has experimented with over the years, taking inspiration from Spanish, French, and even Puerto Rican blood sausage traditions. This recipe produces a crumbly sausage. If you prefer a meatier, more sliceable link, you can add a little day-old bread and/or ground pork to the sausage mixture. When stuffing, you'll need to use a different method from most sausages in this book: The very wet sausage mixture is stuffed one link at a time, using a widemouthed (about ½-inch) funnel.

1¾ teaspoons kosher salt

1 cup Bomba or Calasparra rice

2½ ounces salt pork, finely diced

1 tablespoon olive oil

1 medium Spanish onion, finely chopped

10 medium garlic cloves, coarsely chopped

1 teaspoon granulated onion

¾ teaspoon pimentón (smoked paprika)

¼ teaspoon ground fenugreek (or fennel seeds)

¼ teaspoon ground dried bay leaf

¼ teaspoon dried oregano

¼ teaspoon dried marjoram

¼ teaspoon freshly ground black pepper

¼ teaspoon ground mace

Pinch of cayenne pepper

¼ cup brandy

5 feet large hog casing

1. Bring ¾ cup water and ¼ teaspoon of the salt to a boil in a medium pot. Add the rice, reduce the heat to low, cover, and simmer until tender and the liquid is absorbed, about 18 minutes. Spread onto a rimmed baking sheet to cool to room temperature. Refrigerate until cold, about 1 hour.

2. Bring a medium pot of water to a boil and add the salt pork. Boil for 5 minutes and then drain. Set aside.

3. Heat the olive oil in a 10-inch skillet over medium heat. Add the chopped onion and garlic, and cook, stirring occasionally, for 2 to 3 minutes, until the onion is translucent.

4. Reduce the heat to low and stir in ½ teaspoon of the salt with the granulated onion, pimentón, fenugreek, bay leaf, oregano, marjoram, black pepper, mace, and cayenne. Cook for about 5 minutes.

5. Increase the heat to medium, add the drained salt pork, and cook to slightly render the fat, about 5 minutes. Deglaze the pan with the brandy, allowing it to flame up. Cook for 1 minute, until the scent of alcohol wears off. Remove the pan from the heat, cool to room temperature, and refrigerate until cold, about 1 hour.

6. Prepare the casing (see page 29).

7. In a large bowl, combine the chilled rice, onion mixture, the remaining 1 teaspoon salt, the cilantro, and oregano. Toss well to combine. Add the pork blood and stir to combine. Cover the bowl and refrigerate for 30 minutes.

8. Cut the casing into five 12-inch lengths. Knot one end of each casing, leaving the other end open. Attach the open end of one casing to the spout of a widemouthed funnel and balance

2 tablespoons chopped
  fresh cilantro

1 tablespoon chopped
  fresh oregano

2 cups pork blood
  (see Note)

2 cups hard cider

the funnel on the lip of a quart container or tall Mason jar (this will help keep the casing from rolling around too much as you fill it). Slowly ladle 1½ cups of the sausage filling into the funnel, gently pushing it through the funnel with a plastic straw or skewer. Tie off the top of the casing, leaving 2 inches of unfilled casing. Repeat with the remaining casings. The filling will probably separate a little inside the casing.

9. Combine the cider with 1 quart water in a wide pot. Bring to a boil and reduce to a low simmer. Gently roll each sausage across the counter to reincorporate the blood into the filling. Holding horizontally so the blood and solids don't separate again, lower each sausage into the liquid. Poach until the sausages are cohesive but still have a little puddinglike give to them, about 12 minutes. Let the sausages cool to room temperature, then refrigerate for at least 2 hours.

10. Refrigerate, wrapped well in plastic, for up to 3 days, or freeze for up to 2 months; thaw overnight in the refrigerator before using.

11. The sausage can be enjoyed cold, at room temperature, or hot. Reheat the sausages by gently simmering in a pot of salted water for 5 to 10 minutes. Alternatively, they can be heated on a medium-low grill, or in a 300°F (150°C) oven, until the casing is beginning to crisp, about 10 minutes.

NOTE: *You will need to order pork blood from a butcher shop or slaughterhouse; any extra can be frozen. Do not use coagulated blood in this sausage. Beef blood may be used if pork blood is unavailable.*

LIVERWURST

This smooth, creamy sausage is one of the most popular of all the German wursts. A thick slice sandwiched between two slabs of homemade bread with a round of Bermuda or Vidalia onion is the preferred way to eat it, at least stateside. This recipe is different from those using natural casings, in that the farce is stuffed into a large collagen or plastic casing, or better yet a homemade muslin one (see opposite for instructions on making one). The sausage is cooked in hot water, then chilled before eating. *See photo, page 99.*

1. Cut the meats and fat into 1-inch cubes. Refrigerate the cubes for about 30 minutes to firm them up before grinding.

2. Separately grind the cubes of liver, pork, and fat through the fine disk of a meat grinder. Then mix together and grind again.

3. Sprinkle the dry milk, paprika, salt, pepper, sugar, coriander, marjoram, allspice, cardamom, mace, and onion over the ground meat. Mix well, using your hands.

4. Grind the mixture through the fine disk twice more, freezing the mixture for 30 minutes between grindings. (Or process in a food processor until very smooth.)

5. Stuff the mixture into the prepared casing. It helps to fold the open end of the casing down over itself (like forming a cuff) to get things started, making it easier for the meat mixture to reach the bottom. Pack the meat as firmly as possible.

6. Stitch or tie the open end so it is firmly closed.

7. In a large kettle, bring enough water to a boil to cover the liverwurst by 2 to 3 inches.

8. Put the liverwurst in the boiling water and place a heavy plate on top of it to keep it submerged. When the water returns to a boil, reduce the heat so that the water maintains a temperature of 175°F (79°C). Cook, covered, for 3 hours.

9. Plunge the liverwurst into a bowl of ice water to cool quickly. Drain well, pat dry, and refrigerate, covered, overnight before removing the casing.

10. The sausage can be refrigerated for up to 10 days; do not freeze.

Ingredients

- 1 pound fresh pork liver
- ¾ pound boneless lean pork butt or shoulder
- ¼ pound pork fat
- 3 tablespoons nonfat dry milk powder
- 2 teaspoons paprika
- 2 teaspoons kosher salt
- 1 teaspoon freshly ground white pepper
- 1 teaspoon sugar
- ½ teaspoon ground coriander
- ½ teaspoon dried marjoram
- ¼ teaspoon ground allspice
- ¼ teaspoon ground cardamom
- ¼ teaspoon ground mace
- 1 large sweet white onion, finely diced (about 1½ cups)
- 1 foot large (4-inch diameter) collagen, plastic, or muslin casing

HOW TO MAKE
A MUSLIN CASING

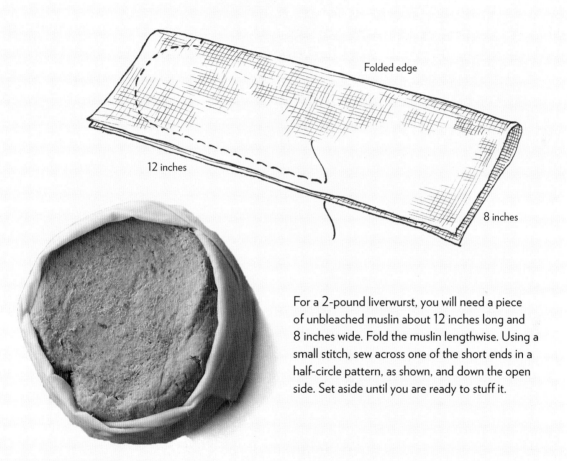

Folded edge

12 inches

8 inches

For a 2-pound liverwurst, you will need a piece of unbleached muslin about 12 inches long and 8 inches wide. Fold the muslin lengthwise. Using a small stitch, sew across one of the short ends in a half-circle pattern, as shown, and down the open side. Set aside until you are ready to stuff it.

Weisswurst, page 145

Thuringer Sausage, page 138

Dijon Mustard, page 293

GERMAN FLAVORS

Kasekrainer, page 153

Liverwurst, page 96

Braunschweiger, page 108

SCRAPPLE

This version of scrapple uses a broad range of seasonings and is typical of recipes from the area around Lancaster, Pennsylvania, where early German immigrants settled. (See opposite page for more background information.)

½ pound pork liver

1½ pounds boneless lean pork butt or shoulder

2 teaspoons kosher salt

¼ teaspoon freshly ground black pepper

1 medium onion, chopped

1¼ cups all-purpose flour

½ cup buckwheat or whole-wheat flour

½ cup yellow cornmeal

1 teaspoon ground coriander

1 teaspoon ground mace

1 teaspoon ground thyme

1 teaspoon fresh marjoram or ½ teaspoon dried

Bacon grease, lard, or canola oil, for cooking

1. Bring a kettle of water to a boil and pour the boiling water over the liver to cover; let stand for 5 minutes. Drain and rinse in cold water.

2. In a large pot, combine the liver, pork shoulder, salt, pepper, and onion. Cover with water and cook over medium heat, covered, until the meat is fork-tender, about 2 hours. Remove the meat from the broth and set aside to let it cool.

3. Measure the broth and add enough water to make 4½ cups of liquid. In a large pot, bring the liquid to a boil.

4. Whisk together the all-purpose flour, buckwheat flour, cornmeal, coriander, mace, thyme, and marjoram in a bowl. Gradually stir the mixture into the boiling liquid. Simmer over low heat, uncovered, for about 30 minutes, stirring frequently.

5. Put the meat through the coarse disk of the meat grinder (or finely chop it). Add the meat to the flour mixture, stirring to combine. The mixture will be thick.

6. Line two 9- by 5-inch loaf pans with plastic wrap, leaving an overhang on both long sides. Spread the mixture evenly between the pans, cover, and refrigerate overnight.

7. Remove the scrapple from the pans, using the plastic overhang, then peel off the plastic and cut the scrapple into slices about ¼ inch thick. The slices can be frozen for up to 3 months, wrapped well in plastic and stored in a freezer bag; thaw overnight in the refrigerator before cooking.

8. To cook, heat some bacon grease in a large skillet until very hot and fry the scrapple slices until golden and crisp, about 5 minutes per side. Serve hot.

ALL ABOUT SCRAPPLE

If you connect scrapple with the Pennsylvania Dutch, you are correct, as most recipes for it originate from those thrifty German immigrants. (The "Dutch," you may recall, are actually *Deutsch*, or German.) As its name suggests, scrapple uses up scraps of meat in a hearty dish that is a combination of meatloaf and cornmeal mush.

Although it isn't stuffed into a sausage casing, scrapple is a fine way to preserve fresh or cooked pork. You can start with either leftover cooked pork or fresh pork. There are probably as many variations as there are Pennsylvania Dutch farmers, who traditionally made scrapple in the fall, at butchering time. The "recipe" spread to communities outside Pennsylvania, too; in Ohio, the sausage was made with hog jowls, feet, liver, and tongue and seasoned with plenty of fresh marjoram.

Properly prepared, scrapple has the flavor of a good pork sausage and the crispness of bacon. That's why it makes a perfect breakfast meat and accompaniment to fried eggs. Purists warn that you should never fry scrapple in anything but lard or bacon grease. Some people like to pour maple syrup over fried scrapple. Others like to sauté a few apple slices in the pan along with the meat, or serve the fried scrapple with fresh applesauce.

Any way you slice it or serve it, scrapple is worth giving a try.

LAP CHEONG

The thin pork sausages (also called *laap ch'eung*) you find in Chinatowns or Chinese groceries are redolent of spices, slightly sweet (thanks to the addition of brown sugar), and delicious in stir-fries and other Asian dishes — or simply sliced and served with Chinese mustard to serve with cocktails. The "smoking" method here relies on the low heat of the oven to cook the sausages through. *See photo, page 84.*

- 3 pounds boneless lean pork butt or shoulder
- ½ pound pork fat
- 3 tablespoons brown sugar
- 1 tablespoon kosher salt
- 2 teaspoons five-spice powder
- 2 teaspoons crushed red pepper
- 1 teaspoon grated orange zest
- 2 tablespoons soy sauce
- 8 feet small sheep or lamb casing

1. Cut the meat and fat into 1-inch cubes. Freeze the cubes for about 30 minutes to firm them up before grinding.

2. Grind the meat and fat through the coarse disk of a meat grinder.

3. In a large bowl, combine the meat mixture, sugar, salt, five-spice powder, crushed red pepper, orange zest, and soy sauce. Mix well, using your hands.

4. Cover with waxed paper and chill in the refrigerator for at least 2 hours or preferably overnight.

5. Prepare the casing (see page 29).

6. Stuff the mixture into the prepared casing, prick air pockets, and twist off into 6-inch lengths. Tie off each link with butcher's twine.

7. Place the sausage on a baking sheet fitted with a wire rack and refrigerate, uncovered, for a few hours or preferably overnight.

8. Preheat the oven to 200°F (93°C).

9. Place the sausages on rack over a roasting pan. Bake for about 5 hours, or until the sausages are firm and nearly dry to the touch. Turn off the oven; let the sausages remain in the oven for 2 hours longer.

10. Discard the fat in the roasting pan. Eat the sausages warm, or let cool and refrigerate for up to 1 week, until ready to use. These sausages also will keep for 2 months in the freezer; thaw overnight in the refrigerator.

PORK RILLETTES

- 2 teaspoons whole black peppercorns
- 2 teaspoons allspice berries
- 1 teaspoon coriander seeds
- 2 whole cloves
- 2 pounds boneless pork butt or shoulder, about 70% lean
- 1 pound pork belly, skin removed and reserved
- ¼ cup kosher salt
- 2 rosemary sprigs
- 2 thyme sprigs
- 2 bay leaves
- 3 garlic cloves, smashed and peeled
- 1 cup dry white wine
- 2 cups chicken stock or water

Fleisher's Craft Butchery first opened its doors in Kingston, New York, in 2004, and ever since it has been an inspiration to up-and-comers intent on adopting the shop's tenet of "butchery as a means for improving and growing a strong food community." (Two such aspiring folks, Susan and Terry Ragasa, are profiled on page 68.) There are now offshoots of Fleisher's original location in New York City (two in Manhattan and one in Brooklyn) and Westport, Connecticut, which is also home to a restaurant. This recipe is inspired by the shop's utterly delectable rillettes — a classic charcuterie spread made by cooking pork in its own fat until meltingly tender. It's perfect for smearing on crusty bread, alone or as part of a charcuterie board (see page 148).

1. Preheat the oven to 275°F (135°C). Grind the peppercorns, allspice, coriander, and cloves with a mortar and pestle or a spice mill until coarse.

2. Cut the pork butt or shoulder into 2-inch cubes and the pork belly into 1-inch cubes. Place in a Dutch oven or other ovenproof pot and add the ground spices and salt, tossing to coat.

3. Add the rosemary, thyme, bay leaves, garlic, and wine, and bring to a boil. Reduce to a bare simmer and cook, skimming the foam from the surface, for 30 minutes.

4. Add the chicken stock, return to a boil, and cover. Transfer to the oven and braise for about 2½ hours, until the pork is falling-apart tender, stirring occasionally.

5. Remove the lid and continue to braise in the oven until the liquid is almost completely evaporated and only fat remains at the bottom of the pot. Remove from the oven and set aside to cool for about 1 hour.

6. Strain the contents of the pot in a sieve, reserving the melted fat. Discard the herb sprigs and garlic cloves. Transfer the pork to a bowl and finely shred with a fork or a potato masher. Add ½ cup reserved fat and mash to combine, adding more as needed to reach a spreadable consistency. Taste and adjust the seasoning as desired.

7. Divide the rillettes among jars or crocks, packing tightly, and top each with a thin layer of reserved fat. Cover tightly and refrigerate until set, at least 4 hours or preferably overnight (and up to 1 week).

8. Let sit at room temperature for about an hour before serving.

Meet the
MAKER

⚜

Jarred
Zeringue
Wayne Jacob's
Smokehouse
La Place, Louisiana

FOR THE LOVE OF ANDOUILLE

"What we make here is a German sausage with a French name. This area has a strong German influence, but everyone speaks French," proclaims Jarred Zeringue, current co-owner (along with Matthew Moreland to the right of Jarred in the photo) of Wayne Jacob's Smokehouse, in La Place, nicknamed "The Andouille Capital of Louisiana." Situated in St. Charles Parish, about 25 miles due west of New Orleans, La Place was once known as the German Coast after the early settlers who congregated on both banks of the Mississippi River — and who brought their own culinary traditions to the French Acadian melting pot. It's against this rich historical backdrop that you can appreciate the andouille that's produced here, and how the food differs from other parts of southern Louisiana. It's also very different from commercial versions of andouille — thicker, smokier, and darker colored, with a coarser texture that's marbled with fat. "People come here and expect the food to be hot and spicy, but it's not like that at all. It's very flavorful, but not extreme. Our andouille is mostly a seasoning meat," offers Jarred, who crumbles it into corn chowder, gumbo, and jambalaya, and even mixes some into ground beef to make an andouille burger. He also makes andouille chips, thinly sliced andouille that's deep-fried to a crisp and served with creole mustard as a bar snack (other offerings include hog head cheese, served with hot sauce and saltines, and boudin balls) — perfect for washing down with Abita craft beer from a nearby town.

Since andouille is an integral part of the county's heritage and ongoing identity, there's not surprisingly a friendly rivalry among the different purveyors. Besides Wayne Jacob's, there's Jacobs World Famous Andouille (started by the same family) and Bailey's World Famous Andouille (you get the idea). "We are the oldest continually operating smokehouse for this type of andouille in the state," boasts Jarred.

Even though Wayne Jacob's has changed ownership over the years, it has always stayed true to the original recipe for andouille (and other favorites, including tasso and beef jerky). So when Zeringue learned that Wayne Jacob's was for sale, he jumped at the chance to buy it himself, adding to his three restaurants in New Orleans. "It was too important of a cultural legacy to let it die, and no one was really looking to get into the restaurant business here since it's such a small town," he says. "We'd been eating this sausage all my life, so I really bought it for selfish reasons." The biggest warning he got from his family and other ardent fans was, "Do not screw it up." Besides, it was a healthy business with a devoted following. "We sell a hundred pounds of andouille a week, with none left over. During the holidays we at least triple that." (This includes shipping to loyal customers across the country.)

Another tradition he has been sure to continue: "Each Tuesday and Thursday we make 200 pounds of cracklins, and those are our busiest days. Locals come in and just hold up one or two fingers to indicate the number of bags they want to buy. They never speak, and when we try to get them to they just laugh." He has also tried to introduce new items to appeal to a new audience, with other types of fresh and smoked sausages (made from pork, duck, turkey, chicken, and deer), boudin (with or without rice), and even prosciutto, in the spirit of using every part of the animal. "We smoke our own salts, peppers, and vegetables, basically anything that can keep us busy and showcase our local ingredients. To be able to layer those smoked items with the meat — it's just a really unique flavor experience." They also grind their own cayenne pepper for the andouille, make their own smoked poblano vinegar, can their own fruit jams and jellies, plus keep bees and hens for honey and eggs. All of these products are for sale in their shop as well as being used in the restaurant.

Although he (and all the other proprietors in town) guards the original recipe under lock and key, Jarred did offer this: "It's just pork that's cut into larger pieces, salt, black pepper, our own cayenne, and garlic. It's the whole philosophy of Cajun cooking: You should treat the food so you get the most flavor out of the simplest ingredients. So many people think if you throw more ingredients on a plate you'll get more flavor. But it's actually just the opposite."

The secret, he says, lies in the smoking. There are four smokehouses behind the restaurant, and it can take 8 to 10 hours to smoke the meats. "But that depends on different things. You have to monitor the temperature of the smokehouse to get the color and flavor that you want in the finished product. Each smoker differs in the right temperature to use and the amount of wood that you use and the time you need to smoke the meat. If it's cold outside, it may take longer, but you can't get the fire too hot or the casings will fall apart. Always aim for low and slow."

TRADE SECRETS

"That low-and-slow philosophy goes for home sausage makers, too: You just can't rush the process. Monitor the temperature of your smoker and adjust as needed. It takes more effort, but the benefit is you'll have delicious sausage that can be left unrefrigerated for a week or more, just like salami."

— Jarred Zeringue

ANDOUILLE

Louisiana's rich culinary landscape is spiked with andouille in all its spicy, robust glory. Don't confuse Cajun andouille with its French cousin of the same name, which doesn't have the characteristic kick from cayenne. (Indeed, in France, andouille is often just a general term for any kind of sausage, while andouiette is made with tripe and chitterlings and is another thing entirely.) In Louisiana, you'd be hard-pressed to find andouille any other way than smoked — typically over pecan wood and sugarcane, which imparts subtle nuances in flavor. You can add pecan (or hickory) wood to your own smoker, too. The recipe here calls for hot smoking, but you could easily use the method described on page 46 for cold smoking. *See photo, page 188.*

- 6 feet beef middle casing (2–3 inches in diameter) or 12 feet medium hog casing
- 5 pounds boneless lean pork butt or shoulder (see Note)
- 1 pound pork fat (see Note)
- ½ cup chopped garlic
- ¼ cup cracked black pepper
- 4 tablespoons kosher salt
- 2 tablespoons cayenne pepper
- 1 tablespoon dried thyme

1. Prepare the casing (see page 29).

2. Cut the meat and fat into 1-inch cubes. Freeze the cubes for about 30 minutes to firm them up before grinding.

3. Grind the meat and fat through the fine disk of a meat grinder.

4. In a large bowl, combine the meat mixture, garlic, black pepper, salt, cayenne, and thyme. Mix well, using your hands.

5. Stuff the mixture into the prepared casing, prick air pockets, and twist off into 12-inch links. Tie off each link with butcher's twine.

6. Place the sausage on a baking sheet fitted with a wire rack and refrigerate, uncovered, for a few hours or preferably overnight.

7. Hot-smoke at 175°F to 200°F (79 to 93°C) for 4 to 5 hours (for beef middles), or 1 to 2 hours (for hog casings), using pecan or hickory wood if possible. The sausage should be firm and should reach an internal temperature of 160°F (71°C). Store in the refrigerator for 5 to 7 days, or freeze and use as needed.

NOTE: *If you want to cold-smoke the sausage, prepare the pork and fat according to the instructions on page 44 to ensure that it is free from trichinosis. Then, reduce the salt to 2 tablespoons and add Insta Cure #1 at the level recommended by the supplier. To cold-smoke, dry the sausage overnight in the refrigerator, then cold-smoke at 70 to 90°F (21 to 32°C) for at least 12 hours, or until the sausage firms up. Cook before eating.*

4 pounds boneless pork butt or shoulder, about 75% lean (see Note)

2 tablespoons paprika

1 tablespoon plus 2 teaspoons kosher salt

1 tablespoon ground coriander

1 teaspoon crushed red pepper

½ teaspoon ground allspice

½ teaspoon ground cinnamon

½ teaspoon ground cloves

7 garlic cloves, minced

½ cup cold water

¼ cup red wine vinegar

4 feet medium hog casing

SMOKED

LINGUIÇA

Linguiça is the basic spicy country sausage beloved in Portugal; it is a close cousin to andouille but without the fiery heat. Similar to Spanish chorizo, paprika is the predominant flavor, only using the sweet variety of the spice instead of the hot. Linguiça shows up in many Portuguese dishes, from soups and stews to pizza and egg dishes. It's as versatile as that. Because authentic versions can be hard to come by, it's worth making a batch (or two) of this sausage for your own supply. *See photo, page 73.*

1. Cut the meat into 1-inch cubes. Freeze the cubes for about 30 minutes to firm them up before grinding.

2. Grind the meat through the coarse disk of a meat grinder.

3. In a large bowl, combine the meat with the paprika, salt, coriander, crushed red pepper, allspice, cinnamon, cloves, garlic, water, and vinegar. Mix well, using your hands.

4. Cover the mixture and refrigerate overnight.

5. Prepare the casing (see page 29).

6. Stuff the mixture into the prepared casing, prick air pockets, and twist off into 10-inch links. Tie off each link with butcher's twine.

7. Place the sausage on a baking sheet fitted with a wire rack and refrigerate, uncovered, for a few hours or preferably overnight.

8. Hot-smoke at 175°F to 200°F (79 to 93°C) for 4 to 5 hours. The sausage should be firm and should reach an internal temperature of 160°F (71°C). Store in the refrigerator for 5 to 7 days, or freeze and use as needed.

NOTE: *If you want to cold-smoke the sausage, prepare the pork according to the instructions on page 44 to ensure that it is free from trichinosis. Then, reduce the salt to 1 tablespoon and add Insta Cure #1 at the level recommended by the supplier. To cold-smoke, dry the sausage overnight in the refrigerator, then cold-smoke at 70 to 90°F (21 to 32°C) for at least 12 hours, or until the sausage firms up. Cook before eating.*

BRAUNSCHWEIGER

This smooth, creamy German sausage from the town of Braunschweig is mildly spiced and has a distinctive flavor all its own. Because it's cured and smoked, it is decidedly different from another renowned pork liver sausage, liverwurst (see recipe, page 96). *See photo, page 99.*

1. Prepare the casing (see page 29).

2. Cut the meats into 1-inch cubes. Freeze for 30 minutes to firm them up before grinding.

3. Grind the meats separately through the fine disk of a meat grinder.

4. In a large bowl, combine the meats, dry milk, salt, sugar, pepper, mustard seeds, marjoram, allspice, ascorbic acid, curing salt, onion, and ice water. Mix well, using your hands. Freeze the mixture for 30 minutes.

5. Grind the seasoned meat mixture through the fine disk of the grinder.

6. Stuff the mixture into the prepared casing, prick air pockets, and twist off into 6- to 8-inch links. Tie off each link with butcher's twine. Do not separate the links.

7. Simmer the sausage in a large pot of water at 180 to 190°F (82 to 88°C) for 1 hour.

8. Remove the sausage from the water, dry thoroughly, and smoke at 150°F (66°C) for 2 hours, or until it reaches an internal temperature of 160°F (71°C).

9. Place the sausage in a large pot of cool water for 30 minutes, then remove and pat dry. Braunschweiger will keep, wrapped well in plastic, for up to 2 weeks in the refrigerator.

NOTE: *We recommend that you use a commercial premixed cure in any recipe for cured sausage. Premixed cures replace the saltpeter in older recipes.*

- 4 feet medium hog casing
- 2½ pounds pork liver, trimmed
- 2½ pounds boneless pork butt or shoulder, about 75% lean
- ¼ cup nonfat dry milk powder
- 1 tablespoon plus 2 teaspoons kosher salt
- 1 tablespoon sugar
- 2 teaspoons freshly ground white pepper
- 1 teaspoon crushed mustard seeds
- ½ teaspoon ground dried marjoram
- ¼ teaspoon ground allspice
- ¼ teaspoon ascorbic acid
- Insta Cure #1 (use supplier's recommended quantity for 5 pounds of meat; see Note)
- 2 tablespoons finely minced onion
- ½ cup ice water

SALAMETTE

MAKES 10 POUNDS

8 pounds prefrozen boneless lean pork butt or shoulder (see Notes)

2 pounds prefrozen pork fat (see Notes)

3 tablespoons plus 1 teaspoon kosher salt

2 tablespoons freshly ground black pepper

1 tablespoon plus 2 teaspoons fennel seeds

1 tablespoon sugar

2 teaspoons anise seed

½ teaspoon ascorbic acid

Insta Cure #2 (use supplier's recommended quantity for 10 pounds of meat; see Notes)

1 tablespoon finely minced garlic

1 cup dry red wine

6 feet medium hog casing

The "ette" in the name is the Italian diminutive, as in "small salami." It is simple to make and to dry (see page 166 for guidelines on drying sausage). *See photo, page 151.*

1. Partially thaw the meat and fat, then cut them into 1-inch cubes.

2. While the meat and fat are still cold, grind them separately through the coarse disk of a meat grinder.

3. In a large bowl, combine the meat, fat, salt, pepper, fennel seeds, sugar, anise seed, ascorbic acid, curing salt, garlic, and wine. Mix well, using your hands.

4. Spread the mixture in a large, shallow pan, cover loosely with waxed paper, and cure it in the refrigerator for 24 hours.

5. Prepare the casing (see page 29).

6. Stuff the mixture into the prepared casing, prick any pockets, and twist off into 4-inch links. Tie off each link with butcher's twine. Do not separate the links.

7. Hang the links in a drying area for 6 to 8 weeks. Test the sausage after 6 weeks by cutting off one link and slicing through it. If the texture is firm enough to suit your taste, the remaining links can be separated and wrapped tightly in plastic wrap for storage in the refrigerator.

NOTES: *Prepare pork according to the instructions on page 44 to ensure that it is free from pathogens.*

The meat is not cooked; please see Dry Sausage Safety on page 43.

We recommend that you use a commercial premixed cure in any recipe for cured sausage. Premixed cures replace the saltpeter in older recipes. Also see page 162.

RULER OVERRULED

In the fourth century, the Christian emperor Constantine the Great tried to ban sausages from the Roman Empire because they were consumed with great gusto at pagan orgies. (He wasn't too crazy about the orgies either.) By this time, however, practically every Italian village had developed its own specialty sausage, and not even the scowls of the emperor could deter sausage lovers from enjoying them.

SPANISH-STYLE CHORIZO

MAKES 10 POUNDS

Dried chorizo is a specialty of Spain, where it is commonly smoked but just as traditionally not; it gets its bold flavor from the addition of pimentón, Spanish smoked paprika — either hot or sweet, or sometimes both. Hungarian paprika can be used instead in a pinch, but the results won't be the same. Here in the United States you can find chorizo that is dried (*curado*), as in this version, or semidried and softer (*semicurado*).

Spanish chorizo is excellent on its own, served as an appetizer or as part of a charcuterie board, or in cooking, where it is integral to many soups and stews as well as certain versions of paella. Don't confuse Spanish chorizo (also called slicing chorizo) and Mexican chorizo (see Fresh Chorizo, page 67) when using in cooking, since they are very different in texture and taste.

1. Partially thaw the meat and the fat, then cut them into 1-inch cubes.

2. While the meat and fat are still cold, grind them separately through the coarse disk of a meat grinder.

3. In a large bowl, combine the meat, fat, salt, pimentón, black pepper, sugar, crushed red pepper, cumin seeds, oregano, fennel seeds, ascorbic acid, curing salt, brandy, vinegar, and garlic. Mix thoroughly, using your hands. Spread the mixture in a large, shallow pan, cover loosely with waxed paper, and cure it in the refrigerator for 24 hours.

4. Prepare the casing (see page 29).

5. Stuff the mixture into the prepared casing, prick air pockets, and twist off into 4-inch links. Tie off each link with butcher's twine. Do not separate the links.

6. Hang the links in a drying area for 6 to 8 weeks. Test the sausage after 6 weeks by cutting off one link and slicing through it. If the texture is firm enough to suit your taste, the remaining links can be separated and wrapped tightly in plastic wrap for storage in the refrigerator.

NOTES: *Prepare pork according to the instructions on page 44 to ensure that it is free from pathogens.*
The meat is not cooked; please see Dry Sausage Safety on page 43.
We recommend that you use a commercial premixed cure in any recipe for cured sausage. Premixed cures replace the saltpeter in older recipes.

- 8 pounds prefrozen boneless lean pork butt or shoulder (see Notes)
- 2 pounds prefrozen pork fat (see Notes)
- 3 tablespoons plus 1 teaspoon kosher salt
- 3 tablespoons pimentón (smoked paprika) or cayenne pepper
- 2 tablespoons freshly ground black pepper
- 2 tablespoons sugar
- 1 tablespoon crushed red pepper
- 1 teaspoon cumin seeds
- 1 teaspoon crushed dried oregano
- 1 teaspoon fennel seeds
- ½ teaspoon ascorbic acid

Insta Cure #2 (use supplier's recommended quantity for 10 pounds of meat; see Notes)

- ¾ cup brandy
- ¼ cup red wine vinegar
- 2 tablespoons minced garlic
- 6 feet medium hog casing

1 pork tenderloin (about 1 pound), silverskin removed

½ cup kosher salt

1 tablespoon light brown sugar

½ tablespoon cognac or other brandy

½ tablespoon freshly cracked black pepper

½ tablespoon herbes de Provence (see page 212)

SAUCISSON OF PORK TENDERLOIN

This recipe was provided by Blake Royer (see profile on page 112), who adapted it from a Jacques Pépin recipe that went viral online, sparking a wave of home cooks (and some professionals, too) to start dry-curing whole tenderloins. Look for the largest tenderloin you can find when making this. You will need to hang the tenderloin in a cool place, such as a cellar or attic where you can open up a window; if you don't have access to either of those, clear out space in a closet and put a fan in there for circulation. As Blake says, you want it to be slightly drafty, but don't overdo it.

1. Trim 2 to 3 inches from the narrow end of the tenderloin, or as needed to create a uniform thickness for even curing.

2. In a small brining bag or large resealable plastic bag, combine the salt and sugar. Put the tenderloin inside and blow a little air into the bag. Pinch it closed and toss it around to coat the meat very well. Seal the bag and refrigerate for at least 12 hours.

3. Remove the bag from the refrigerator and take the pork from the bag. Wipe dry with paper towels, then rub all over with the cognac and sprinkle with the pepper and herbs.

4. Wrap the tenderloin in muslin or cheesecloth, tie up like a roast with butcher's twine, and hang in a cool, slightly drafty place for 5 to 6 weeks, until the meat has the consistency of a salami — not too soft on the inside and not too hard on the outside. Wrap in plastic for storage in the refrigerator.

Blake Royer

*Home Charcuterie
Aficionado*

Toronto, Ontario

THE HOBBYIST

"I would say I have a passion for charcuterie in all its forms. Duck prosciutto was the first project I ever did. Eventually I did get into salami and dry-cured sausage, and there's a whole level of intensity and science behind those, so it took me a while to get up the courage to do them," says Blake Royer, who for a period of five years was one half of the duo behind the well-regarded blog *The Paupered Chef.* Although the blog ceased operations in 2012 when Blake and his partner, Nick Kindelsperger, decided it was time to move on (to paying jobs, so no hard feelings!), the content is still on the site and is as relevant today as it was then.

"We were living in Manhattan with a couple of friends, in a really tiny apartment, with this tiny little stove. At the time, there was this wave of interest in chefs and good food, and a bunch of us started blogs and we fed off each other's energy. We had a fearless, experiment-driven approach and weren't afraid to look stupid." What made the blog so compelling was the pair's passionate exploration of all things food related, propelled by their youthful energy and proximity to the incredible diversity of New York City's food scene.

It was also Blake's enthusiasm for his sausage-making enterprises that captured the attention of amateurs and professionals alike. "It's intimidating but it doesn't have to be." What Blake recommends for aspiring home sausage makers is a graduated approach: First try simple cured meats like bacon or guanciale (his first foray), then whole-muscle curing, and gradually move into dry-cured sausages. He recalls curing saucisson of pork tenderloin using a recipe by Jacques Pépin in the beginning. "The great thing is that you're working with a sausage-shaped muscle, and a muscle is inherently sterile inside, so you can get a sense for what's it like to air-cure a cut of meat. This was the project that gave me the courage to make the leap to home-cured sausage." (For the recipe, see page 111.)

But first he had to master the basics of making fresh sausage and how to properly stuff the casings, which he did with a simple funnel and the handle of a wooden spoon ("nothing fancy but it did the trick"). He recommends starting with sheep's casing for cured sausages, since they are more slender than hog casings and will be easier to monitor. "Don't sweat the stuffing process: It's better to underfill than overfill, since you can always squeeze out excess air by twisting the casings." To educate himself about the art of dry curing, he said he watched a lot of *River Cottage* episodes, and, having graduated from cooking school himself, was able to spend some time in professional butcher shops. "I did a couple of stages at The Fatted Calf in Napa Valley, which at the time was doing a ton of experimenting with charcuterie. I also spent a few days at this fantastic place in Madison, Wisconsin, called Underground Meats. They had some really interesting recipes." (See page 126.)

When experimenting at dry-curing at home, using an old refrigerator found on the street, Blake says he struggled the most with regulating the humidity. "My issue was always that the environment was too dry and the casing would become too brittle, meaning it was no longer permeable and the inside would rot. That's also true in whole-muscle curing." That's why he encourages beginners to master how to get the humidity right by experimenting with a couple of pork tenderloins, which are relatively inexpensive. "What you want is to get a nice even color and dry texture all the way through; if it's stiff and hard on the outside and sort of pink on the inside, then most likely your humidity levels are too low. I got pretty geeky and bought humidity sensors and all that. But a pan of salty water will go a long way if the fridge is not too huge." Blake says that temperature, unlike humidity, is easy to control. "A basement is probably the best place to do it, but if you don't have one then put it in a dark room, like a closet. Ambient room temperature is usually okay."

Then there was the issue of mold. "In the beginning, I was really terrified about what was good mold versus bad. What's great about a whole muscle is that the inside of the meat is always sterile, so you only need to worry about the outside. With the pork tenderloin, at first I would wipe away any mold with a brine solution, which protects it and keeps it from coming back. With sausage, I started inoculating the meat with cultures that feed on the natural sugars in the meat and create an acidic environment that will prevent spoilage. I would recommend getting the good white mold into your curing chamber, and it becomes the resident mold."

Another factor to consider is the amount of time that goes into dry curing. "It's on a different time scale than most cooking, where you see the results right away or a few hours later. This is very different. It also incorporates a lot more science and there's a lot more to learn, in a good way." Blake describes dry curing as a fun lifetime skill and more than just a hobby. "We're renovating our house in the spring, and one of the projects is to build a real curing chamber in the basement, not too big but with all the proper controls." With any luck he'll start blogging about his exploits again, too.

TRADE SECRETS

"Start with the highest-quality meat, especially if you will be dry curing the sausage. It just makes no sense to use anything but.

"You're not doing it at home because you're looking for a cheap salami. You're looking to craft something and put your heart into it and learn, and what would be the purpose in not buying the absolute best-quality meat you could?

"There's the caveat that you might screw up a few times, but once you get the hang of it, it's going to taste better, especially pork-based charcuterie.

"The quality of the fat will determine the quality of the difference between the flavor, and the flavor of pasture-raised pork and commodity pork is night and day. That's a no-brainer."

— Blake Royer

FRENCH GARLIC SAUSAGE

MAKES 10 POUNDS

Despite its name, this classic saucisson is much more nuanced than it sounds. Just look at the lengthy list of dried herbs and spices in the recipe, which also calls for a full cup of cognac (*vive la France!*). It will become a must-have on any of your charcuterie boards — and also makes a flavorful addition to cassoulet and choucroute.

1. Partially thaw the meat and fat, then cut them into 1-inch cubes.

2. While the meat and fat are still cold, grind them separately through the coarse disk of a meat grinder. Mix together the meat and fat, and freeze for 30 minutes.

3. Grind the meat mixture through the fine disk of the meat grinder.

4. Mix in the salt, garlic, and cognac.

5. In a food processor or blender, process the sugar, paprika, white pepper, bay leaf, ascorbic acid, basil, cinnamon, cloves, mace, oregano, sage, thyme, cayenne, and summer savory until you have a fine powder.

6. Mix the powdered herbs and spices along with curing salt into the meat mixture with your hands, blending thoroughly.

7. Spread the meat mixture in a large, shallow pan, cover loosely with waxed paper, and cure in the refrigerator for 24 hours.

8. Prepare the casing (see page 29).

9. Stuff the mixture into the prepared casing, prick air pockets, and twist off into 6- to 7-inch links. Tie off each link with butcher's twine; do not separate the links (see Notes, opposite).

10. Hang the sausage to dry for 8 to 12 weeks, until it is firm. Test the sausage after 8 weeks of drying by cutting off one link and slicing through it. If the texture is firm enough to suit your taste, the remaining sausage can be cut down and wrapped tightly in plastic wrap for storage in the refrigerator.

- 7 pounds prefrozen fresh ham (see Notes)
- 3 pounds pork fat (see Notes)
- 3 tablespoons plus 1 teaspoon kosher salt
- 3 tablespoons minced garlic
- 1 cup cognac or other brandy
- 2 tablespoons sugar
- 1 tablespoon paprika
- 1 tablespoon freshly ground white pepper
- 1 teaspoon crushed dried bay leaf
- ½ teaspoon ascorbic acid
- ½ teaspoon dried basil
- ½ teaspoon ground cinnamon
- ½ teaspoon ground cloves
- ½ teaspoon ground mace
- ½ teaspoon dried oregano
- ½ teaspoon ground sage

½ teaspoon dried thyme

¼ teaspoon cayenne pepper

¼ teaspoon dried summer savory

Insta Cure #2 (use supplier's recommended quantity for 10 pounds of meat; see Notes)

4 feet large beef casing

NOTES: *Prepare pork according to the instructions on page 44 to ensure that it is free from trichinosis. See Dry Sausage Safety on page 43.*

We recommend that you use a commercial premixed cure in any recipe for cured sausage. Premixed cures replace the saltpeter in older recipes.

Depending on the beef casing, it may be easier to cut the casing into 6- to 7-inch lengths, tie each length at one end, stuff the casing, then tie the other end.

SAUCISSON, A LOVE STORY

The French adore their saucisson so much so that they eat 100,000-plus tons of it each year. Besides saucisson sec and saucisson à l'ail, there are myriad other options to pack your picnic basket with — probably as many varieties as there are regions. Among the more noteworthy is rosette de Lyon, hailing from what is considered the culinary capital of the country (and that's saying something). This country-style, rose-colored dry-cured sausage, made from leg of pork, is exceptionally rich and coarse in texture. Or consider another type of charcuterie, jambon de Bayonne, a dry-cured ham from southwest France that is the French answer to Italian prosciutto and Spanish jamon Iberico — and which bears the coveted Protected Designation of Origin by the European Union.

sirloin

short loin

rib

chuck

head

round

flank

plate

brisket

Merguez, page 32

BEEF, LAMB, AND VEAL SAUSAGES

A cook turns a sausage, big with blood and fat, over a scorching blaze without a pause, to broil it quick.

— Homer, *The Odyssey*

We don't know exactly what kind of sausages Ulysses and his cohorts were eating, but we do have an inkling into how they were prepared. Grilling sausages outdoors, whether in the backyard or over a campfire, still brings out the adventurer in all of us. The beef, lamb, and veal sausages in this chapter come to us from all over the world — Homer's Mediterranean isles, Central Europe's farms, English pubs, and our own American Midwest, among other points on the world atlas.

While pork is the meat thought of most often for sausage, many delicious and distinctive sausages are traditionally made from beef, lamb, and veal (even goat). Beef chuck is an excellent cut for making sausage; it is relatively high in fat (similar to pork shoulder) and economical. Leaner cuts, including sirloin and eye of round, may require the addition of extra fat to keep the sausage juicy; flank steak and brisket decidedly so. For all-beef links, use suet as the added fat. Lamb and veal sausages are naturally tender and palatable, and amenable to a variety of seasonings.

ENGLISH BANGERS

MAKES 2 POUNDS

Bangers is British slang for the popular sausages made of meat — traditionally pork, here tweaked with beef — and bread crumbs (or ground rusks, as Brits would use). The origin of the name came during World War I, when food shortages forced sausage producers to fill out the products with meat scraps, cereal grain, and water; as a result, the sausages would hiss and pop (or bang!) when cooked over open fires in the trenches. You can find bangers in pubs all over the British Isles, usually as part of bangers and mash, with onion gravy an optional indulgence. Feel free to follow suit, or simply grill or panfry them and serve with grainy mustard.

1. Prepare the casing (see page 29).

2. Cut the meat and fat into 1-inch cubes. Freeze the cubes for about 30 minutes to firm them up before grinding.

3. Grind the meat and fat through the fine disk of a meat grinder.

4. In a large bowl, combine the meat, fat, bread crumbs, salt, pepper, mace, thyme, coriander, nutmeg, and egg yolks. Mix well, using your hands. Freeze for about 30 minutes.

5. Grind the seasoned meat mixture a second time through the fine disk.

6. Stuff the mixture into the prepared casing, prick air pockets, and twist off into 4-inch lengths.

7. Place the sausage on a baking sheet lined with a wire rack and refrigerate, uncovered, for a few hours or preferably overnight.

8. Cut the links apart. Refrigerate, wrapped well in plastic, for 2 to 3 days, or freeze for up to 3 months; thaw overnight in the refrigerator before using.

9. Cook as desired (see methods on pages 38 to 40) to an internal temperature of 160°F (71°C).

INGREDIENTS

- 6 feet small hog casing
- 2 pounds boneless lean beef chuck
- 4 ounces pork fat
- 2 cups fresh bread crumbs
- 1 tablespoon kosher salt
- 1 teaspoon freshly ground black pepper
- ½ teaspoon ground mace
- ½ teaspoon dried thyme
- ¼ teaspoon ground coriander
- ¼ teaspoon freshly grated nutmeg
- 2 large egg yolks, beaten

BARE BANGERS

Some pubs skip the casing and serve skinless bangers. To make this variation, shape the meat mixture into logs, roll each log in beaten egg and then in dried bread crumbs, and chill on a baking sheet until firm. Panfry the sausages until golden, about 15 minutes.

ROSEMARY AND MUSTARD BEEF SAUSAGE

MAKES 2 POUNDS

- 4 feet medium hog casing
- 2 pounds boneless beef chuck blade roast, well marbled
- 4 garlic cloves, minced
- 2 teaspoons yellow mustard seeds
- 1 tablespoon minced fresh rosemary leaves
- 2 teaspoons kosher salt
- 1 teaspoon freshly ground black pepper
- 1 teaspoon sugar
- ½ teaspoon crushed red pepper (optional)
- 1 tablespoon Dijon mustard

What makes this sausage a standout is its coarse texture and the choice combination of yellow mustard seeds, fresh rosemary (emphasis on fresh), and minced garlic. Dijon mustard is also added for extra flavor; serve more mustard alongside the sausages after grilling or roasting.

1. Prepare the casing (see page 29).

2. Cut the meat into 1-inch cubes. Freeze the cubes for about 30 minutes to firm them up before grinding.

3. Grind the meat with the garlic, mustard seeds, and rosemary through the coarse disk of a meat grinder.

4. In a large bowl, combine the meat mixture, salt, black pepper, sugar, crushed red pepper (if using), and mustard. Mix well, using your hands.

5. Stuff the mixture into the prepared casing, prick air pockets, and twist off into 4-inch lengths.

6. Place the sausage on a baking sheet lined with a wire rack and refrigerate, uncovered, for a few hours or preferably overnight.

7. Cut the links apart. Refrigerate, wrapped well in plastic, for 2 to 3 days, or freeze for up to 3 months; thaw overnight in the refrigerator before using.

8. Cook as desired (see methods on pages 38 to 40) to an internal temperature of 160°F (71°C).

KISHKE

Originally made by stuffing a cow's intestine with a seasoned mincemeat filling, kishke — the word is Yiddish for "intestine"—is considered a delicacy in Eastern European Jewish cooking. The name is derived from the Polish *kiszka* and Ukrainian *kishka*, and the word has become part of American slang for "guts" (as in "It takes a lot of kishke to clean out a cow's intestine!"). Kishke is often added to cholent, a hearty, slow-cooking stew traditionally served on the Sabbath, but it is also good on its own.

1. Prepare the casing (see page 29). Cut the casing into 8-inch lengths and sew one end of each length with cotton thread.

2. Cut the meat into 1-inch cubes. Freeze the cubes for about 30 minutes to firm them up before grinding.

3. Grind the meat through the fine disk of a meat grinder.

4. In a large bowl, combine the meat, onions, salt, pepper, and egg yolks. Sprinkle with ⅓ cup matzo meal and work all the ingredients together with your hands, adding the remaining matzo meal as needed until the mixture is the consistency of meatloaf.

5. Stuff the mixture into the prepared casings, allowing room for expansion. Sew closed the open end of each sausage.

6. Bring a large pot of water to a boil. Cook the sausages in the boiling water for 5 minutes. Drain.

7. Preheat the oven to 350°F (175°C).

8. Roast the sausages in a roasting pan in the oven, basting frequently with the pan juices, for about 1½ hours, or until well browned. After cooking, they may be sliced and served warm or added in chunks to cholent. Remove the thread before serving.

3 feet medium beef casing

2 pounds boneless beef chuck blade roast, well marbled

2 small onions, finely chopped

2 teaspoons kosher salt

1 teaspoon freshly ground black pepper

2 egg yolks

⅓–½ cup matzo meal

LAMB, ROSEMARY, AND PINE NUT SAUSAGE

MAKES 3 POUNDS

- 4 feet sheep or small hog casing
- 2½ pounds boneless lean spring lamb
- ½ pound lamb fat
- 2 tablespoons pine nuts, lightly toasted
- 1 tablespoon minced fresh rosemary leaves
- 1 teaspoon kosher salt
- 1 teaspoon freshly ground black pepper
- 2 garlic cloves, minced

The delicate flavor of spring lamb can make a delicious sausage, especially with characteristic Mediterranean flavors. It's an excellent reason to procure fresh lamb from a farm (or farmers' market) during the brief season. Roast or grill the sausage and serve with creamy polenta and a sprightly salad of fresh-picked spring herbs and baby greens. *See photo, page 122.*

1. Prepare the casing (see page 29).

2. Cut the meat and fat into 1-inch cubes. Freeze the cubes for about 30 minutes to firm them up before grinding.

3. In a large bowl, combine the meat, fat, pine nuts, rosemary, salt, pepper, and garlic. Mix well, using your hands.

4. Grind the mixture through the fine disk of a meat grinder.

5. Stuff the mixture into the prepared casing, prick air pockets, and twist off into 3-inch links.

6. Place the sausage on a baking sheet lined with a wire rack and refrigerate, uncovered, for a few hours or preferably overnight.

7. Cut the links apart. Refrigerate, wrapped well in plastic, for 2 to 3 days, or freeze for up to 3 months; thaw overnight in the refrigerator before using.

8. Cook as desired (see methods on pages 38 to 40) to an internal temperature of 160°F (71°C).

Lamb, Rosemary, and Pine Nut Sausage, page 121

Tabbouleh, page 303

Luganega, page 66

Baba Ghanoush, page 302

Venison and Lamb Cevapi, page 199

Moroccan Goat Sausage, page 128

MEDITERRANEAN FLAVORS

Tzatziki, page 302

Pimentón Lentil
Patties, page 258

Hummus with
Sumac, page 302

LAMB, GINGER, AND DRIED APRICOT SAUSAGE

MAKES 3 POUNDS

Crystallized ginger and dried fruit tip the balance of flavors to the sweeter side, which the lamb carries off beautifully. Serve the sausage with a nutty (fruit-free) rice pilaf and wilted greens.

1. Prepare the casing (see page 29).

2. Cut the meat and fat into 1-inch cubes. Freeze the cubes for about 30 minutes to firm them up before grinding.

3. In a large bowl, combine the meat and fat cubes with the salt, pepper, apricots, ginger, and lemon juice. Mix well, using your hands. Freeze the mixture for about 30 minutes.

4. Grind the mixture through the fine disk of a meat grinder.

5. Stuff the mixture into the prepared casing, prick air pockets, and twist off into 3-inch lengths.

6. Place the sausage on a baking sheet lined with a wire rack and refrigerate, uncovered, for a few hours or preferably overnight.

7. Cut the links apart. Refrigerate, wrapped well in plastic, for 2 to 3 days, or freeze for up to 3 months; thaw overnight in the refrigerator before using.

8. Cook as desired (see methods on pages 38 to 40) to an internal temperature of 160°F (71°C).

Ingredients:

- 4 feet sheep or small hog casing
- 2½ pounds boneless lean lamb
- ½ pound lamb fat
- 1 tablespoon kosher salt
- 1 teaspoon freshly ground black pepper
- 2 tablespoons finely chopped dried apricots
- 1 tablespoon finely chopped crystallized ginger
- 2 tablespoons fresh lemon juice or white wine

NO-STICK CHOPPING

It can be tricky to chop sugar-coated candied ginger, but there are a couple of tips to prevent sticking: Coat your sharp chef's knife with oil or cooking spray and dip it in flour before you begin, and wipe clean and repeat as needed. Sprinkling the cut surfaces of the ginger with more sugar will keep those exposed surfaces from resticking; once you've cut the pieces into matchsticks, sprinkle with sugar before turning and cutting crosswise into fine dice. Some people prefer to use kitchen scissors, cleaning them under warm water as needed. Others use a mini chopper with additional sugar, which can be strained out through a sieve after chopping.

North African Lamb Tagine,
page 348

MAKERS

⚓

Jonny Hunter and Charlie Denno

Underground Meats

Madison, Wisconsin

Jonny Hunter

OPEN-SOURCE SAUSAGES

"I very much identify as someone from Madison. The culture is all about agricultural wonderfulness and sustainability. The restaurants have amazing relationships with the farmers, which are less than five miles away, and these farms can totally support the city," explains Jonny Hunter, founder of Underground Meats and The Underground Food Collective, which also includes Underground Butcher, the retail shop where the meats and charcuterie are sold, Forequarter Restaurant, and Underground Catering.

Jonny traces their beginnings to when they were catering and doing pop-up events. "We began working with farmers directly and realized that there was this complicated regulatory system involved with processing the animals, and we didn't want to have to rely on a processor that wouldn't adhere to our standards. We were able to open our own plant and devise the HACCP (Hazard Analysis Critical Control Points) food safety plan with the support of the state." They also had a successful Kickstarter campaign, raising money from investors who supported their mission to provide the first open-source HACCP plan, which they did (see opensourcefoodsafety.org). "We did it not only to become USDA certified, but for the chance to share the plan with other plants. When we were trying to get started, no one else would share information with us. It costs literally thousands of dollars to do the necessary studies and product testing, and these other plants didn't want to train their competition. We decided to create a transparent, concise HACCP model for small producers who might otherwise be priced out of the market. This effort definitely helped us out a lot with our reputation."

Jonny says that the idea for Underground Meats emerged from three things: There wasn't any high-quality processing nearby and they were excited to find a niche; they wanted to understand all elements of the food chain; and they wanted to create markets for farmers and other vendors. "We thought our effect on the local agricultural economy would be greater by also becoming a processor."

Underground Meats opened in 2009, and then the butcher shop and restaurant (called Underground Kitchen, which has since ceased operations due to a fire) opened to provide places for the meat to go. "Charlie Denno joined in 2012 and had such a strong interest in what we were doing that he was able to take over running Underground Meats after about a year. He has been integral to its success," he says.

Charlie's personal experience with butchery started about seven years before he joined Underground Meats. "I was living with a good friend of mine who was a chef, and we wanted to explore nose-to-tail butchery, so a local farmer agreed to raise a pig for us and we taught ourselves how to make fresh sausages. The next year we got two pigs and ventured into dry-cured salami, and eventually we got ten pigs in a single year and gave our friends and family all our products. Then we started a sort of charcuterie share." Later, during his stint as the head sausage maker at a regional deli chain, he learned the commercial side of the business. "I took the opportunity to gain an understanding of the food safety programs that are an essential part of being a USDA-inspected operation. I also learned what not to do, like buying commodity pork. It was the antithesis of what they were doing at Underground Meats, which I was already aware of. Our sourcing is impeccable."

Besides helping to write and implement the HACCP plan for the Underground Meats processing plant, Charlie is responsible for locating farms and animals that are raised to their standards and then figuring out what to do with them. "We try and be our own stocker and larder while also keeping the food chain local to this little pocket of the Midwest, and that helps drive the food scene here as well."

On its website, Underground Meats sells a line of dry-cured salami that includes classic varieties like finocchiona; less familiar ones such as German-style aahlewurst, made with rum and coriander; and others with a twist — like pepperoni that's spicy with ghost chiles. At Underground Butcher, the retail shop where they sell their meats, they have a full line of charcuterie as well as seasonal sausages made with local ingredients, such as venison sausage seasoned with northern prickly ash berries, Wisconsin's only native citrus. (See page 128 for their recipe for Moroccan Goat Sausage, another specialty product.)

Charlie says that experimentation is highly encouraged at the shop — an approach that should be adopted at home, too. "Sometimes the results are horrible, or we get mixed results. Jonny is not afraid of allowing us to make mistakes if it's in the spirit of progress." Or, even better, if the mistake winds up being fortuitous. "I was making smoked beef and pork sausages with chunks of local aged cheddar one day and put them in the smoker, then went home and forgot all about them. The next day, the sausages were so shriveled that I just tossed them into a tub in the walk-in fridge to let them cool off before they could be thrown away. Well, turns out a fellow worker decided to taste one, loved it, and now we end up using the very same process to make our snack sticks." How's that for a happy accident!

Charlie Denno

MOROCCAN GOAT SAUSAGE

According to Charlie Denno of Underground Meats (see profile, page 126), who contributed the recipe, this sausage is inspired by *ras el hanout,* a North African blend of spices that roughly translates to "head of the shop" (as in "head of the class"). The precise blend varies from region to region (see opposite for one version) but often consists of cardamom, cumin, coriander, cinnamon, ginger, turmeric, and peppers. He swaps out the mace and anise for cumin and cayenne, for a fiery kick.

"We like to use goat because it is a highly sustainable resource and we love its funky gaminess. We buy male dairy goats from select farmers who care for these animals up until they are harvested (it's a fact of goat farming that males are often euthanized and their remains discarded). We like to think of this as a win-win for us and the farmer. We get great meat that we feel is underutilized in this country, and they get a revenue stream out of what would normally be discarded."

He adds that the recipe has a "gang of spices" but doesn't yield a particularly hot sausage. "If you are timid about eating goat, this recipe may very well change your opinion. The gamey nature of the meat is quelled by warming spices and the brightness of parsley and lemon." Goat is very lean; for a richer sausage, replace up to 30 percent with short ribs or other fatty beef. *See photo, page 122.*

Ingredients

- 10 feet lamb casing
- 5 pounds boneless goat meat
- ¼ cup kosher salt
- 2 tablespoons dried oregano
- 1 tablespoon plus 1 teaspoon ground cumin
- 1 tablespoon ground cinnamon
- 1 tablespoon ground coriander
- 1 tablespoon cayenne pepper
- 1 tablespoon freshly ground black pepper
- 2 teaspoons garlic powder

1. Prepare the casing (see page 29).

2. Cut the meat into 1-inch cubes. Mix the meat cubes together with the salt, oregano, cumin, cinnamon, coriander, cayenne, black pepper, garlic powder, cardamom, ginger, nutmeg, cloves, and turmeric. Chill for at least 2 hours.

3. Meanwhile, toast the pine nuts in a dry skillet until golden brown. They can burn quickly, so remove them from the pan once they have started to color. Set aside.

4. In a small bowl, combine the wine, lemon juice, and raisins. Cover and set aside.

5. Grind the meat mixture through the medium disk of a meat grinder. Freeze for 30 minutes.

6. Grind the meat through the medium disk of the grinder a second time.

7. Mix the soaked raisins, pine nuts, and parsley into the meat mixture with your hands until well combined and a little sticky.

2 teaspoons ground
  cardamom

2 teaspoons ground
  ginger

1½ teaspoons freshly
  grated nustmeg

1 teaspoon ground
  cloves

1 teaspoon ground
  turmeric

⅓ cup pine nuts

¼ cup white wine

2 tablespoons fresh
  lemon juice

½ cup golden raisins

¼ cup chopped fresh
  flat-leaf parsley

8. Stuff the mixture into the prepared casing, prick air pockets, and twist off into 6-inch lengths. Alternatively, make long coils instead of twisting into links.

9. Place the sausage on a baking sheet fitted with a wire rack and refrigerate, uncovered, for 1 day.

10. Cut the links apart. The sausage can be refrigerated for 2 to 3 days, or frozen for up to 3 months; thaw overnight in the refrigerator before using.

11. Cook as desired (see methods on pages 38 to 40) to an internal temperature of 160°F (71°C).

RAS EL HANOUT

Do as they do in northern Africa and improvise with whatever spices you have on hand — and have an affinity for. And remember to always buy spices (preferably whole, in bulk, and grind as you go) from a store with high turnover, and replenish whenever they lose their fragrance.

To make it: Toast 1½ teaspoons coriander seeds and ¾ teaspoon cumin seeds in a dry skillet over medium heat, tossing, until aromatic, 2 to 3 minutes. Let cool, then grind with ½ teaspoon crushed dried chile peppers in a mortar and pestle (or a spice mill) to a fine powder. Transfer to a small bowl. Add 1¼ teaspoons ground cinnamon, 1 teaspoon paprika, and ½ teaspoon each ground cardamom, ground ginger, and ground turmeric. Mix to combine. Store in a sealed jar at room temperature for up to a month. Makes about 3 tablespoons.

Meet the **MAKER**

⚜

Cathy Barrow

Food Writer and
Cookbook Author

Washington, D.C.

THE PRESERVATIONIST

Soon after starting a blog (under the guise of *Mrs. Wheelbarrow's Kitchen*), Cathy Barrow came to be a respected expert on many aspects of home preservation, tapping into the renewed interest in putting up jams, jellies, and pickles. (She's written an award-winning book on the subject since then, called *Mrs. Wheelbarrow's Practical Pantry*, published in 2014.) But what really put her on the map was Charcutepalooza — a year-long celebration of sausage making that she cocreated with Kim Foster.

The year was 2011, sausage making was experiencing a revival, and the web-based event struck a chord with followers, bloggers, and chefs and other food cognescenti across the globe. Like so many successful ventures, this one involved just the proper ratio of serendipity and chutzpah. "I was a virtually unknown blogger with a very small and very loyal readership. Kim also had a blog at the time called *Yummy Mummy*. I was trying to figure out a hook for my blog and how to get more traffic, and one Sunday morning in late December Kim tweeted, 'It's so cold in my basement I could hang meat.' I tweeted back, 'You should hang a duck breast and you'll have duck prosciutto in 7 days.' Then I told her how to do it in 140 characters. She replied, 'Oh man we should do a year of this where you teach us all how to make charcuterie and we'll call it charcutepalooza.'"

That was the serendipitous part; Cathy provided the chutzpah, relying on her past experience as a marketing director for nonprofits to put together a plan that would pull together sponsors, discounted meats from D'artagnan, a charcuterie class in Gascony where the winner would eventually be announced, and a travel agency — appropriately called Truffled Pig — that funded all flights to France. She had also designed the "curriculum" of monthly challenges for 12 months. "This all happened within three days. Food52 came on board about three weeks later, and then the *Washington Post* ran a story about us on the front page of the food section and it took off from there. Michael Ruhlman participated, the former charcuterie teacher from the CIA was participating. There were about 400 bloggers that were doing this all over the world, and at least 200 of them were very active. I would post a challenge each month and they would respond and post photos. We did increasingly complex challenges, including gallatines where you bone out an entire chicken, plus pâtés and a lot of other highly complex preparations. Everyone was really into it."

As for Cathy's personal know-how, she viewed charcuterie as just another aspect of preserving, which she had been doing for years. "I had made sausage, I had made duck prosciutto, I'd made pâté, but I didn't know a lot about it otherwise. I learned it on the job. I would figure it out and then I would put up a challenge. I think that struck a chord with a lot of people, too, since it wasn't like I was an expert, I was just a home cook figuring it out. I'm a good teacher and I could write it out and give them some instructions."

Cathy says the whole experience has reverberated in her kitchen to this day. For starters, she learned how to make smaller quantities of things. "And that's hard; a lot of these recipes start with ten pounds of meat or an entire pork shoulder, so I guess over the years I've gotten better at making discrete recipes that I know I can get through myself or share with friends. The amusing part is that my husband is a vegetarian and it's just the two of us, so I don't do a lot of big projects. The beauty of charcuterie is that I can garnish vegetarian meals I make for my husband with this tiny bit of my pancetta or prosciutto or whatever else I have on hand. For example, I often make shakshuka for him and then add some sausage for me." She also found it a huge relief to no longer feel obliged to stuff all sausages, instead making bulk sausage that she can take out as needed. "That was a big aha moment."

Perhaps the biggest takeaway from the year-long endeavor was that she gained an appreciation for buying whole cuts of meat and learning to break them down herself. "Only then can you fully understand that there's value and there's also variety. You can make pork steaks, sausage, salami, all from one shoulder. I can break down poultry like nobody's business. But the bigger cuts, sometimes it's not pretty, but I figure it's going in the grinder, for heaven's sake."

She always buys meat directly from the source, either the farmers themselves or a trusted butcher in town. "I don't eat much meat, and what I eat I want well raised and properly butchered. Over the years I've just found that I know the farmers that I trust where I consistently bring things home and say this is the best pork chop or chicken or whatever. I also visit most of the farms, and that makes all the difference. I want to know where it's from and treat it with tender, loving care and have a really great meal."

After all, for Cathy, charcuterie is just a way of life: "I think anybody who has studied in Europe, which I did, comes back completely converted. I learned pig butchery and charcuterie in France, where they've been doing it this way for hundreds of years. They use every last bit, and nothing is wasted. There's artistry in that."

TRADE SECRETS

"Whenever I am teaching charcuterie or writing about this subject, I always try to bring people back to the reality of the smart, economical, no-waste way of consuming an animal that you raised for a long time, hundreds of years ago. People start raising pigs in early spring and they'll have babies all the way through to August, and then they are going to be slaughtering pigs starting in November through February or March, when they start raising the babies again. That's when I buy my pork.

"Also, initially I thought I had to buy these big packages with 100 feet of casings, and it's a lot. Now I have a really good butcher that I just love and she sells me 10 feet of casings."

— Cathy Barrow

MERGUEZ

Originating in North Africa, this spicy lamb sausage's popularity has since spread to the Middle East. This recipe was inspired by one that Cathy Barrow has used for years — and which was one of two recipes (along with chorizo) she used in the May challenge of the year-long Charcutepalooza, which focused on sausage making (see profile, page 130). Her recipe won a wild card on Food52 and eventually made its way into one of their cookbooks. Cathy makes hers in bulk — or in coils for parties — but here we call for stuffing the sausage into casing. *See photo, page 116.*

1. Prepare the casing (see page 29).

2. Cut the lamb and fat into 1-inch cubes. In a large bowl, sprinkle the lamb and fat with the garlic, salt, paprika, oregano, coriander, cumin, anise seed, cinnamon, cayenne, and black pepper. Mix well to coat the meat and fat with the spices. Chill, covered, for at least 2 hours, or preferably overnight.

3. Grind the meat mixture through the small disk of a meat grinder.

4. Add the wine and water to the ground meat mixture. Mix with your hands until well combined and a little sticky.

5. Stuff the mixture into the prepared casing, prick air pockets, and twist off into 10-inch lengths if using lamb casings, or 6-inch lengths if using hog casings.

6. Place the sausage on a baking sheet fitted with a wire rack and refrigerate, uncovered, for 1 day.

7. Cut the links apart. The sausage can be refrigerated for 2 to 3 days, or frozen for up to 3 months; thaw overnight in the refrigerator before using.

8. Cook as desired (see methods on pages 38 to 40) to an internal temperature of 160°F (71°C).

20 feet lamb casing, or 10 feet medium hog casing

4 pounds boneless lamb shoulder

1 pound pork fat

8 garlic cloves, minced

3 tablespoons kosher salt

2 tablespoons paprika

1 tablespoon dried oregano

2 teaspoons ground coriander

2 teaspoons ground cumin

2 teaspoons anise seed

2 teaspoons ground cinnamon

2 teaspoons cayenne pepper

1½ teaspoons freshly ground black pepper

¼ cup red wine

¼ cup water

BOCKWURST

2 feet small hog casing

1¾ pounds boneless veal

¼ pound pork fat

1 teaspoon kosher salt

¾ teaspoon ground cloves

½ teaspoon freshly ground white pepper

¼ cup finely minced onion

2 teaspoons finely chopped fresh flat-leaf parsley

2 teaspoons minced fresh chives

1 cup milk

1 large egg, beaten

Bock means "buck" or "he-goat" in German — perhaps the original meat in this mild-flavored veal sausage. Bockwurst is traditionally seasoned with parsley and chives, as here, and served in the spring, washed down with full-bodied, hoppy bock beer. It tastes best when poached and served piping hot.

1. Prepare the casing (see page 29).

2. Cut the meat and fat into 1-inch cubes. Freeze the cubes for about 30 minutes to firm them up before grinding.

3. Grind the veal and fat separately through the fine disk of a meat grinder.

4. In a large bowl, combine the veal, fat, salt, cloves, pepper, onion, parsley, chives, milk, and egg. Mix well, using your hands. Freeze for 30 minutes.

5. Grind the seasoned mixture through the fine disk of the meat grinder.

6. Stuff the mixture into the prepared casing, prick air pockets, and twist off into 3- to 4-inch lengths.

7. These sausages can be refrigerated, covered, for up to 2 days.

8. Bring a large pot of water to a boil. Add the sausages, reduce the heat, and poach the sausages in the simmering water for 20 to 25 minutes.

9. Drain the sausages, pat dry, and cut the links apart. Refrigerate for up to 2 days or freeze for up to 2 months; thaw overnight in the refrigerator. To serve, reheat gently until warmed through.

Bryan Bracewell

Southside
Market & Barbeque

Elgin, Texas, and Bastrop, Texas

ALL ABOUT THE SAUSAGE

In the Lone Star State, beef equals barbeque equals sausage. They all go together. You can travel far and wide in Texas (that means 660 miles from east to west and 790 miles from north to south) and never go hungry, what with all the barbeque joints dotting the vast landscape. Yet Austin and the surrounding Hill Country are where you'll find the highest concentration of places that make and sell hot links — Texas's own brand of fiery all-beef sausages. Everyone seems to have his or her own favorite haunt, and for many folks that would be Southside Market & Barbecue, self-described as "The Oldest Barbeque Joint in Texas."

Legend has it that Southside was started by a local butcher named William J. Moon, who offered door-to-door delivery of his meats and eventually opened the original Southside location in 1886. Over the years, Southside changed hands, but the original recipe for its signature sausage, called Elgin hot guts (so named because of the hog casings, or "guts"), was preserved, according to Bryan Bracewell, the current owner. "My grandparents bought Southside in 1968, so I'm the third generation to take over the business in the past 48 years," he says. "From the beginning in 1882 to 1968, no recipe was ever written down. It was just passed down to a guy who knew how to make it. So when my granddad bought the business, he spent a couple months shadowing this guy and documenting everything so it would be consistent each day. He made him weigh everything on a scale. Originally the sausage was mostly beef but occasionally it would have pork, since the sausage was a way to use up whatever didn't sell at the butcher shop that day. But my granddad changed it to being made solely with beef from that point forward."

As for the "hot" part of the sausage, there's a story behind that as well: "It was always called Elgin hot guts, because for the time it was considered spicy. The original recipe was called Original, but my granddad wanted to make it more family friendly (i.e., less spicy). So he took some of the spice out of the sausage and put it in the sauce. But the Original name had stuck by then, so he kept that even though the formula changed. Then by the 125th anniversary, in 2007, we started to lose the name hot guts. It wasn't really spicy and people felt it was misleading. So I went back to the original and called it 1882 Hot Recipe. And now we're sort of reviving hot guts, not on the menu board but just among old-timers and regular customers."

Which is all just to say that you can order the Original beef sausage (which is the original minus the spices) or the 1882 Hot sausage (which is the *real* original), either in the market or from their online store.

All of their sausages are still ground, seasoned, and stuffed on site, five days a week. "We make it fresh and cook it fresh, every day. If we have to freeze sausage, we sell it frozen or mostly just end up taking it home ourselves. On an average Saturday, any time of year, we sell about 1,400 pounds of sausage (plus 1,000 pounds of brisket). On holidays we plan for 2,000 pounds," says Bryan.

"When I got out of school in 1998, we were selling 95 percent fresh and 5 percent smoked. There had always been a smoker at the plant, but it just wasn't used that much since the preference was for fresh sausage. But sausage consumption had changed and people wanted a quick meal that they could throw on the grill. So my wife joined the company and went on the road and sold our smoked sausages to the grocery stores in the region. Today, our retail business is 85 to 90 percent smoked and 10 to 15 percent fresh. We have a USDA-inspected plant out back of the shop in Elgin for the retail business." The sausage is smoked, low and slow, over Texas post oak wood for retail sales and the restaurant, but Bryan says they are still known as "the fresh sausage place" among locals, who sometimes buy up to 100 pounds at a time for family reunions or other gatherings.

On the days leading up to July 4, by far the busiest time of year, you'll find Bryan smoking sausages out back. "I don't get to cook near as much as I'd like to, but I do think it's important for me to sweat along with the crew when it's especially busy. I like to hang out with the guys and show them that I love them and am right there with them."

Even though Southside is also known for its smoked brisket, ribs, and other smoked meats, they have a saying: We've got sausage and stuff to go with sausage. "Sausage has always been king. It's the center of the plate. We're proud of it. It's the reason we're on the map. We're all about the sausage."

TRADE SECRETS

"It's real simple for me: When I was studying food science as an undergraduate at Texas A&M University, I was pretty intimidated to learn that there was this whole science behind making sausage. And our sausage does have science behind it. But the reality is that you can make delicious sausage by just keeping it simple. Simple cut of meat, simple spices, simple process. Don't overthink it."

— Bryan Bracewell

Southside's original slaughterhouse

VEAL SALTIMBOCCA SAUSAGE

MAKES 2 POUNDS

Saltimbocca, which means "jumps in the mouth" in Italian, refers to the utter deliciousness of the Roman specialty that bears the name. (As in, "I want this dish to jump into my mouth!") Here, the components of the classic dish — veal, prosciutto, sage, and white wine — are combined in a sausage that you will also want to devour, *presto pronto*.

1. Prepare the casing (see page 29).

2. Cut the veal into 1-inch cubes. Freeze for 30 minutes to firm them up before grinding.

3. Grind the veal through the fine disk of a meat grinder. Refrigerate.

4. In a large bowl, combine the prosciutto, sage, salt, pepper, and wine. Add the veal and mix well, using your hands. Freeze for 30 minutes.

5. Grind the meat mixture through the fine disk of the meat grinder. Freeze for 30 minutes.

6. Stuff the mixture into the prepared casing, prick air pockets, and twist off into 5-inch links.

7. Bring a large pot of water to a boil. Add the sausages, reduce the heat, and poach the sausages in the simmering water for 20 to 25 minutes.

8. Drain the sausages, pat dry, and cut the links apart. Refrigerate for up to 2 days, or freeze for up to 2 months; thaw overnight in the refrigerator. To serve, reheat gently until warmed through.

4 feet small hog casing

2 pounds boneless veal shoulder

4 ounces prosciutto, cut into thin strips

15 fresh sage leaves, chopped

2 teaspoons kosher salt

½ teaspoon freshly ground white pepper

½ cup dry white wine

ALL-BEEF SUMMER SAUSAGE

MAKES 3 POUNDS

3 feet large beef casing

3 pounds boneless beef chuck blade roast, well marbled

4 garlic cloves, minced

1 tablespoon kosher salt

2 teaspoons brown sugar

1 teaspoon mustard seeds

1 teaspoon freshly ground black pepper

Insta Cure #1 (use supplier's recommended quantity for 3 pounds of meat; see Notes)

1 cup water

¾ teaspoon liquid smoke

Consider this an easy-to-make (meaning only cooked, not smoked) alternative to more traditional summer sausage, typically made with pork fat and smoked low and slow (see Thuringer Sausage, page 138). Purists can make the sausage the traditional way: Hot-smoke the sausages (omit liquid smoke) until they reach an internal temperature of 160°F (71°C). Summer sausage makes an excellent snack or hors d'oeuvre, sliced and served with sharp cheese and rye or pumpernickel bread. *See photo, page 72.*

1. Prepare the casing (see page 29).

2. Cut the meat into 1-inch cubes. Freeze the cubes for about 30 minutes to firm them up before grinding.

3. Grind the meat through the fine disk of a meat grinder.

4. In a large bowl, combine the ground meat and the garlic. Mix well, using your hands.

5. In a smaller bowl, combine the salt, sugar, mustard seeds, pepper, curing salt, water, and liquid smoke. Stir until blended. Add the spice slurry to the meat mixture. Mix well, using your hands.

6. Stuff the mixture into the prepared casing, prick air pockets, and twist off into 6-inch lengths.

7. Place the sausage on a baking sheet lined with a wire rack and refrigerate, uncovered, for a few hours or preferably overnight.

8. Preheat the oven to 200°F (93°C). Transfer the sausage (on the rack and baking sheet) to the oven and cook for 4 hours, or until the internal temperature reaches 160°F (71°C). The meat will remain bright red even when fully cooked. (Or smoke the links in a smoker as described on page 48.)

9. Let cool before separating the links. The sausages can be refrigerated, covered, for up to 3 weeks.

NOTES: *We recommend that you use a commercial premixed cure in any recipe for cured sausage. Premixed cures replace the saltpeter in older recipes.*

The processing in this recipe may not fully cook the beef. See Dry Sausage Safety on page 43.

THURINGER SAUSAGE

The earliest record of Thuringer bratwurst (or *rostbratwurst*, as it is more formally called in Germany) dates back to 1404, and the oldest recipe dates back to the early seventeenth century. For it to be a true Thuringer wurst, it must be produced in the Thuringen region using only ingredients from that region. But you can — and should — replicate the sausage no matter where you live. This version follows tradition and smokes it slightly for a semidry result; it will keep longer than fresh sausage, but since it is not completely dried, it should be consumed within a couple of weeks. *See photo, page 98.*

1. Partially thaw the pork fat and cut it into 1-inch cubes. Cut the beef into 1-inch cubes, then freeze the cubes for about 30 minutes to firm them up before grinding.

2. Grind the fat and meat separately through the fine disk of a meat grinder.

3. In a large bowl, combine the fat, meat, salt, pepper, sugar, paprika, caraway seeds, celery seeds, coriander, mace, mustard seeds, ascorbic acid, nutmeg, and curing salt. Mix well, using your hands.

4. Spread the mixture in a large, shallow pan, cover loosely with waxed paper, and cure it in the refrigerator for 24 to 48 hours.

5. Prepare the casing (see page 29).

6. Stuff the mixture into the prepared casing, prick air pockets, and twist off into 4- to 6-inch links. With butcher's twine, tie two separate knots between *every other* link, and one knot at the beginning and another at the end of the stuffed casing.

7. Hang the sausage in a cold place for 1 to 2 days (or place on a baking sheet fitted with a wire rack and refrigerate, uncovered, for 1 to 2 days).

8. Bring the sausage to room temperature, then cold-smoke at 90°F (32°C) for 12 hours.

9. Hang to dry for another day or two before eating. Make sure the sausage is kept cool.

NOTES: *Prepare pork fat according to the instructions on page 44 to ensure that it is free from trichinosis. Also, see Dry Sausage Safety on page 43.*

We recommend that you use a commercial premixed cure in any recipe for cured sausage. Use the commercial premixed cure at the level recommended by the supplier. Premixed cures replace the saltpeter in older recipes.

Ingredients

- 1 pound prefrozen pork fat (see Notes)
- 4 pounds boneless lean beef
- 2 tablespoons plus 1 teaspoon kosher salt
- 1 tablespoon freshly ground white pepper
- 1 tablespoon sugar
- 2 teaspoons paprika
- ½ teaspoon pulverized caraway seeds
- ½ teaspoon ground celery seeds
- ½ teaspoon crushed coriander seeds
- ½ teaspoon ground mace
- ½ teaspoon crushed mustard seeds
- ¼ teaspoon ascorbic acid
- ¼ teaspoon freshly grated nutmeg
- Insta Cure #1 (use supplier's recommendation for 5 pounds of meat; see Notes)
- 4 feet medium hog casing

KOSHER SALAMI

8 pounds boneless lean beef chuck

2 pounds beef fat

3 tablespoons plus 1 teaspoon kosher salt

1½ tablespoons sugar

1 tablespoon coarsely ground coriander seeds

1 tablespoon freshly ground white pepper

1½ teaspoons coarsely crushed white peppercorns

½ teaspoon ascorbic acid

Insta Cure #2 (use supplier's recommended quantity for 10 pounds of meat; see Notes)

1 cup dry white wine

1½ teaspoons minced garlic

4 feet large beef casing

This recipe makes real kosher salami *if* (and only if) you have access to kosher-butchered beef. If not, then just like kosher dill pickles, it's the flavor that counts. Since this is an all-beef recipe, use blade-cut chuck, which has about the right proportion of lean to fat. Trim away all fat when cubing the meat and precisely measure the amounts. Before you begin this or any other smoked sausage recipe, read the How to Smoke instructions beginning on page 46. *See photo, page 150.*

1. Cut the meat and fat into 1-inch cubes. Freeze the cubes for about 30 minutes to firm them up before grinding.

2. Grind the meat through the fine disk of a meat grinder into a large bowl. Grind the fat through the coarse disk of the meat grinder into the bowl with the meat.

3. Mix together the salt, sugar, coriander, ground and crushed white pepper, ascorbic acid, curing salt, wine, and garlic, and pour the mixture over the meat and fat. Mix well, using your hands.

4. Spread the mixture in a large, shallow pan, cover loosely with waxed paper, and cure it in the refrigerator for 24 to 48 hours.

5. Prepare the casing (see page 29).

6. Stuff the mixture into the prepared casing, prick air pockets, and twist off into 8- to 9-inch links. Tie off each link with butcher's twine. Do not separate the links.

7. Hang the sausage in a cold place to dry for 1 week.

8. Wipe the sausage dry and smoke at 120°F (49°C) for 8 hours.

9. Increase the smoking temperature to 150 to 160°F (66 to 71°C) and smoke for 4 hours longer.

10. Because the smoking aids in the drying process, the salami should be ready to eat after about 3 weeks of additional drying in a cold place (see Notes). When dried, cut the links apart, wrap the salami in plastic, and refrigerate.

NOTES: *We recommend that you use a commercial premixed cure in any recipe for cured sausage. Premixed cures replace the saltpeter in older recipes.*

The processing in this recipe may not fully cook the beef. See Dry Sausage Safety on page 43.

Loukanika, page 152

COMBINATION SAUSAGES

Sausage usually carries the label of its birthplace even when it is imitated in foreign countries which have forgotten that the frankfurter was invented in Frankfurt and baloney in Bologna.

— Waverley Root, *The Food of France*

Some of our favorite traditional sausages are artful combinations of pork, beef, veal, and lamb. This potpourri of proteins makes sense given sausage's beginnings as a frugal way for farmers and other people to make something tasty, nourishing, and versatile out of whatever scraps were left over at butchering time.

There's another reason as well, and it is the same reason cooks have relied on the trio of pork, veal, and beef in making meatballs and meatloaf: The combination of tastes and textures makes for a wonderfully nuanced result. This is especially true of leaner cuts of meat, which can be too tough or stringy without being bolstered by another.

In Europe, innovations in charcuterie historically centered on some of the important areas of salt deposits — the mountains of eastern France; the area around Salzburg (literally, "Salt Town"), Austria; the Alsace-Lorraine territory in northern France; and several regions within Eastern Europe and Poland. The domestic animals of each area found their way into sausages that reflected local foodways and crops.

Frankfurters, Vienna sausages, Genoa salami, Bavarian summer sausage — all of these and their many cousins are common in our supermarkets today, far from their points of origin. When we make our own versions, going back to the original recipes, we also make a connection to our sausage-making forebears.

BRATWURST

If you mention bratwurst to a German, you're likely to be asked, "Which one?" as there are more than 40 different kinds of brats (it rhymes with *trots*). Every region and many cities and towns have their own versions, with different seasonings, sizes, textures, and methods of making (some are fresh, others semidry, still others are only smoked). This one is inspired by Nürnberger *rostbratwurst*, from the city of (you guessed it!) Nürnberg. There, the locals stipulate that the links must be made only with pork and be 7 to 9 centimeters (3 to 4 inches) long, weigh 20 to 25 grams (less than 1 ounce), and contain marjoram, mace, and no more than 35 percent fat.

Like many a foodstuff adopted here in the United States, liberties are taken; this one combines pork and veal and swaps in allspice for the mace. Grill it so the outside turns brown and almost crunchy, according to tradition; panfrying is another good option. Serve on a generous heap of sauerkraut (or three to a roll, as dictated in Nürnberg), with horseradish on the side.

- 3 feet small hog casing
- 1½ pounds boneless lean pork butt or shoulder
- 1 pound boneless veal shoulder
- ½ pound pork fat
- 1 teaspoon kosher salt
- 1 teaspoon freshly ground white pepper
- ½ teaspoon crushed caraway seeds
- ½ teaspoon dried marjoram
- ¼ teaspoon ground allspice

1. Prepare the casing (see page 29).

2. Cut the pork, veal, and fat into 1-inch cubes. Freeze the cubes for about 30 minutes to firm them up before grinding.

3. Grind the pork, veal, and fat separately through the fine disk of a meat grinder.

4. Mix the ground meats and fat together, freeze for 30 minutes, and grind again.

5. In a large bowl, combine the meat mixture, salt, pepper, caraway seeds, marjoram, and allspice. Mix well, using your hands.

6. Stuff the mixture into the prepared casing, prick air pockets, and twist off into 3- to 4-inch lengths.

7. Place the sausage on a baking sheet fitted with a wire rack and refrigerate, uncovered, for 1 to 2 days.

8. Cut the links apart. Refrigerate, wrapped well in plastic wrap, for 2 days, or freeze for up to 2 months; thaw overnight in the refrigerator.

9. Cook as desired (see methods on pages 38 to 40) to an internal temperature of 160°F (71°C).

WHEN IN SHEBOYGAN . . .

In Sheboygan, Wisconsin, the self-styled Bratwurst Capital of the World, bratwurst making, cooking, and eating are a serious business. Each of the meat markets in town has its own recipe for bratwurst. In order to put on a brat fry, as it's called, true bratwurst lovers make a pilgrimage to their favorite spot for brats and to a bakery for crusty hard rolls, the traditional holder for the cooked meat.

Bratwurst is always grilled slowly over a charcoal fire that is not too hot (coals should be covered with gray ash, not glowing). However, this is never referred to as grilling. It is called frying out, and the grill itself is called a fryer, as is the person who is doing the frying. The bratwurst is never pricked during grilling, lest any of the precious juices and flavors seep out. It takes about 20 minutes, with lots of turning, to produce a perfectly cooked brat. (It's traditional for the fryer to drink a beer — or two — during this process.)

After the brats are done, some Sheboyganites simmer them briefly (5 to 10 minutes) in a brine, as suggested by Miesfeld's Meat Market, a local institution, but this is entirely optional. If you want to give the method a try, pour the contents of two cans of beer into a pot, add half a stick of butter along with a sliced peeled onion, and bring to a simmer. (In the city of Milwaukee, one hour to the south of Sheboygan, locals prefer to simmer the bratwurst in this mixture *before* frying, but this is just not done in Sheboygan.)

Here, bratwurst is always served hot with brown mustard, a dab of ketchup, dill pickles, and raw onions ("the works") piled into a buttered hard roll. Two brats to a roll is called a double. Single brats are for wimps. Have lots of cold ones on ice for washing it all down.

FRESH KIELBASA

The ingredients and even the spelling and pronunciation of *kielbasa* are as variable as the vagaries of spring weather, the time of year when kielbasa is traditionally made. What passes for kielbasa in one area might be regarded as inauthentic in another — understandable given that even its exact roots are up for grabs (Polish or Ukrainian, depending on your allegiance). This version uses a blend of lean pork, beef, and veal. The best way to eat kielbasa? Grilled over a charcoal fire, then tucked into a hard roll slathered with spicy brown mustard.

6 feet large hog casing

3 pounds boneless lean pork butt or shoulder

1 pound boneless lean beef chuck

½ pound boneless veal shoulder

½ pound pork fat

2 tablespoons paprika

1 tablespoon freshly ground black pepper

1 tablespoon dried marjoram

1 tablespoon dried summer savory

2½ teaspoons kosher salt

½ teaspoon ground allspice

3 garlic cloves, minced

1. Prepare the casing (see page 29).

2. Cut the pork, beef, veal, and fat into 1-inch cubes. Freeze the cubes for 30 minutes to firm them up before grinding.

3. Grind the meats and fat together through the coarse disk of a meat grinder.

4. In a large bowl, combine the meat mixture, paprika, pepper, marjoram, summer savory, salt, allspice, and garlic. Mix well, using your hands.

5. Stuff the mixture into the prepared casing, prick air pockets, and twist off into long links (18 to 24 inches is traditional).

6. Coil the links and place on a baking sheet fitted with a wire rack. Refrigerate, uncovered, for 1 to 2 days.

7. Cut the links apart and refrigerate for 2 to 3 days, or freeze for up to 2 months; thaw overnight in the refrigerator.

8. Cook as desired (see methods on pages 38 to 40) to an internal temperature of 160°F (71°C).

SUMMER SAVORY

Not as familiar here as in Europe, summer savory has been described as tasting like mint with a kick from black pepper. It is called the "bean herb" in Germany since its flavor marries so well with beans and is often used in soups and stews. If you can't find summer savory in fresh or dried form, substitute thyme, which has a stronger flavor, or better yet thyme mixed with a touch of mint or sage.

WEISSWURST

MAKES 5 POUNDS

6 feet medium hog casing

1¾ pounds boneless lean pork butt or shoulder

1¾ pounds boneless veal shoulder

1½ pounds pork fat

¾ cup nonfat dry milk powder

½ cup minced fresh flat-leaf parsley

Zest of 2 lemons

¼ cup kosher salt

2 teaspoons granulated onion

2 teaspoons freshly ground white pepper

½ teaspoon ground mace

¾ cup crushed ice (about 6 ounces)

To the uninitiated, weisswurst (literally "white sausage") calls for an instruction manual on how it should be served and enjoyed: First, Bavarians typically shun the sausage after breakfast (or rather, after the peal of the church bells at noon). Second, they never eat the skin, preferring to snip off either end and suck out the insides. But there's a less daunting way to enjoy these mild-mannered links: Using a fork and knife, cut the sausage in half lengthwise, then split each half down the middle and use the knife to peel off the skins, slathering the sausage meat with mustard as you go. Pretzels are the accompaniment of choice in this neck of the German woods. *See photo, page 98.*

1. Prepare the casing (see page 29).

2. Cut the pork, veal, and fat into 1-inch cubes. In a large bowl, combine the meats and fat with the dry milk, parsley, lemon zest, salt, granulated onion, pepper, and mace. Mix well, using your hands. Freeze for 30 minutes.

3. Grind the meat mixture through the medium disk of a meat grinder. Freeze for 30 minutes.

4. Mix the crushed ice into the ground meat mixture. Grind through the medium disk of the meat grinder. Freeze for 30 minutes.

5. Transfer the mixture to the bowl of a heavy-duty stand mixer fitted with the paddle attachment. Beat until smooth and emulsified (see page 156).

6. Stuff the mixture into the prepared casing, prick air pockets, and twist off into 3-inch lengths.

7. Cut the links apart. Refrigerate, wrapped well in plastic, for 2 to 3 days, or freeze for up to 3 months; thaw overnight in the refrigerator before using.

8. Cook as desired (see methods on pages 38 to 40) to an internal temperature of 160°F (71°C); poaching is recommended.

Carolina Story

*Straw Stick &
Brick Delicatessen*

Washington, D.C.

FOR THE LOVE OF CHARCUTERIE

Like many others who have devoted their professional lives to charcuterie, Carolina Story — co-owner with her husband, Jason Story, of Straw Stick & Brick Delicatessen (more on the name later) — says she discovered her passion for all things cured meats while traveling. "My family is from Columbia, South America, but we had relatives in France, Spain, and Italy, and every year my parents would take us somewhere. That's where and when I fell in love with charcuterie. Salami has always been my favorite, but I just love all kinds of cured meats. I also love history, and when you combine food with history . . . for me it was just the most beautiful thing."

The couple met in a wine class while students at The Culinary Institute of America (CIA) and bonded over a mutual love of charcuterie. "I ate a whole lot of it but I didn't make it. Jason, however, was really talented at making it." Turns out Jason had honed his craft while working for Brian Pulcyn at his restaurant in Detroit, Michigan, where Jason grew up. (Along with Michael Ruhlman, Pulcyn literally wrote the book on charcuterie — *Charcuterie: The Craft of Salting, Smoking & Curing* — that has been the inspiration for many of those people profiled in this book.) Carolina credits Jason as being the expert behind the preparation, although she also makes sausages and salami. "I'm very much the artist and my role is to play around with the ingredients and methods, but Jason is very precise. And with salami, you really can't deviate from the recipe and you have to be very methodical and meticulous. So it's really perfect for him because that's how he is in life. I'm very whimsical and he's more precise. So we balance each other out."

Because they were both trained chefs, they viewed the shop as a means of justifying all their experimentation. "When we first opened we made everything. We used the business to make every recipe we were curious about. Of course we eventually had to streamline the product line. Our repertoire is over 132 items and we base what's offered on the popularity of the products. But we had a slogan when we first started, 'Old World Craftsmanship,' and while we don't say it so much anymore, it's still true. The beauty for us is that we are applying that craft with ingredients that are local to us."

In that spirit, they source whole animals from nearby farms, break them down in the shop, and have the goal of using the entire animal in the most delicious way. When deciding on which products to make, Carolina said that it took quite a bit of studying, and she discovered a talent for identifying the flavor profiles and appreciating the recipes and doing the research to learn how each item is served. "When it comes to charcuterie, different countries have different preferences and quality standards. In Spain, the moister and softer, the better. In Italy, it's more about items that are fermented and larger diameter and just a little bit soft. In France, it seemed as if it was all about making it as dry as possible, really hard and dry. We became committed to making the product as authentic as we can to the country of origin of the recipe."

Even customers help the cause by bringing in different meats from their travels. "I don't ask them; they just do it because they know I like to taste everything firsthand before trying to make it myself." And setting up shop in Washington, D.C., meant that they would have customers from all over the world. "What I find especially fulfilling is when a new customer stops by and they are so excited to find the food that is prepared the way that they remember it from their home country."

Now about the name: The shop was originally called The Three Little Pigs Charcuterie, after Carolina's favorite childhood story. But in 2014, they were forced to change the name for being too similar to an establishment in New York City (that one had the French translation). So a customer suggested Straw Stick & Brick, which provided a wonderful link to the tale and the shop's past namesake. (They also swapped the word *charcuterie* for *delicatessen*, mostly to avoid having to keep explaining the meaning of *charcuterie* but also because it seemed to resonate with customers who were looking for more than just salumi. As chefs, the couple wanted to be able to provide people with delicious ways to enjoy the products — hence, they offer a blackboard menu of sandwiches.)

In the end, the couple viewed the whole ordeal as being a positive experience. "People really like our shop for the customer service and the products we offer, so we didn't need to stick with the original name. After all, we're still the little shop in the red brick building with the bright blue doors."

TRADE SECRETS

"When it comes to preparing a charcuterie board, let go, get inspired. Don't overstress about making it look perfect. The perfection actually comes from the whimsy. For example, I think of the board as a Christmas tree and treat the olives as ornaments. Sometimes I'll start with one method and then about halfway through I'll shift it over and it is what it is when it's done. Each board is different.

"People develop their own style depending on their personalities. One staff member loves motorcycles, so every board ends up looking like a motorcycle chain. Another is a graphic designer, and she'll roll up the meats and arrange them in the shape of something, like a fish. That's what's so fun. It shouldn't be daunting. The art is all about letting go."

— Carolina Story

CHARCUTERIE BOARD 101

The best way to showcase your sausage-making efforts is to present them, *tous ensemble*, on a board or platter with assorted accompaniments. Think of the sausages as the stars and the other items as super-talented backup singers — while essential to the experience, they are really there to enhance the sausages' flavor and texture. In other words, you'll want to include items that are tart, sweet, tangy, and bitter but also cool, crisp, creamy, crunchy — everything that sausage is not.

Straw Stick & Brick has a bit of a cult following in and around Washington, D.C., for its distinctive charcuterie boards, which are always made to order, never in advance. According to Carolina Story (see profile, page 146), "The reason why charcuterie is served the way it is, on these boards, is that it is something very special. But also, historically speaking, the whole point of making charcuterie — all those hams and bacon and salami — is to make sure that you and your family have enough to eat throughout the year. That's why you are served these little slivers of meat: You want to share all that love that you've put into making the meats with all your friends. And that's what makes it elegant."

Here, Carolina shares tips and techniques for preparing your own memorable boards:

PRESENTATION: My style is very natural. I want each piece of meat to show its beauty without manipulating it. I like to see the way the paper-thin slices fall into place. Everyone at the shop has the opportunity to create boards as well. I set the standard but I do try to have the staff feel a sense of ownership in what they do, no matter what position they are in. I just give them the pricing structure and quantity and encourage them to practice their art. It's so wonderful to see each person's unique expression as well as how proud he/she feels in what's been done. That's the best part: When you make a charcuterie board and you see how great it looks, you are so proud of yourself. I always feel so proud of myself when I see it from afar.

QUANTITY: First I need to know how the customer intends to serve the board for me to be able to gauge how much charcuterie to recommend. Will the board be the main dish at an event, or served as an appetizer? And if it's an appetizer, what else, if anything, is going to be served? As a general rule, I recommend 2.5 ounces of meat per person. Because charcuterie is shaved so thinly, it takes about 14 slices of salumi to reach an eighth of a pound of meat per person, and I figure that 14 slices of meats, when supplemented with cheese and bread, would be enough to make each person happy. I find that our customers enjoy charcuterie as an indulgence, but that doesn't mean they want *too* much meat (a few leftovers are okay). They also don't want to have too little for guests to feel satisfied. That's where my formula has proven successful.

VARIETY: The assortment of meats tends to be the same whether the board is for two people or twenty. I feel that guests should have a nice variety of flavor profiles. So unless otherwise requested, I usually recommend three or four salumi (salami and other sausages), one pâté, and one or two cured meats such as prosciutto. I have found this balance to be the one that garners the most positive feedback.

SALUMI SELECTION: I always start with the salami, my personal favorite and also a crowd favorite. Although many people are connoisseurs these days, the vast majority are used to types of salumi that are readily available at major grocery stores. Because of this, I try to start with more relatable options such as Genoa salami or hot soppressata. Then I can venture out to something a little less common, such as a spreadable sausage. My typical board will include two hard salamis, one or two smoked sausages, and one spreadable sausage, such as liverwurst. Avoid too many smoked meats, whose flavor can overpower anything else on the board.

ACCOUTREMENTS: I always keep a fully supplied larder, and charcuterie boards are the perfect application for those house-made items, including pickled vegetables, fruit jams, chutneys, and so on. We include our caramelized onion mustard with all of our boards. It's sweet and has a nice kick to it that pairs well with our smoked meats and some of our sweeter salumi, as does our fig jam, apricot chutney, and cherry mustard. In addition, olives and cornichons add a nice acidity to the boards. The yummiest ones are the oil-cured or Castelvetrano olives.

CHEESE: I love cheese with charcuterie. I usually recommend only offering two or three varieties, because I don't want the spread to become a cheese plate as opposed to a charcuterie board. I like to include three cheeses that have a range of textures: one creamy, one semisoft, and one hard. For larger events with more than 20 guests, I recommend offering only hard cheeses since they are easiest to serve (in cubes) and also to pick up (with provided toothpicks). Although soft cheeses are delicious, they tend to get messy with large groups that might have had too many cocktails. Consider instead cloth-bound cheddars, Spanish manchego, and all the many Italian and French soft-ripened cheeses — these are the ones that pair best with the meats that we offer and also with a great variety of salami.

FINAL CONSIDERATION: I like to serve the meats (and cheeses) at room temperature, to allow them to express all the subtle flavors and complexities.

CHARCUTERIE

Kosher Salami, page 139

Pickled Pearl Onions, page 286

Mortadella, page 90

Grainy Mustard, page 293

'Nduja Americana, page 88

Pickled Bell Peppers, page 286

Pepperoni, page 170

Salamette, page 109

Mortadella Mousse, page 284

Liverwurst, page 96

Tomato Chutney, page 295

LOUKANIKA

Redolent of cumin, garlic, and orange zest, loukanika (also spelled lukaniko) is a specialty of Greece. There, the mountainous terrain is ideal for raising lamb and sheep (hence feta cheese), and this sausage makes delicious use of the tougher cuts of lamb by mixing it with lean pork and pork fat. Retsina, a resinous Greek wine, is traditionally added as well; because good examples of this wine can be hard to find here (and it is, as they say, an acquired taste), you can use Metaxa, a Greek liqueur made from brandy that's perfect for sipping with the sausage, too. *See photo, page 140.*

1. Prepare the casing (see page 29).

2. Cut the pork, lamb, and fat into 1-inch cubes. Refrigerate the cubes for about 30 minutes to firm them up before grinding.

3. Heat the olive oil in a skillet over low heat, and sauté the onion and garlic just until softened, about 5 minutes. (Do not let the garlic burn or it will be bitter.) Refrigerate the mixture for 30 minutes.

4. In a large bowl, combine the onion mixture with the pork, lamb, fat, coriander, salt, cumin, pepper, thyme, and orange zest. Mix well, using your hands.

5. Grind the mixture through the fine disk of a meat grinder.

6. Add the wine to moisten the mixture and knead well, using your hands.

7. Stuff the mixture into the prepared casing, prick air pockets, and twist off into 6-inch links.

8. Place the sausage on a baking sheet fitted with a wire rack and refrigerate, uncovered, for a few hours or preferably overnight.

9. Cut the links apart. Refrigerate, wrapped well in plastic, for 2 to 3 days, or freeze for up to 3 months; thaw overnight in the refrigerator before using.

10. Cook as desired (see methods on pages 38 to 40) to an internal temperature of 160°F (71°C).

Ingredients:

- 6 feet medium hog casing
- 2 pounds lean pork
- 1 pound boneless lamb leg or shoulder
- ½ pound pork fat
- 2 tablespoons olive oil
- 1 large onion, chopped
- 4 garlic cloves, chopped
- 1 tablespoon ground coriander
- 1 tablespoon kosher salt
- 1 teaspoon ground cumin
- 1 teaspoon freshly ground black pepper
- 2 teaspoons chopped fresh thyme or 1 teaspoon dried
- Zest from 1 orange
- ½ cup retsina or Metaxa

KASEKRAINER

- 8 feet medium hog casing
- 1 tablespoon vegetable oil
- ½ cup minced onion
- 2 pounds boneless lean pork butt or shoulder
- 2 pounds boneless beef, about 70% lean
- 1 tablespoon minced garlic
- ¼ cup nonfat dry milk powder
- 3 tablespoons salt
- 1 tablespoon dextrose or granulated sugar
- 1 tablespoon ground black pepper
- 1 tablespoon dry mustard
- 1 tablespoon paprika
- 1 teaspoon ground coriander
- 1 teaspoon cayenne pepper
- Insta Cure #1 (use supplier's recommended quantity for 4 pounds of meat; see Note)
- 1 cup crushed ice (about 7 ounces)
- 7 ounces Gruyère cheese, diced into ¼-inch cubes

Every country has its street-cart food culture, and for Austria it's the kasekrainer — plump pork sausages chock-full of chunks of cheese. Emmenthaler is the classic choice, but other melting cheese such as Gruyère and Dutch Gouda are options. When cooked, the sausage turns wonderfully crisp on the outside, melty and creamy on the inside. In other words, this sausage, with roots in neighboring Slovenia, is utterly delicious. This recipe is courtesy of Jeremy Stanton of The Meat Market (see profile, page 154). *See photo, page 99.*

1. Prepare the casing (see page 29).

2. Heat the oil in a 6-inch skillet. Sauté the onion until softened and starting to brown. Set aside to cool.

3. Cut the pork and beef into 1-inch cubes. Freeze for 30 minutes.

4. In a large bowl, sprinkle the garlic, dry milk, salt, dextrose, black pepper, dry mustard, paprika, coriander, cayenne, and curing salt over the meat. Mix well, using your hands.

5. Mix the crushed ice into the meat mixture. Grind through the medium disk of a meat grinder. Freeze for 30 minutes.

6. Transfer the meat mixture to the bowl of a stand mixer. Beat with the paddle attachment until smooth and emulsified.

7. Fold the onion and cheese into the meat mixture.

8. Stuff the mixture into the prepared casing, prick air pockets, and twist off into 6-inch lengths.

9. Place the sausage on a baking sheet fitted with a wire rack and refrigerate, uncovered, for 1 day.

10. Cut the links apart. The sausage can be refrigerated for 2 to 3 days, or frozen for up to 3 months; thaw overnight in the refrigerator before using.

11. Cook as desired (see methods on pages 38 to 40) to an internal temperature of 160°F (71°C).

NOTE: *We recommend that you use a commercial premixed cure in any recipe for cured sausage. Premixed cures replace the saltpeter in older recipes.*

Jeremy Stanton

The Meat Market

Great Barrington,
Massachusetts

THE LOCALIST

If you're ever in the vicinity of Great Barrington, in the Southern Berkshires of Massachusetts, around lunchtime, head over to The Meat Market for their house-made liverwurst sandwich, topped with pickled onions, lettuce, and mustard on rye bread. You're likely to see the owner, Jeremy Stanton, in the open production kitchen, where the shop's many meats, sausages, and charcuterie are crafted by a well-choreographed cadre of butchers.

Before opening The Meat Market in 2011, Jeremy, a classically trained chef, worked in the food industry for over 20 years. His first memory of sausage was when he was working on the line at a restaurant in Stockbridge: "We sourced the sausage from an Italian guy who had a little place on South Main up in North Adams and would bring the sausage down in the back of his Buick. He made a hot Italian sausage with pickled cherry peppers and there was just something so delicious about that piquant flavor." Later, after he had opened a pasta company that garnered a lot of local food awards, he had another epiphany when he was introduced to Barbara Del Molino, the matriarch of a family who had immigrated to Great Barrington from Italy and had been making sausage for years in their basement. "It was unbelievable! I converted the pasta shop into a sausage shop for a few days, gathering together all these people with Barbara to make an amazing amount of sausages."

Once he got the butchery bug, he worked for Stafford Premium Meats, a slaughterhouse that was managed by the New England Heritage Breeds Conservancy and funded by the Cabbage Hill Foundation. "I was hired to develop a line of value-added products using local meats, and sausage fits into that category. While there I discovered that the best way to have an impact on local agriculture is for people like me to run small shops that could ultimately supply the community with great-quality products."

With that goal in mind, he launched Fire Roasted Catering — and, some 10 years after that, he opened The Meat Market. "It was all about supporting local farmers, and my friends who wanted to be farmers. I had started the catering company for the same reason — basically, to buy whole animals, which is the most affordable way to buy meat from local farmers. It's better for the farmers, too, since they don't have to break down the animal and sell the cuts individually or deal with the trimmings." Now he works with over 20 local farms, including one run by his brother, Sean, who raises pigs and chickens on the same property where the brothers grew up.

Jeremy says that he learned fairly quickly that making sausage relies on knowing where the meat is coming from. "We buy only local pigs that are on a non-GMO diet and raised sustainably — meaning out in the open, with access to fresh water and sunshine and proper foraging, and with kind handling all the way through the process." As Jeremy sees it, shopping at butcher shops like his, the customers will also know where their food is coming from, and can trust that it's from a high-quality source. "I have this relationship on your behalf. You're going to pay more, but we'll give you really good advice. So come to your butcher shop and ask a few questions and you'll get some very good information." Such as, when it comes to fresh sausages, "Many people think that it's how much fennel seed or paprika that goes into the sausage, but it really boils down to how much salt, the quality of the meat, and the processing. If you get those three things right, you could make sausage with just salt and it would be amazing."

Grinding is also important, he adds, and they do it all in the walk-in refrigerator. "The key is to keep the meat cold enough, and that's often the biggest problem for home sausage makers. Once that fat starts to melt, you'll have smearing, which can also happen when you have a dull blade and when you are running your meat back through for a second grind." Jeremy swears by his old-fashioned piston stuffer for cased sausages, since you can control the pace. He uses an old commercial refrigerator for the fermentation cabinet — which doubles as display case, with its glass doors — that has a small electric kettle to regulate the temperature and humidity levels, plus a fan to keep the air moving. "This case does really well when it's full of fresh product, which helps to raise the humidity level."

That's generally no problem, since The Meat Market has an extensive charcuterie program where, in addition to a rotating roster of fresh, smoked, and dry-cured sausages, they make whole-cured muscles and whole hams. "We also have an annual event called Sausage Fest where we transform the shop into a crazy scene with live music and a bunch of different sausages for tasting. We've had that for five years straight, but now that we've opened up the restaurant, called Camp Fire, next door, that's going to change everything, in a good way." He plans to do more events since he'll have more space. "*Forbes* magazine recently ranked our burger as the best in America, so I guess we should have a Burger Fest next." No doubt there'll be some sausages on deck at that, too.

TRADE SECRETS

"The key is don't use your KitchenAid grinder as the stuffer. There's a worm drive in your grinder, and it pushes the chunks through the blade and your knife and they cut it into pieces. When you run those pieces back through to stuff the casings, it smears the fat into layers and it heats it up, and what you get is almost an emulsified sausage with all these small layers. Get a five-pound manual canister stuffer instead. Also, in sausage, salt content by weight is critical; there's a formula for the salt for a reason. There's a technique for binding the protein for a reason. These things are not there by accident; they are there intentionally, historically."

— Jeremy Stanton

EMULSIFICATION
EXPLAINED

It's a sticky mess: You're beating meat and fat, after all, in your beloved (and costly) stand mixer, and the motor is heating up the farce like nobody's business. But you're determined to make hot dogs, bologna, mortadella, or whatever sausage is on your mind — the ones that have that inimitable ultra-fine texture that can only be achieved by a process that is not pretty but is doable — and by people across all spectrums.

"To make the emulsification, you are basically taking the fat and wrapping it in a protein balloon, like when making mayonnaise," explains Jeremy Stanton of The Meat Market (see page 154). "The proteins get developed into sheets that turn into balloons, and if you go too far the walls break down and the fat leaks out." In other words, in a proper emulsification, the meat and fat are evenly dispersed for a smooth, uniform texture.

Most people are familiar with emulsifications from making hollandaise sauce or homemade mayonnaise, whereby you've mixed some liquid (vinegar or wine) with an emulsifying agent (usually mustard or egg) and then ever so slowly whisked in oil, at first drop by drop and then in a steady stream, until the consistency is luscious and velvety smooth. That is, until the sauce "breaks" (meaning the protein balloons have burst), resulting in an unsightly mess. Unlike hollandaise, which is capable of being repaired by some fancy footwork (for the ultimate cooking recovery), a broken emulsion in sausage is not salvageable; you just have to scrap the whole mixture and start all over. Otherwise, as Michael Ruhlman describes it, "It will feel a little like eating clumps of soggy ground-up newspaper."

For the vast majority of people who find that prospect unappetizing, here's a formula that's guaranteed to succeed — with a bit of practice. We suggest using a stand mixer over a food processor, for the more powerful motors. Having said that, the more powerful mixing action of the stand mixer motor can cause the ingredients to heat up, thereby impeding emulsification. The easy answer: Freeze everything — ingredients and equipment — for at least an hour before you get started.

SWEDISH POTATO SAUSAGE

4 feet medium hog casing

1 pound boneless very lean beef

½ pound boneless lean pork butt or shoulder

½ pound pork fat

5 large russet potatoes

1 large yellow onion, coarsely chopped

2 teaspoons kosher salt

½ teaspoon freshly ground black pepper

½ teaspoon freshly ground white pepper

¼ teaspoon ground allspice

¼ teaspoon ground mace

¼ teaspoon freshly grated nutmeg

1 garlic clove, minced

Chicken broth, for cooking sausage

On December 13, otherwise known as St. Lucia's Day, Swedes make *potatis korv*, a sausage with potatoes and onions, as part of the holiday feast — as do the many descendants of the Swedish immigrants who settled the upper Midwest. In the original recipes, the potatoes and onions are left uncooked and ground with the meat by hand; here, the potatoes are briefly boiled to make that difficult task easier to do (of course, if you are using an electric grinder, you may opt to skip the cooking step and just cut the raw potatoes into chunks). Once stuffed, the links are tied, end to end, to form a ring, and then poached in a flavorful broth before serving, warm or chilled, with assorted mustards and some lingonberry jam.

1. Prepare the casing (see page 29).

2. Cut the beef, pork, and fat into 1-inch cubes. Freeze the cubes for about 30 minutes to firm them up before grinding.

3. Grind the meats and fat separately through the fine disk of a meat grinder. Freeze until you are ready for step 6.

4. Peel the potatoes and boil in a pot of lightly salted water until just beginning to soften but still quite firm in the centers, about 10 minutes. Let cool.

5. Cube the cooled potatoes and mix with the onion. Pass the mixture through the fine disk of the grinder.

6. In a large bowl, combine the ground meats and fat and the potato mixture. Add the salt, black pepper, white pepper, allspice, mace, nutmeg, and garlic. The mixture will be sticky, so dip your hands in cold water, then mix well, using your hands.

7. Stuff the mixture into the prepared casing, prick air pockets, and twist off into 12-inch links. With butcher's twine, tie two separate knots between each link and one knot at each end. Separate the links by cutting between the two knots, then bring the ends of each link together and tie to form a ring.

8. Bring enough chicken broth to cover the rings to a simmer in a pot. Poach the rings in the broth for 45 minutes. Serve warm, or refrigerate and serve cold. The sausages may be refrigerated for up to 3 days, or frozen for up to 2 months; thaw overnight in the refrigerator.

HOT DOGS
AMERICA'S FAVORITE SAUSAGE

Even professional sausage makers often steer clear of hot dogs in their regular rotation. For one, it's hard to find that one style that pleases everyone — then there's the whole business of achieving the right texture, no easy feat for the faint of heart. Just ask Brent Young of The Meat Hook (see profile, page 208), who is also a partner at Ripper's, a hot dog haven in Rockaway Beach: "It's not about your skill level, it's that you don't have $50,000 to embark on the project of making hot dogs." What he is referring to is the commercial equipment that he found he needed to rely on to achieve hot dog perfection. Besides, this is one sausage you can find in every supermarket throughout the land, and in a great many options, too. Herewith are just some of the ways in which you'll find hot dogs across the land.

Who eats the most hot dogs? In 2015, Americans spent over 2.5 billion dollars on hot dogs in supermarkets, and Americans eat around 150 million hot dogs on Independence Day alone!

➤ In the 1929 Disney animated short film, Mickey Mouse runs a hot dog stand and says his first spoken words, "Hot dogs! Hot dogs!"

➤ There are several conflicting stories about the invention of the hot dog, but one likely origin was in the late 1800s, when German immigrant Charles Feltman sold sausages in rolls at Coney Island, New York.

➤ In 2007, Joey Chestnut broke the hot dog–eating contest world record by consuming 66 hot dogs and buns in only 12 minutes.

➤ When King George VI and his wife, Elizabeth, visited the United States in 1939, the first time a British monarch had set foot in the United States, President Franklin Roosevelt served hot dogs and beer at their state dinner. Those were the king and queen's first hot dogs and helped improve their image with the American public and its support for aiding the British in World War II.

➤ The longest hot dog in the world was 668 feet and weighed 260 pounds, made in 2011 in Paraguay for the country's 200th birthday.

➤ Neil Armstrong and Buzz Aldrin ate hot dogs on their way to the moon, during the *Apollo 11* mission in 1969.

THE TOP 10 HOT DOG–CONSUMING AMERICAN CITIES

1. LOS ANGELES	6. ATLANTA
2. NEW YORK CITY	7. DETROIT
3. PHOENIX	8. CHICAGO
4. PHILADELPHIA	9. WASHINGTON, D.C.
5. BOSTON	10. TAMPA

According to consumption stats from the National Hot Dog and Sausage Council and the Nielson Company.

HOT DOG STYLES AROUND THE UNITED STATES

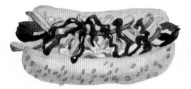

THE CLEVELAND POLISH BOY is a grilled kielbasa topped with coleslaw, french fries, and barbecue sauce.

A SCRAMBLED DOG, found in Georgia, is chopped up and covered in chili, pickles, and onions and piled high with oyster crackers.

THE HALF-SMOKE, from Washington, DC, is a spicy hot dog–sausage hybrid, topped with mustard, onions, and chili.

MAINE RED SNAPPER, so named for its atomic red (thanks to food coloring) hot dog, in its ultra-snappy lamb casing, is tucked into a top-split roll (the same for the iconic lobster roll) and slathered with the state's other claim to fame: Raye's Down East Schooner yellow mustard.

THE SEATTLE-STYLE hot dog is a grilled frank, split in half, served with cream cheese and onions in a bun.

TEXAS DOG keeps it on the simple side with a grilled frank (the thicker kind) cradled in a soft bun and loaded with salsa, melted yellow cheese, and sliced jalapeños (beef chili optional) — often with tortilla chips and more melted cheese alongside.

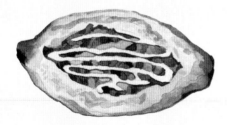

SONORAN HOT DOGS, from Arizona, are wrapped in bacon, served in a bolillo roll, and topped with pinto beans, grilled onions, tomatoes, jalapeño sauce, mayo, mustard, and cheese — usually followed by a nap (and then an antacid).

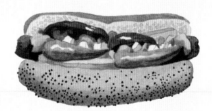

CHICAGO DOGS feature a steamed frankfurter and poppy-seed bun, and are "dragged through the garden" with chopped onions, mustard, relish, a dill pickle spear, tomato slices, and celery salt. Ketchup is unacceptable!

BOUDIN BLANC

To most of the world, boudin blanc (French for "white pudding") is a delicate sausage with a fine, almost creamy texture. It bears little resemblance to the boudin blanc of Louisiana's Cajun country — a sturdy construction of pork, rice, onions, and red pepper noted for its darker color and its heat. This is the traditional French preparation and is worthy of a holiday feast. It's quite a revelation.

- 4 feet medium hog casing
- 1 pound boneless, skinless chicken breast
- 1 pound boneless veal shoulder
- ½ pound pork fat
- 3 large yellow onions, sliced
- 1 cup milk, plus more for cooking
- ¾ cup dried bread crumbs
- 1 teaspoon kosher salt
- ¼ teaspoon ground allspice
- ¼ teaspoon freshly grated nutmeg
- ¼ teaspoon freshly ground white pepper
- 1 tablespoon chopped fresh chives
- 1 tablespoon chopped fresh flat-leaf parsley
- 2 large whole eggs
- 2 large egg whites
- 1 cup heavy cream

1. Prepare the casing (see page 29).

2. Cut the chicken, veal, and fat into 1-inch cubes. Freeze the cubes for about 30 minutes to firm them up before grinding.

3. Grind the pork fat through the fine disk of a meat grinder.

4. In a large skillet over medium heat, slowly melt half of the ground fat. Add the onions and cook slowly, covered, until the onions are translucent, 15 to 20 minutes. Let cool.

5. In a medium saucepan, bring the milk to a boil and add the bread crumbs. Cook, stirring constantly, until the mixture thickens enough to stick to the spoon, about 5 minutes. Let cool.

6. Grind the chicken and veal together through the fine disk of the grinder.

7. In a large bowl, combine meat mixture, onions, the remaining ground fat, salt, allspice, nutmeg, pepper, chives, and parsley. Mix well, using your hands. Freeze for 30 minutes.

8. Grind the seasoned mixture through the fine disk of the grinder.

9. In the bowl of a standing mixer fitted with the paddle attachment, beat the mixture until it is thoroughly blended. Beat in eggs and egg whites, then add the bread crumb mixture. Continue beating and add the cream, a little at a time.

10. Stuff the mixture into the prepared casing, prick air pockets, and twist off into 4-inch links. Refrigerate, covered, for 1 to 2 days.

11. Place the sausage in a large pot. Cover with a mixture of half milk and half water. Bring the liquid to a simmer and poach the sausage gently for 30 minutes, until cooked to an internal temperature of 160°F (71°C).

12. Drain the links and let cool before cutting them apart. Pat dry, then refrigerate, covered, for up to 3 days. To serve, grill or panfry until just heated through.

- 5 pounds boneless lean beef from chuck, round, or shank
- 3 pounds prefrozen boneless lean pork butt or shoulder (see Notes)
- 2 pounds prefrozen pork fat (see Notes)
- 3 tablespoons plus 1 teaspoon kosher salt
- 2 tablespoons whole black peppercorns
- 1½ tablespoons sugar
- 1 tablespoon freshly ground white pepper
- 1 teaspoon ground cardamom
- 1 teaspoon ground coriander
- ½ teaspoon ascorbic acid
- Insta Cure #2 (use supplier's recommended quantity for 10 pounds of meat; see Notes)
- 2 teaspoons minced garlic
- 1 cup best-quality brandy or red wine
- 4 feet large beef casing

GENOA SALAMI

In America, this sausage is named for the Ligurian town on the Adriatic coast where it originated, where it is actually known as *salami di Sant'Olcese*. It is also typically made with pork and veal, but we swapped in beef (easier to find here) for the veal (found in abundance there). Whole peppercorns and red wine are what define this type of salami, along with the larger-than-usual diameter (4 inches)— just the right size for layering on sandwiches, with fresh basil (another Ligurian staple), tomatoes, and buffalo-milk mozzarella. *See photo, page 72.*

1. Cut the beef, pork, and fat into 1-inch cubes. Refrigerate the cubes for about 30 minutes to firm them up before grinding.

2. Grind the meats and fat separately through the coarse disk of a meat grinder.

3. Mix the meats and fat together and freeze for about 30 minutes, then grind the mixture through the fine disk of the meat grinder.

4. In a large bowl, combine the meat mixture, salt, peppercorns, sugar, white pepper, cardamom, coriander, ascorbic acid, curing salt, garlic, and brandy. Mix well, using your hands.

5. Spread the mixture in a large, shallow pan, cover loosely with waxed paper, and cure in the refrigerator for 24 hours.

6. Prepare the casing (see page 29).

7. Stuff the mixture into the prepared casing, prick air pockets, and tie off into 12-inch links with butcher's twine. Do not separate the links.

8. Hang the links to dry in a cold place for 8 to 12 weeks. Test the sausage after 8 weeks by cutting off one link and slicing through it. If the texture is firm enough to suit your taste, the remaining sausage can be cut down and wrapped tightly in plastic for storage in the refrigerator.

NOTES: *Prepare pork according to the instructions on page 44 to ensure that it is free from trichinosis.*

The meat is not cooked; please see Dry Sausage Safety on page 43.

We recommend that you use a commercial premixed cure in any recipe for cured sausage. Premixed cures replace the saltpeter in older recipes.

SALUMI VS. SALAMI

Variety is the slice of life: All these types of salumi start the same way, as raw meat that undergoes aging and curing until it is totally transformed — and entirely edible!

Don't be confused by the similarity in spelling: salumi and salami are similar but distinct. *Salumi* (or *salume*, in the singular form) is the Italian term for any dry-cured meat — and is the Italian equivalent of *charcuterie* in France. There are two types of salumi: whole-muscle cured meats, or "cold cuts," and dried sausages (aka salami).

Common types of salumi (usually made with neck, leg, or shoulder) fall along the lines of prosciutto, speck, and culatello; coppa (aka capocollo); guanciale (pig's jowl); bresaola (beef); pancetta; lardo; and mocetta (made from wild game).

Of all the various kinds of dried sausages, probably more are labeled as salami than anything else. The term encompasses many different sizes and shapes of highly spiced dried sausages. Some are short and fat; others, quite long. Some have a distinctive smoky flavor, while in others wine is the dominant flavor. Common types of salami include finocchiona; soppressata; mortadella; salsiccia salame; lucanica; Toscano; Calabrese; cotechino; cacciatore (or cacciatorini); sanguinaccio; and mostardella, among countless others.

Many types of salumi are prohibited from being imported in the US, but fortunately there's been a resurgence of Italian salumerias (a sort of deli) that are picking up where old-world traditions left off. (See page 86 for a profile of one such producer.)

5 pounds prefrozen boneless lean pork shoulder (see Notes)

2 pounds prefrozen pork fat (see Notes)

3 pounds boneless lean beef

3 tablespoons plus 1 teaspoon kosher salt

1 tablespoon freshly ground black pepper

2 teaspoons freshly ground white pepper

2 teaspoons sugar

1 teaspoon ground coriander

½ teaspoon ascorbic acid

Curing salt (use supplier's recommended quantity for 10 pounds of meat; see Notes)

1 teaspoon minced garlic

1 cup dry white wine

10 feet medium hog casings

DRIED

CACCIATORINE

A peppery, "hunter-style" salami (*cacciatore* means "hunter" in Italian), this sausage is speckled with bits of fat and is traditionally made in a short, stubby shape — just the right size for carrying in hunters' pockets for a portable snack or lunch. It is most often made with pork or wild boar; here, a combination of pork and beef makes for a delightful update.

1. Partially thaw the pork and fat and cut them into 1-inch cubes. Meanwhile, cut the beef into 1-inch cubes and freeze for 30 minutes.

2. Grind the beef, pork, and fat separately through the coarse disk of a meat grinder.

3. In a large pan, mix together the beef, pork, and fat. Add the salt, black pepper, white pepper, sugar, coriander, ascorbic acid, curing salt, garlic, and wine. Mix well, using your hands.

4. Spread the mixture in a large, shallow pan, cover loosely with waxed paper, and cure in the refrigerator for 24 hours.

5. Prepare the casing (see page 29).

6. Stuff the mixture into the prepared casing, prick air pockets, and tie off into 4-inch links with butcher's twine. Do not separate the links.

7. Hang the sausage to dry in a cold place for 8 to 12 weeks. Test the sausage after 8 weeks by cutting off one link and slicing through it. If the texture is firm enough to suit your taste, the remaining sausage can be separated. Wrap tightly in plastic and refrigerate for up to 3 weeks.

NOTES: *Prepare the pork according to the instructions on page 44 to ensure that it is free from trichinosis.*

The meat is not cooked; please see Dry Sausage Safety on page 43.

We recommend that you use a commercial premixed cure in any recipe for cured sausage. Premixed cures replace the saltpeter in older recipes.

Paul Bertolli
Fra' Mani
Handcrafted Foods
Berkeley, California

THE CURE-ATOR

"We are known for our high-pH, dry-cured, silky, and lightly tangy salami. There's nothing much in it but meat and salt and wine, but also microorganisms and surface cultures," says Paul Bertolli, founder of Fra' Mani — and a renowned former chef and highly regarded authority on the art of dry curing. In hindsight, it seems inevitable that he would end up where he is today: His grandfather was a pork butcher who emigrated from Italy and set up shop in Chicago, and Paul himself grew up in an Italian neighborhood in San Francisco, where his first job was at a butcher shop and the gourmet store next door that carried sausages from all over the world. "I was cutting meat and selling sausage when I was around 14 years old. I was well aware of what was going on in San Francisco with legacy Italian salami companies and I loved the product."

After deciding on a culinary career during college, he gained acclaim as the chef at Chez Panisse for 10 years and then as chef and co-owner of Oliveto, in Oakland, for another 10 years. All the while he had been making salumi in his cellar and, eventually, put it on his own menu. "People really loved it and they wanted to buy it but I wasn't licensed to sell it." He had also developed a whole-hog program at Oliveto, procuring a hog every week from Niman Ranch, and wrote *Cooking by Hand,* with a significant chapter on curing. "Toward the end of my restaurant career, I had seen where the commodity meat industry was going, with the confinement of animals and the degradation of the meat supply. There were four big companies who owned all the meat and controlled in a vertical way how the meat was being raised and processed. I decided to make a difference in that whole scene."

Paul began trotting his "old-world-style, slow-processed salami" to buyers around the country, and they (literally) ate it up. "I saw an opportunity to spark a renaissance in this food category. We were at risk of losing the wisdom and skills of older butchers, and there was this mechanization in technology, with products made by machines instead of machines being at the service of the products. I wanted to turn that around. We were going to do it the old way, which is still the best way."

Paul purchases all meat from small family farms in the Midwest that raise their animals on open pasture, without antibiotics. "The meat is freshly slaughtered — we never freeze it before we make the salami. That's because you want meat with microbiological integrity — you don't want contaminants in the meat and that can happen easily during extended storage. The cold chain needs to be maintained all the way to the place you plan to make it. For us it comes from a distance, but it's on cold-refrigerated trucks and it's slaughtered on Thursday and it arrives on Monday morning, when we start working." (According to Paul, there are simply not enough local suppliers to fill Fra' Mani's high demands.)

The "old way" also applies to the way the products are made, and his methods were inspired by his time living in Italy in the early 1980s. "I followed these *norcini* (pig butchers) around Tuscany when I was working and living there in the country-side. The norcini would come in the winter when it was time to kill the pigs and they would break down the animals and make all kinds of different things in a way that was very rustic. They had a ladder on which the pig would be splayed, and they used an axe to cut down the spine and knives the rest of the day to do everything else. That was my first experience of working with butchers in the old country." When he returned to the United States, he took courses at the state-of-the-art Iowa State University Meat Laboratory, where they have a strong meat science program. "I wanted to marry an old-world and new-world understanding of how to do this. I felt I needed to know whether I could make that delicious salami I had in Vincenza here, under the guidelines of USDA inspection. I knew I had things to learn."

He started with six types of dry salami (now expanded to eight) and also pancetta, and there was such an immediate demand for his products that he had to come up with a plan to fulfill that, beyond what was possible in his already sizable plant. "I fell back on the whole-hog concept that had been in my mind all along. How do we honor this being and make products out of the entire animal?" Hence the cotto salami, capicolla, cooked sausages, hams, and galantines that Fra' Mani produces today. Mortadella, another of these products, is one of their biggest sellers. "We don't use any trims, we don't load it with water, it's not full of fat. It's basically belly and shoulder and snout and other primal or subprimal cuts of the hog."

Fra' Mani has a diverse customer base that includes high-end restaurants, hotels, food-service operators, small chain groceries, mom-and-pop shops, cheese stores, and some large groceries in every state (except for North Dakota). Even though that sounds like a massive undertaking, he says the USDA still considers Fra' Mani to be a small-scale operation. "We still make our salami one at a time, so it is in limited supply. It's still a handcrafted product."

TIPS AND TECHNIQUES FOR
DRY CURING AT HOME

"To boil it down, it all starts with the pig," says Paul Bertolli (see profile, page 164). After that, it's up to science as well as the person making it. Here, Paul offers a glimpse into the methods he uses to produce his award-winning salami, from start to finish, with time-tested tips along the way. Home salumi makers, take note.

Preparing the Sausage

THE MEAT-TO-FAT RATIO: You don't just make salami out of any cut of meat. In fact, the way in which you can diversify primarily the appearance and to a certain extent the flavor of the salami is to use different cuts of meat. I favor meat from the working muscles of the animal, meaning from the leg or shoulder, and we try to remove as much of the sinew as possible, as that can interfere with the textural integrity of the product.

Essentially you are working with a proportion of 80 percent fat to 20 percent lean. If you want a product that's richer and softer, you can use a higher proportion of fat in the mix. We only use plate fat, which is the fat that's highest on the hog in the loin over the coppa. It's very hard, white, and pure. That's the favored fat that we use.

THE "FACE" OF THE PRODUCT: It's very important to have sharp blades and plates and very cold meat, about 26 to 28 degrees. You just cut with better accuracy. What you want in dried salami is to see visible particles of meat and fat that defines the "face" of the product. We grind the meat and in order to vary the face we may use some chopped meat, using a large bowl chopper that you can make random cuts with. Then you can mix that with a more uniform

ground product for a mosaic face that I find very appealing and mimics what I was doing with these butchers in Tuscany years ago when they were chopping all the meat by hand.

THE SEASONINGS: I don't flavor the salami with anything other than salt and pepper, a tiny bit of wine, and a little bit of garlic. The rest of it comes from the fermentation process itself. The fermentation organisms are added to this mixture of meat and fat, and we also add the curing agent, either celery powder or juice, which is a natural source of nitrite. You need nitrite to fix the pigment in the flesh of the meat; otherwise it would be gray or brown. Nitrite also has a prophylactic effect against certain kinds of pathogenic bacteria. And since salami is never cooked, there are certain hurdles you have to cross for food safety.

THE CASINGS: Once you've added all the seasonings and curing agents and microorganisms, the meat mixture is stuffed into casings. We use only natural hog casings. For dried salami, the best product is the natural skin: It's permeable, mold likes to live on it, it dries evenly with the product as it develops, it's natural, and it shrinks with the product. Stuffing is a very important step, and it's all done by hand. By that I mean we use a machine, but the person doing the stuffing has to be savvy about the tension that needs to be held in the casings so you get proper bind. Casings are fussy and no one casing is exactly like the other. Some of them have a thinner wall, so you have to be able to gauge that.

THE LINKS: Once all the meat is stuffed, the salami is tied off by hand. We generally use hemp twine, except linen is used for finer sausage with less weight. Each type of salami has its own tying sequence, which helps in identifying it but also allows the salami to hold up while hanging, anywhere from 40 to 100 days. So each salami is trussed and looped and hung on a rack and put into a fermentation chamber.

Curing the Salami

PHASE ONE: Pork is about 65 percent water. What you are trying to do in drying salami is to get rid of roughly half of the water by the end of the process in such a way that the meat dries and ages very evenly and consistently. It's all about managing water and air.

The fermentation organisms are first going to denature the proteins so that the water that's bound in the cell walls or fibers of the meat starts to drip off. The first phase of the process is called dripping. The product will stay in a warm room — we use a very low-temperature fermentation, which is roughly warm room temperature for us. You can go faster, but you won't be unlocking the enzymes that are pathways to aroma and flavor. It's sort of like wine: You can make a hot, fast fermentation that's fruity, but it's not something you want to age. Fermentation happens over the course of about seven days while the salami loses 10 to 12 percent of water.

During this same time, we inoculate the casings with penicillin mold and yeast. There are also native yeasts in the plant that we can't control or get rid of, so these salamis also take on a surface culture (sort of like Roquefort in a cheese cave) that starts to grow in those first seven days, when you'll start to see a light dusting on the outside.

PHASE TWO: Once that happens, the salami is moved to a room that's a little cooler. In the arc of time that marks fermentation to maturation to aging, you are basically moving from warmer to cooler and also from wetter to dryer environments and then back to one that's a little more moist. It's almost

100 percent humidity in the fermentation room. To manage the air during this stage, you use temperature and air movement to moderate the humidity of the room so that it's dryer outside the product. You use the atmosphere as a pump, as we say. As soon as the moisture inside the product moves to the surface, the fans and evaporators will turn on and together they move that moisture out of the room.

What you are trying to create is a condition whereby the moisture is moving from the core of the product to the surface, and from the surface out of the room, at the same rate. In other words, the dispersion rate equals the evaporation rate. This is how you get salami that is perfectly dry. By that I mean the outside surface is not dark or dry, and the salami is basically supple all the way through.

That's the art of drying. Machines don't know how to do this; that's what I do. That's what I had to learn the first four years when I was using my Italian equipment. How do I manage this equipment to get that sausage dried properly? It's very challenging. The equipment is supposed to be perfectly even in the rooms, but they're not. You've got a boiler running water through the circulatory system of the plant and this heats up the room, you've got fans that turn on air, and you've got condensate lines that pull water from the room through the evaporators.

You have to learn to master this equipment so that it's doing the job you want it to do and turn out the product you want to make. There's a microprocessor in every room that tells us what phase of drying we are in, and we have tools that measure the humidity and temperature and air movement. We even have probes that measure the pH in each room.

But these tools can only report the conditions. I have to be the one to decipher that there's a big load of moisture in the room and devise a way to change the parameters.

PHASE THREE: Once the salami has reached the proper state, it moves out of the fermentation chamber into drying — or aging — rooms. We're looking for that target of water loss but at the same time we're creating the conditions whereby the surface culture can start to do its thing. Mold is important for shielding the product from light so it doesn't oxidize, and it conditions the protein so the fungi on the surface can get into the product and aromatize it. Mold also mitigates the drying process: When environmental conditions cause small amounts of moisture to move from the inside to the surface, the mold will consume that moisture to thrive and grow there.

At the same time, there's all this biochemical stuff happening that has to do with what has been unlocked in the meat by fermentation. These are very complex chemical transformations that create the particular kind of aroma and flavor you are looking for, and that continue throughout the aging process.

The cultures we used at the beginning, which play a part in breaking down the proteins and fat, help to create the particular "bite" or texture that the salami will have. We put a small amount of corn sugar in the mixture before fermentation, and the good bacteria (like in cheese or yogurt) begin to metabolize that sugar and create lactic acid to produce that agreeable tang in the salami.

It's this combination of what happens chemically in the process of aging with the fixed acids that are there from the fermentation and then the effect of the mold ripening and the loss of moisture in the meat and the concentration of flavor. You have to have all of these things in balance. The one thing depends upon the other in order to produce all that desirable aroma and flavor and acidity.

The aging time depends on the diameter of the product as well as the grind and the casing, which can vary in thickness along the intestines. For the most part, we age our salami anywhere from 40 days for our salamette (2½ inches in diameter) to 100 days for our larger-format products (5 inches in diameter).

Ultimately, because so much of what happens inside the casing is out of our control, all we can continue to do is make sure the product dries evenly and ages long enough to develop the flavors that we think are the hallmark of our product.

The Art of Drying

The biggest complaint I hear from home sausage makers is that their salami turned out really dry on the outside but is soft on the inside, and that tells me this person doesn't understand what this product needs to feel like through its various stages. That's what I learned when I was in Italy, where they didn't have machines in some of these small *agritourismi*, and when making salami in my cellar: If more moisture was needed, I'd throw water on the floor; for more air movement, I'd open a window; if it needed to be drier, I'd turn on lights to help evaporate moisture. This is the sense that you get in your body and in your hands and this is what you want to make your machines do.

For me, the equipment that I bought for the plant, including the grinder and chopper, is meant to mimic the work of the hand. The machinery that humidifies and evaporates and moves air is more like the wind and the light. I can tell in an instant whether it's too dry or too wet or the fan is coming on too hard or we need to move a rack that's in the back of the room 360 degrees because it's only molding on one side.

There's so much handwork involved in paying attention to how the product is evolving and making little tweaks to achieve the right atmosphere. That's what distinguishes my product from others, or at least I'd like to think so.

VEAL SALAMI

5 pounds boneless
lean veal

3 pounds prefrozen
boneless pork (see
Notes)

2 pounds prefrozen
pork fat (see Notes)

3 tablespoons plus
1 teaspoon kosher salt

2 tablespoons sugar

1 tablespoon freshly
ground black pepper

1 tablespoon freshly
ground white pepper

1 teaspoon crushed
anise seed

½ teaspoon freshly
grated nutmeg

½ teaspoon ascorbic
acid

Curing salt (use sup-
plier's recommended
quantity for 10 pounds
of meat; see Notes)

½ cup brandy

½ cup dry vermouth

4 feet large beef casing

The seasonings in this delicious salami, made with a combination of veal and pork, are both spicy and sweet and are accented by brandy and dry vermouth.

1. Cut the veal into 1-inch cubes. Freeze the cubes for about 30 minutes to firm them up before grinding.

2. Partially thaw the pork and fat and cut into 1-inch cubes.

3. Grind the veal, pork, and fat separately through the coarse disk of a meat grinder.

4. Mix the veal and fat together in one bowl and freeze the mixture, along with the pork in a separate bowl, for 30 minutes.

5. Grind the veal mixture using the fine disk of the grinder.

6. In a large pan, combine the veal mixture with the ground pork. Add the salt, sugar, black pepper, white pepper, anise seed, nutmeg, ascorbic acid, curing salt, brandy, and vermouth. Mix well, using your hands.

7. Spread the mixture in a large, shallow pan, cover loosely with waxed paper, and cure it in the refrigerator for 24 hours.

8. Prepare the casing (see page 29).

9. Stuff the mixture into the prepared casing, prick air pockets, and tie off into 6- or 7-inch lengths with butcher's twine. Do not separate the links.

10. Hang the salami to dry in a cool place for 8 to 12 weeks. Test the sausage after 8 weeks by cutting off one link and slicing through it. If the texture is firm enough to suit your taste, the remaining sausage can be separated. Wrap tightly in plastic and refrigerate for up to 3 weeks.

NOTES: *Prepare pork according to the instructions on page 44 to ensure that it is free from trichinosis.*

The meat is not cooked; please see Dry Sausage Safely on page 43.

We recommend that you use a commercial premixed cure in any recipe for cured sausage. Premixed cures replace the saltpeter in older recipes.

PEPPERONI

Pepperoni sausage is sometimes referred to as a stick of pepperoni because that's just about what it resembles. There are many different varieties of this American favorite, some decidedly hotter than others, but most, if not all, rely on a beef and pork combination (and paprika for the trademark red hue). All are quite pungent. Pepperoni come in different sizes, the most common being about an inch in diameter. Some commercial packers also make what they call pizza pepperoni, which is about twice the diameter of regular pepperoni and is not as dry. This type is better able to withstand the high temperature of a hot oven without becoming too crisp. If you intend to use your pepperoni primarily as a topping for pizza, you might want to experiment with the drying time, too. *See photo, page 151.*

3 pounds boneless lean beef chuck, round or shank

7 pounds prefrozen boneless pork butt or shoulder, including fat (see Notes)

3 tablespoons plus 1 teaspoon kosher salt

3 tablespoons paprika

2 tablespoons cayenne pepper

1 tablespoon crushed anise seed

1 tablespoon sugar

½ teaspoon ascorbic acid

Insta Cure #2 (use supplier's recommended quantity for 10 pounds of meat; see Notes)

1 teaspoon very finely minced garlic

1 cup dry red wine

6 feet small hog casing

1. Cut the beef into 1-inch cubes. Freeze the cubes for about 30 minutes to firm them up before grinding.

2. Partially thaw the pork and cut into 1-inch cubes

3. Grind the beef and pork separately through the coarse disk of a meat grinder.

3. In a large pan, combine the meats, salt, paprika, cayenne, anise seed, sugar, ascorbic acid, curing salt, garlic, and wine. Mix well, using your hands.

4. Spread the mixture in a large, shallow pan, cover loosely with waxed paper, and cure in the refrigerator for 24 hours.

5. Prepare the casing (see page 29).

6. Stuff the mixture into the prepared casing, prick air pockets, and twist off into 10-inch links. With butcher's twine, tie two separate knots between *every other* link, and one knot at the beginning and another at the end of the stuffed casing. Cut between the double knots. This results in pairs of 10-inch links.

7. Hang the pepperoni by a string tied to the center of each pair; let dry for 6 to 8 weeks. Test the sausage after 6 weeks by cutting off one link and slicing through it. If the texture is firm enough to suit your taste, the remaining sausage can be cut down and wrapped tightly in plastic for storage in the refrigerator for several months.

NOTES: *Prepare the pork according to the instructions on page 44 to ensure that it is free from trichinosis.*
The meat is not cooked; please see Dry Sausage Safety on page 43.
We recommend that you use a commercial premixed cure in any recipe for cured sausage. Premixed cures replace the saltpeter in older recipes.

SMOKED COUNTRY-STYLE SAUSAGE

- 4 feet small hog casing
- 3 pounds boneless pork butt or shoulder, about 75% lean
- 2 pounds boneless beef chuck, about 75% lean
- ¼ cup nonfat dry milk powder
- 1 tablespoon plus 1 teaspoon kosher salt
- 1 tablespoon paprika
- 1 tablespoon sugar
- 2 teaspoons mustard seeds
- 2 teaspoons freshly ground white pepper
- ¼ teaspoon ascorbic acid

  Insta Cure #1 (use supplier's recommended quantity for 5 pounds of meat; see Note)
- ½ cup ice water

Sometimes you will see something similar to this sausage in the fresh meat case at the grocery, called smoked country links. It is delicious both as a breakfast sausage or as part of an hors d'oeuvres selection.

1. Prepare the casing (see page 29).

2. Cut the pork and beef into 1-inch cubes. Freeze the cubes for about 30 minutes to firm them up before grinding.

3. Grind the pork through the fine disk of a meat grinder. Grind the beef through the coarse disk of the grinder.

4. In a large bowl, combine the meats. Add the dry milk, salt, paprika, sugar, mustard seeds, pepper, ascorbic acid, curing salt, and ice water. Mix well, using your hands.

5. Stuff the mixture into the prepared casing, prick air pockets, and twist off into 2- to 3-inch links. Tie off each link with butcher's twine.

6. Smoke for 2 hours at 180 to 190°F (83 to 88°C).

7. Bring a large pot of water to a boil. Add the sausages and reduce the heat to maintain a water temperature of 180 to 190°F (83 to 88°C). Simmer the sausages for 30 minutes. The sausages should rise to the top when they are done and the internal temperature on an instant-read thermometer should be 160°F (61°C).

8. Place the sausages in a bowl of cold water for 30 minutes. Remove, dry thoroughly, and refrigerate, covered, for up to 2 weeks.

NOTE: *We recommend that you use a commercial premixed cure in any recipe for cured sausage. Premixed cures replace the saltpeter in older recipes.*

SMOKED CHEDDAR SUMMER SAUSAGE

Fans of summer sausage will appreciate this version, which has cheddar cheese running through it — meaning it's all the better for slicing and using to top crackers (no extra cheese required). Smoking makes the flavor that much deeper. *See photo, page 2.*

1. Prepare the casing (see page 29).

2. Cut the beef, pork, and fat into 1-inch cubes. In a large bowl, combine the meats, fat, garlic, salt, sugar, pepper, mustard seeds, and curing salt. Mix well, using your hands. Freeze for 30 minutes.

3. Grind the meat mixture through the medium disk of a meat grinder. Mix in the water and cheese. Freeze for 30 minutes.

4. Stuff the mixture into the prepared casing, prick air pockets, and twist off into 12-inch lengths. Cut the links apart.

5. Smoke at 180 to 190°F (83 to 88°C) for 2 hours, or until the internal temperature is 160°F (71°C).

6. Place the sausage in a bowl of cold water for 30 minutes. Remove, dry thoroughly, and refrigerate, wrapped in plastic, for up to 2 weeks.

NOTE: *We recommend that you use a commercial premixed cure in any recipe for cured sausage. Premixed cures replace the saltpeter in older recipes.*

- 4 feet medium hog casing
- 2½ pounds boneless beef chuck
- 1¼ pounds boneless pork shoulder
- 6 ounces pork fat
- 2 tablespoons minced garlic
- 1 tablespoon plus 1 teaspoon kosher salt
- 1 tablespoon brown sugar
- 1 tablespoon freshly ground black pepper
- 1 tablespoon mustard seeds

  Insta Cure #1 (use supplier's recommended quantity for 3¾ pounds of meat; see Note)
- 1½ cups water
- 12 ounces cheddar cheese, cut into ¼-inch cubes

- 4 feet medium hog casing
- 3 pounds boneless pork butt or shoulder with fat
- 2 pounds boneless beef chuck, trimmed
- ¼ cup nonfat dry milk powder
- 1 tablespoon plus 1 teaspoon kosher salt
- 1 tablespoon paprika
- 1 tablespoon sugar
- 2 teaspoons freshly ground white pepper
- ½ teaspoon ground celery seeds
- ½ teaspoon ground coriander
- ½ teaspoon dried marjoram
- ½ teaspoon freshly grated nutmeg
- ½ teaspoon dried thyme
- ¼ teaspoon ascorbic acid
- Insta Cure #1 (use supplier's recommended quantity for 5 pounds of meat: see Note)
- 1 tablespoon minced garlic
- ½ cup ice water

SMOKED KIELBASA

For most Americans, this is the real-deal kielbasa: hot-smoked and with that signature ring shape you find in every grocery store — and in Polish butcher shops such as Pekarski's (see profile, page 174). It has a wonderful snap when cooked on a hot grill — or panfried or roasted — and is ideally served with rye bread and brown mustard, plus a frosty bock beer.

1. Prepare the casing (see page 29).

2. Cut the pork and beef into 1-inch cubes. Freeze the cubes for about 30 minutes to firm them up before grinding.

3. Grind the pork through the coarse disk of a meat grinder. Grind the beef through the fine disk.

4. Mix the meats together in a large bowl. Add the dry milk, salt, paprika, sugar, pepper, celery seeds, coriander, marjoram, nutmeg, thyme, ascorbic acid, curing salt, garlic, and ice water. Mix well, using your hands.

5. Stuff the mixture into the prepared casing, prick air pockets, and twist off into 8- to 10-inch links. Tie off each link with butcher's twine.

6. Spread the mixture in a large, shallow pan, cover loosely with waxed paper, and cure in the refrigerator for 1 day.

7. Smoke at 180 to 190°F (83 to 88°C) for 2 hours, or until the internal temperature is 160°F (71°C).

8. Place the sausage in a bowl of cold water for 30 minutes. Remove, dry thoroughly, and refrigerate, wrapped in plastic, for up to 2 weeks.

NOTE: *We recommend that you use a commercial premixed cure in any recipe for cured sausage. Premixed cures replace the saltpeter in older recipes.*

Mike Pekarski

*Pekarski's, Home of
New England's Farm
Made Sausage*

South Deerfield, Massachusetts

CHIP OFF THE OLD
BUTCHER BLOCK

There's an unassuming little shop on a two-lane highway in South Deerfield, Massachusetts, that has what many consider the best kielbasa around, period. Mike Pekarski, the owner and grandson of the founders, still mixes the spice blend himself and is there 6 days a week, 52 weeks a year. It all started back in 1948, when Pekarski's was a custom slaughterhouse for local farmers before making sausages and smoking meats themselves from whole cuts of pork, and never trimmings (still the policy today). The shop closed shortly after New Year's Day in 1981, and after that his grandfather used the building as his workshop.

Meanwhile, "I was floundering at UMass and finally convinced my dad to get back in the sausage business in 1991," recalls Mike. They renovated the original building, keeping the hefty iron doors intact, and then more recently installed two new smokehouses. "These chambers are larger than what my grandfather used, and allow for smoking more meat at the same time — up to 600 pounds — which is pretty essential given that we sell upwards of 800 pounds of kielbasa each day in the weeks leading up to Easter." That's on top of the 300-plus cured hams they begin preparing months in advance of the holiday.

Pekarski's has also expanded its offerings, from three basic sausages — kielbasa, breakfast sausage, and Italian sausage — to 50-plus options they make today, including andouille, chorizo, and the newest addition, farmer's sausage, a close kin to French garlic sausage. "When I am creating a new sausage, I will look at all the different recipes for that sausage, which can be almost completely different, and see what's the common strand of DNA among them," Mike says, "and then I'll put our foundation of meat-to-fat ratio and spices behind it to make the taste distinctly ours."

It's that foundation — the fat-to-meat ratio and proprietary blend of spices — that keeps their many fans coming back each year. That and the commitment to keeping to the original recipe that was developed by his grandfather over 50 years ago. And with just six employees, such adherence to tradition can be a time-consuming undertaking. But Mike has invested in upgrades to help with efficiency. "This is our new grinder, which can hold 250 pounds of meat. It lets us accomplish in two or three hours what used to take six, and the same goes for the stuffer." There's no shortcut to producing the wonderful flavor of their smoked meats, however. "First they are left to cure in a refrigerated curing room and then are moved into one of our two smokehouses for as long as eight hours," explains Mike while opening one of those original massive doors to reveal a rolling rack holding 600 pounds of kielbasa suspended over a fire smoldering below the grates in the floor. "We burn only select hardwoods such as hickory, apple, and rock (sugar) maple to give our smoked meats their particular flavor."

That makes Pekarski's one of the few remaining smokehouses on the East Coast that doesn't rely on liquid smoke or chips — and a destination for their legions of fans, who will think nothing of driving over 300 miles for their favorite kielbasa.

GARLIC RING BOLOGNA

MAKES 5 POUNDS

Often described as the missing link between old-world sausage and modern-day hot dogs, garlic ring bologna is decidedly different from the "baloney" we all grew up eating, presliced and processed to the max. It's also thicker, and more garlicky, than your typical frank, with a brighter, pinker interior. Ring bologna can be sliced and eaten right from the package, with cheese and crackers, or cooked with beans according to tradition.

1. Cut the pork, veal, and fat into 1-inch cubes. Freeze the cubes for about 30 minutes to firm them up before grinding.

2. Grind the pork, veal, and fat separately through the fine disk of a meat grinder.

3. In a large bowl, combine the meats, fat, salt, mustard seeds, pepper, allspice, marjoram, ascorbic acid, curing salt, and garlic. Mix well, using your hands. Freeze the mixture for 30 minutes.

4. Prepare the casing (see page 29).

5. Grind the seasoned meat mixture through the fine disk of the grinder.

6. Stuff the mixture into the prepared casing, prick air pockets, and twist off into 18-inch links. Be careful not to overstuff, or when you form the rings, the casing may burst. Tie double knots between the links with butcher's twine, then cut the links between the knots. Bring the tied ends of each link together and tie securely, forming a ring.

7. Hang the sausage in a cool drying area for 2 hours, or refrigerate, uncovered, on a baking sheet fitted with a wire rack for 8 to 10 hours.

8. Hot-smoke at 110 to 120°F (44 to 49°C) for 2 hours.

9. Bring a large pot of water to a boil. Add the sausages and reduce the heat to maintain a water temperature of 180 to 190°F (83 to 88°C). Simmer the rings for 30 minutes. The sausages should rise to the top when they are done and reach an internal temperature of 160°F (61°C) on an instant-read thermometer.

10. Place the sausages in a bowl of cold water for 30 minutes. Remove, dry thoroughly, and refrigerate, wrapped in plastic, for up to 2 weeks.

NOTE: *We recommend that you use a commercial premixed cure in any recipe for cured sausage. Premixed cures replace the saltpeter in older recipes.*

- 2 pounds boneless pork butt or shoulder, about 70% lean
- 1 pound veal hearts
- 2 pounds pork fat
- 1 tablespoon plus 1 teaspoon kosher salt
- 2 teaspoons crushed mustard seeds
- 2 teaspoons freshly ground white pepper
- 1 teaspoon ground allspice
- 1 teaspoon dried marjoram
- ¼ teaspoon ascorbic acid

  Insta Cure #1 (use supplier's recommended quantity for 5 pounds of meat; see Note)
- 4 garlic cloves, very finely minced
- 4 feet medium hog casing

METTWURST

MAKES 5 POUNDS

- 4 feet medium hog casing
- 3 pounds boneless lean beef chuck
- 2 pounds boneless pork butt or shoulder with fat
- 1 tablespoon plus 1 teaspoon kosher salt
- 1 teaspoon freshly ground white pepper
- ½ teaspoon ground allspice
- ½ teaspoon ground celery seeds
- ½ teaspoon freshly grated nutmeg
- ½ teaspoon grated fresh ginger
- ¼ teaspoon ground marjoram
- ¼ teaspoon ascorbic acid
- Insta Cure #1 (use supplier's recommended quantity for 5 pounds of meat; see Note)

Although you can find firm versions of mettwurst, it's the soft, spreadable chubs that have gained a loyal following over the years (and we mean hundreds). It makes a great sandwich spread or toast topper, straight from the package. Like many other smoked sausages, it is fully cooked but requires refrigeration before eating.

1. Prepare the casing (see page 29).

2. Cut the beef and pork into 1-inch cubes. Freeze the cubes for about 30 minutes to firm them up before grinding.

3. Grind the beef and pork separately through the fine disk of a meat grinder.

4. Mix the ground meats together in a large bowl. Add the salt, pepper, allspice, celery seeds, nutmeg, ginger, marjoram, ascorbic acid, and curing salt. Mix well, using your hands.

5. Stuff the mixture into the prepared casing, prick air pockets, and twist off into 6-inch links. Tie off each link with butcher's twine.

6. Spread the mixture in a large, shallow pan, cover loosely with waxed paper, and cure it in the refrigerator for 1 day.

7. Hot-smoke at 110 to 120°F (44 to 49°C) for 2 hours, and then raise the temperature to 150°F (66°C) and smoke for 2 hours longer.

8. Bring a large pot of water to a boil. Add the sausages and reduce the heat to maintain a water temperature of 180 to 190°F (83 to 88°C). Simmer the sausages for 30 minutes. The sausages should rise to the top when they are done and the internal temperature on an instant-read thermometer should be 160°F (61°C).

9. Place the sausages in a bowl of cold water for 30 minutes. Remove, dry thoroughly, and refrigerate, wrapped in plastic, for up to 2 weeks.

NOTE: *We recommend that you use a commercial premixed cure in any recipe for cured sausage. Premixed cures replace the saltpeter in older recipes.*

Bison Sausage with
Jalapeño, page 185

GAME SAUSAGES

After going 12 miles we were fortunate enough to find a few willows, which enabled us to cook a dinner of jerked elk.

— Journals of Meriwether Lewis and William Clark

You don't have to be a hunter yourself — or even live where hunting is a hobby or avocation (like most city dwellers) — to have access to game meats these days; many species of game are raised on licensed ranches and preserves (see the profile of Fossil Farms on page 186), or you can buy truly *wild* game from certain purveyors (see the profile of Broken Arrow Ranch on page 192). Both types of operations make their meat available for sale to restaurants, butcher shops, and the general public. (See the Resources on page 355.)

Game meat is typically leaner, and gamier tasting, than meat from domesticated animals, even those that are pastured and grass-fed. It is also usually more expensive, so you'll want to make sure to treat it with care when using in sausages — it would be a shame to go to the trouble of procuring the elk, venison, wild boar, or other game meat (even if the only hunting you do is online) and then wind up with tough links. Heed the guidelines on the following pages to make the most of your quarry.

Rules of the Game

Just think: In less than 250 years, we've gone from being a people whose survival relied entirely on hunting for food (learning the essential techniques from the Native Americans) to a time when most Americans wouldn't know where to begin if they had to track, kill, dress, and prepare their own meat. So when you are privileged enough to have a cut of venison or wild boar, elk, rabbit, or even alligator, you'll need to know exactly how to prepare it. Here are some basic guidelines to help you on your way.

Age matters. "Young is tender, old is tougher." This adage is true whether we are talking about domestic animals or about the bear you bagged yesterday. Certainly steaks, chops, and roasts are most people's first choices. But an animal has a fixed number of those prized parts. The rest — especially in a muscular older animal — is sausage material. A corollary to this rule is that the older the meat, the more developed the flavor. In other words, game gets gamier with age.

Dress for success. If you hunt for your own meat, be sure to field dress it properly. The size and species of the animal, as well as weather conditions (especially temperature), all play a role in determining the proper procedures to follow. If you are inexperienced or unsure of the right way to go about it, be sure to consult one of the many references that deal with hunting and butchering game, or contact your local Cooperative Extension Service.

Lean and mean. In big game, especially, the strongest flavor is found in the fat of the animal. Unless you really relish the strongest of gamey flavors, it is best to trim the fat from the game meat and replace it in a recipe with pork or beef fat, both of which tend to be mild tasting. And note: Because wild animals are active and develop their muscles, their meat is usually quite lean. To make sausage, you will definitely need to add fat.

Handle with care. Wild boar can be carriers of trichinosis, and wild hare can carry tularemia, a serious bacterial disease. Anytime you handle raw game meat, wear rubber gloves for safety, and be sure to maintain a scrupulously clean kitchen to avoid any cross contamination. See page 44 for a refresher course on the importance of cleanliness.

Hung out to dry. Just as domestic beef is aged or "hung," many hunters age wild game to improve flavor, texture, and tenderness. (For food safety guidelines, see page 22.)

WILD BOAR

Wild boar, also known as feral pigs or razorbacks, are wreaking havoc across the country, with an estimated six million hogs nationwide — and nearly half of them in Texas alone. They cause an estimated $2 million in damages each year to the agricultural community, digging up fields, devouring crops, tearing down fences, and otherwise causing harm to ranchers and farmers as well as natural wildlife. In Texas, they do particular harm to native species and the 400 stream segments in the state that are infected with bacteria from the pigs' fecal matter. That's why Texans are allowed (actually encouraged) to hunt wild hogs year-round. For consumers, this is good news; wild boar is utterly delicious, so it's a win-win for farms and foragers alike.

Trotter Transformation

If you desire the unique flavor of wild boar but don't have the wherewithal to procure the meat, try this shortcut: Marinate pork in a flavorful brine. You'll need a large roasting pan and space in the refrigerator to hold it. You'll also need to turn the meat twice a day for four days during the transformation process, so plan accordingly.

- 5 pounds fresh ham (not cured or smoked), boned and skinned, with fat
- 4 cups ruby port wine, or enough to cover
- 1 tablespoon whole juniper berries
- 1 tablespoon whole black peppercorns
- 3 bay leaves (fresh or dried)
- 3–4 fresh sage leaves
- 3–4 thyme sprigs

Put the fresh ham flat in a roasting pan just large enough to accommodate it. Pour the wine over the meat, then sprinkle in the juniper berries, peppercorns, bay leaves, sage, and thyme. Cover and marinate in the refrigerator for 4 days, turning the ham twice a day. At the end of 4 days, pat the meat dry, remove any spices clinging to it, and use as you would fresh boar meat (i.e., in some amazing sausage!).

Cristiano Creminelli

Creminelli Fine Meats

Salt Lake City, Utah

THE MAESTRO

Many aspiring sausage makers (including some of those profiled elsewhere in this book) first felt the calling to make charcuterie when they were visiting Italy and stumbled upon a *salumificio* in some ancient town, just when they were breaking down a whole pig, the way it's been done there for centuries. If they were in the Piedmont region, in northwest Italy, they may very well have stumbled upon the Creminelli *famiglia* doing just that.

But here's a tale in reverse: Cristiano Creminelli, whose family has been making charcuterie in Italy since the 1600s, left the fold to pursue his passion in the United States. "I'm the black sheep of my family. One of my favorite travel destinations was always the United States because it is just so different from Italy. Before I decided to move here, I had visited about ten times and I started thinking it would be fun to do something here in the same way I did there."

It remained a far-fetched idea until he met Chris Boller, now his business partner, during the 2006 Olympics in Turin, when his family's business was one of the suppliers to House America. "I told Chris my idea during a casual conversation and I didn't think it would go anywhere. But Chris called me some months later and said he had done a little research and thought it was a good time to pursue it. So I told my wife I was just going to the United States to check out the situation, and I ended up staying." She happily joined him soon after.

Not that leaving his hometown was all that easy. "In Italy, when you move from one village to another, just two or three miles away, people are shocked, and I actually moved to the other side of the planet." What allowed him to take the leap was that he knew he had all this expertise, as well as the desire to keep the legacy alive. (He also credits the ability to Skype with his father every day over a meal — lunch for dad, dinner for son.) He spent a full year looking for the right location and settled on Utah. Having some connections there was a big draw (especially since he wasn't at all proficient in English), as was the dry climate that provided the ideal curing environment. "Plus the farms were raising the pigs in the way that I wanted. Salami is made with 99.9 percent meat, and the source of the meat is super, super, super important. I can also call the farmer and say I need the meat tomorrow, which is obviously a huge benefit."

He said he was blessed with an outpouring of support from the very beginning. A local business let him use the basement while he looked for a permanent spot, and when he tried out a few simple salami with friends and potential clients, the responses were encouraging. "That really made my day. It's wonderful to make money, but more than anything I love seeing people try new food that they actually like." After all, when he arrived here there were very few options available, and mostly mass produced, plus there was an audience that was ready and willing for new discoveries. "For a food producer like me, being in the United States is like being in the center of the world. People are really aware of what they want to eat, and they are also open to tasting new food. I find myself in the perfect place to be."

There was a catch, however. Having grown up practicing the kinds of natural controls used in curing salami in his region, Cristiano had to adjust his methods to meet USDA requirements, and here he provides a unique perspective on what it's like to be a producer in the United States as opposed to Italy. In the end, he says he views this as an opportunity to be part of a symbiotic relationship with the USDA where they can feed each other information and, hopefully, help to effect changes in the way producers are regulated here. Even though he isn't allowed to have windows in his processing facility, he can try to replicate the effect of a fresh breeze through accepted means. To maintain consistency, he relies on a combination of qualitative and quantitative measures from batch to batch, and he has cultivated an instinct to hit the sweet spot. "You can do it by machine, but I'm a huge fan of the human touch and making adjustments to make sure the salami is at the proper pH levels and the right temperature. I prefer to rely on people, and I've had the same people working for me for a long time and they know the way I work."

Creminelli Fine Meats currently offers 12 different varieties of salami on its website and for distribution. "But I'm always experimenting. Right now I have 10 different salami hanging in my curing room. If I find a wonderful ingredient I will try to re-create the flavor in salami, and most recently that was mezcal. I am always trying to find something new. That's what provides me with the most happiness." Some of these salami will be released as special-edition recipes available on their website (see page 355) — or Cristiano will just share them with some fortunate friends.

(see page 355)

TRADE SECRETS

"I tell everybody that temperature is the most important factor: The colder the meat, the better the texture. Be sure you have really cold meat because then you can season it however you want. For dry curing, spices are important, but 75 percent of the flavor comes from the fermentation. Even a very small change in the pH will have a huge effect on the final product."

— Cristiano Creminelli

CRAFTED WITH KINDNESS

ITALIAN-STYLE WILD BOAR SAUSAGE

This recipe was provided by Cristiano Creminelli (see profile, page 182), who says he still enjoys making fresh sausage as a hobby. "In northern Italy, wild boar is a classic alternative to regular pork, and is often roasted on the grill or cured into salami." Wild boar (or *cinghiale*) has a slightly sweeter, nuttier taste than domesticated pork, and is also leaner. For that reason, you may have to add more pork fat to achieve the proper fat-to-lean ratio. "For the best results, I recommend using Barbera in the sausage, although any bold, fruity red wine will work. I also suggest grilling the sausage, which pairs very well with a Brunello or Chianti. *Salute!*"

- 4 feet small hog casing
- 1½ pounds boneless lean wild boar meat
- 10 ounces skinned fresh pork belly
- 6 tablespoons full-bodied red wine, such as Barbera
- 3½ teaspoons sea salt
- 1 teaspoon freshly ground black pepper
- 1 teaspoon minced garlic
- ½ teaspoon freshly grated nutmeg

1. Prepare the casing (see page 29).

2. Cut the boar and pork belly into 1-inch cubes. Mix together and freeze for 30 minutes.

3. Grind the meat mixture through the medium disk of a meat grinder.

4. In a large bowl, combine the ground meat with the wine, salt, pepper, garlic, and nutmeg. Mix well, using your hands.

5. Stuff the mixture into the prepared casing, prick air pockets, and twist off into 4-inch lengths.

6. Place the sausage on a baking sheet fitted with a wire rack and refrigerate, uncovered, for 1 day.

7. Cut the links apart. Refrigerate, wrapped well in plastic, for 2 to 3 days, or freeze for up to 3 months; thaw overnight in the refrigerator before using.

8. Cook as desired (see methods on pages 38 to 40) to an internal temperature of 160°F (71°C).

BISON SAUSAGE WITH JALAPEÑO

MAKES 1 POUND

- 1 pound boneless bison chuck, shoulder or stew meat
- 3 ounces pork fat
- ¼ cup bison or beef broth, well chilled
- 1 jalapeño chile, seeded and minced
- 1 teaspoon kosher salt
- 1 teaspoon minced garlic
- ¼ teaspoon paprika
- ¼ teaspoon ground black pepper
- Pinch of ground coriander
- Pinch of dried marjoram
- Pinch of dried rubbed sage
- Pinch of ground dry mustard
- Pinch of dried thyme

If you want a truly American-style sausage, here's your answer. Bison (or buffalo) meat is more readily available than ever these days, but you may need to check with your butcher to buy it before the meat has been ground. If you can only find ground bison/buffalo meat, mix that with the ground pork fat and other ingredients in step 3. Here, the sausage is formed into patties, but you can stuff the mixture into small hog casing if desired, scaling up the ingredients to make five pounds. Serve with cornbread and black-eyed peas, or use as a filling for tacos or in a hearty chili. *See photo, page 178.*

1. Cut the bison and pork fat into 1-inch cubes. Freeze for 30 minutes to firm them up before grinding.

2. Grind the meat and fat through the medium disk of a meat grinder.

3. In a large bowl, combine the ground meat and fat with the broth, jalapeño, salt, garlic, paprika, pepper, coriander, marjoram, sage, dry mustard, and thyme. Mix well, using your hands.

4. Form into thin patties, using a scant ¼ cup for each one. Thinner is better (bison is very lean, and you want the patties to cook through quickly before they have a chance to dry out).

5. Cook as desired (see methods on pages 38 to 40) to an internal temperature of 160°F (71°C).

BISON REBOUND

Native to North America, bison (or buffalo) did indeed once roam free, with an estimated 30 million living across the country. By 1889, however, these iconic animals had been hunted to just over 1,000 in population. Thanks mostly to the efforts of a few private ranchers — and a small herd that managed to escape "The Great Buffalo Hunt" in a remote valley of Yellowstone National Park — there are a reported 500,000 bison nationwide today, but only 5,000 of these are truly wild and free of disease, with another 30,000 wild bison in conservation areas. The rest are typically the result of crossbreeding with cattle and are being raised as livestock — in other words, farmed bison is what we eat today.

⚜

Lance Appelbaum

Fossil Farms

Boonton Township, New Jersey

CONSCIENTIOUS SOURCING

In a state known for its tomatoes and dairy cows, you'll also find ostrich and bison roaming free over a vast plot of land in Sussex County, New Jersey, about an hour's drive from New York City. That property, aptly named Roaming Acres, is the farming half of an operation that was created back in 1997 by two brothers, Lance and Todd Appelbaum, to fill a void for natural raised meats. The other half is Fossil Farms, a source of sustainably raised exotic meat and game (and other, more familiar meat products) procured from small family-owned and -operated farms across the country.

The idea for the business struck while on a ski vacation in 1995, when the brothers tried ostrich and bison for the first time. As Lance recalls, "I was absolutely floored with how clean and how good these meats tasted. Over dinner we decided that Todd would raise the animals and I would sell them as healthy alternatives to red meat to anyone who was willing to listen." They started their farm with ostrich, then added bison before moving on to other naturally sustainable meats. All the while, they began sourcing meat from other farms that share their same commitment to sustainability. "I like that we are a farm and also a distributor, so to say, because we have direct knowledge of our animals' environment as well as what consumers today are looking for, which is 100 percent all-natural meat with no antibiotics, no hormones, no steroids. We raise animals strictly on ethics and quality."

Fossil Farms, then, provides a service to both farmer and consumers. "A lot of small farms look to us to get their products to market because we are experts in that field. We can also help our customers by letting them know where their products are coming from."

Their requirements for their source farms are stringent. "We have to understand the mechanics of their operation. We have to approve the feed they supply their animals with. My brother grinds his own feed on the farm so we know exactly what the animals are eating. We want to be able to have the same knowledge about the animals we source. We want to be able to tell our customers exactly what the animals were raised on, how long they were raised on it, and why they were raised on it. That's a very important factor for us." They also require the animals to have more room than required by any definition of free-range. At Roaming Acres and their partner farms, the animals spend their days in open fields, where they have ample room to roam and access to fresh water. "We really instill in the source farms the idea that if you allow the animals the chance to live freely, it will pay off in dividends. That's pretty powerful to us."

Helping to educate these small farms and improve their processes and profitability is a big part of their misson. "Just because they've been doing something for so many generations doesn't mean it's always right. We can let them know what's worked on our own farm or other farms that we work with. We can also let them know what the market wants to see based on these old-time techniques."

In the same way, Fossil Farms aims to educate consumers by exposing them to alternative meats. "Most people start with bison and then ostrich before branching out to more exotic options like alligator," says Lance. The same goes for more conventional meats. "We have a very diverse clientele and want to be able to provide something for everyone, so we butcher the whole animal and use every last part."

Besides selling whole animals or cuts of meat, Fossil Farms (and Roaming Acres) produce their own cured meats. They also have an extensive line of jerky. As for sausages: "What's amazing is what's *not* on the market," says Lance. "We make a traditional pork Toulouse sausage using an old-style recipe and our own Berkshire pork. We also make a sausage with a mix of wild boar, which has a nutty flavor to it, dried cranberries, and Shiraz wine. It's unique, and at the end of the day it's a lot of fun to come up with new flavor profiles." And rewarding, too. "Sometimes we'll walk into a store in Albuquerque and we'll spot our products on the shelf, and I'm amazed to see the results of something we started over 19 years ago, out of my parents' basement. There wasn't the same concern with ethical farming practices back then, and people said it wouldn't work." Now Fossil Farms has come to be a trusted source of humanely raised meats for chefs and home cooks alike.

The only downside of all this success, for Lance, is that he doesn't have more time to make sausage at home these days. "But I just bought a new smoker and I can't wait to start playing around with making jerky and smoking my own meats."

TRADE SECRETS

"You don't have to do much if you start with the best ingredients, especially meat. But I also say that you have to watch the temperature. Make sure you are always working in a cold environment. Maybe the garage is not the best way to go."

— Lance Appelbaum

REGIONAL AMERICAN FLAVORS

Andouille, page 106

Louisiana Crayfish
Sausage, page 229

Buffalo Chicken
Sausage, page 210

Bison Sausage with
Jalapeño, page 185

Turkey and Cranberry
Sausage, page 221

Clam Dogs, page 238

Tartar Sauce, page 238

QUAIL LINKS

Here's a small-batch, no-casing sausage that's easy enough for anyone to make. Quail has a mild flavor that's slightly more assertive than chicken; wild quail will be even more so. A ready-to-cook quail weighs 3 to 7 ounces (including the giblets), so you'll need three to five birds to get the pound of meat needed for this recipe. If you happen to have a larger supply of quail (say, after a weekend of hunting), you can easily scale up the recipe. You can find farm-raised quail at specialty shops, gourmet grocers, and online. Because these sausages are not stuffed in casings, we recommend panfrying them rather than grilling.

- 1 pound boneless quail meat (including giblets)
- ¼ pound chicken fat
- 1 teaspoon kosher salt
- ½ teaspoon ground ginger
- ¼ teaspoon freshly ground black pepper
- ¼ teaspoon dried sage
- ¼ teaspoon dried thyme

1. Freeze the quail meat and chicken fat for 15 minutes.

2. Grind the meat and fat through the fine disk of a meat grinder.

3. In a medium bowl, combine the meat mixture, salt, ginger, pepper, sage, and thyme. Mix well, using your hands.

4. Grind the seasoned mixture through the fine disk of the grinder.

5. Wet your hands with cold water. Taking about 1 tablespoon of meat at a time, form little logs of sausage.

6. Place the sausage on a baking sheet fitted with a wire rack and refrigerate, uncovered, until firm, about 30 minutes.

7. Cook as desired (see methods on pages 38 to 40) to an internal temperature of 165°F (74°C); panfrying is recommended.

BIRDS OF A FEATHER

You could easily swap out the quail with other game birds, including dove or partridge (quail's cousin). Dove has a taste that's more full-bodied (since it's a red-meat bird), while partridge is similar to chicken and other poultry. Squab, which is farmed in the United States, is another option; each 1-pound bird will yield about ½ pound of meat. This culinary favorite has a richer flavor that's more akin to dark chicken meat and a velvety smooth mouthfeel.

ELK SAUSAGE WITH CAPERS AND WINE

MAKES 5 POUNDS

5 feet medium hog casing

4 pounds boneless elk meat, trimmed and cubed

1 pound beef fat

2 tablespoons kosher salt

2 teaspoons cayenne pepper

2 teaspoons freshly ground black pepper

1 teaspoon crushed anise seed

2 tablespoons small capers, rinsed and drained

2 garlic cloves, finely chopped

¼–⅓ cup dry red wine

Elk is an assertive meat and calls for assertive seasonings, such as the cayenne, anise seed, and capers in this distinctive sausage. Whenever possible, buy capers that are packed in salt; when rinsed and drained, they'll actually be less salty than those packed in brine, and they'll always have a superior flavor. These sausages are best when roasted or grilled.

1. Prepare the casing (see page 29).

2. Cut the meat and fat into 1-inch cubes. Freeze the cubes for about 30 minutes to firm them up before grinding.

3. Grind the meat and fat together through the coarse disk of a meat grinder.

4. In a large bowl, combine the ground meat mixture, salt, cayenne, black pepper, anise seed, capers, and garlic. Mix in ¼ cup of the wine, then add a bit more if needed to achieve a stuffing consistency.

5. Stuff the mixture into the prepared casing, prick air pockets, and twist off into 4-inch links.

6. Place the sausage on a baking sheet fitted with a wire rack and refrigerate, uncovered, for a few hours or preferably overnight.

7. Cut the links apart. Refrigerate, wrapped well in plastic, for 2 to 3 days, or freeze for up to 3 months; thaw overnight in the refrigerator before using.

8. Cook as desired (see methods on pages 38 to 40) to an internal temperature of 160°F (71°C).

Meet the
MAKER

Chris Hughes

Broken Arrow Ranch

Ingram, Texas

THE HUNTER AND GATHERER

If you've eaten wild venison or boar in a restaurant lately, chances are the meat was provided by Broken Arrow Ranch — the only fully inspected source of truly free-range wild game, based out of Ingram, in the Texas Hill Country. In Texas, these animals are so overpopulated they have been determined to pose a continuing threat to the native species. This is especially problematic on the many private ranches, some of them as large as national parks. That's where Broken Arrow steps in, working with the ranchers to help control the invasive animal population to a sustainable level — and to make the most of that delicious wild game meat.

Chris Hughes — the current, second-generation owner — explains that Broken Arrow was the brainchild of his father, Mike, who had purchased property in the area after retiring from a career in professional offshore diving and learned of the overpopulation problem (already an issue in 1983). "He also knew that venison was a popular menu item and yet it had to be imported from New Zealand. He thought there was a good potential for a market and a good potential for supply. He put those two together and that's how the company was started."

What makes Broken Arrow unique in the meat industry is the use of field harvesting. Because these are wild animals, as opposed to being raised as a meat source or agricultural product, they would normally have to be transported over long distances for processing, which is a very stressful situation for them. "These deer and antelope exist on large tracts of private land, anywhere from 1,000 up to 240,000 acres. We had to figure out how to get to these animals and then how to harvest them without causing them all that stress. What we came up with was the concept of a field harvest, where we go to the ranch with a mobile processing unit and our government meat inspector. We hunt for the animals on the ranch and then harvest them one at a time using the most humane method available, anywhere." After the initial on-site processing and inspection, the animal is brought back to Broken Arrow's fixed facility for butchering and aging.

Chris says they have forged relationships with more than 100 ranches around the state. "It's a win-win-win business model: The ranchers win because they get their population back down to a sustainable level, plus they've been paid because the animals have value; the animals win because they've lived a natural, wild life and when they are harvested it's done with the utmost respect; and consumers win because they get to enjoy high-quality, truly wild game meat."

There's a reason that Broken Arrow is the only game in town (or even the country), though some people have tried every now and then. "It's a very complex process. There are a lot of moving parts to it, it's very expensive to carry out, and as a consequence our meats are relatively expensive to purchase. The reality is that you spend a lot of time and expense driving to these ranches, often as far as 200 miles each way, and you can hunt that ranch for eight hours and end up with 30 to 40 deer, 12 deer, or zero deer. That's just part of it."

At Broken Arrow, aging is a two-part process: First they'll dry-age the whole carcass by hanging it for three to five days to allow some of the moisture in the meat to evaporate, concentrating the flavors. This is when the natural enzymes in the muscles start breaking down the connective tissue. Then they butcher the animal into primal cuts, vacuum-pack these to stop the evaporation, and do what's called a "wet age." During this important process, the evaporation is stopped but that break-down of connective tissue is allowed to continue, so the meat becomes exceptionally tender. "You can dry-age beef for 30 days because there's a lot of fat on it, and that fat gives it flavor. But venison is so lean and there's almost no marbling, so if you dry-aged it for that long it would just taste dry. That's why we do more of a wet-aging technique."

Besides selling these cuts to restaurants (which represents 80 percent of all sales) and home cooks through their website, Broken Arrow produces six differ-ent types of sausages. "We've made sausage since the beginning. With any kind of butchery you need to use the whole animal. Plus, we love sausage, and I never sell anything I wouldn't want to eat myself. Down here we have a strong German influence and we worked with an old German butcher to come up with our recipes." Broken Arrow makes Italian links and summer sausages in either venison or wild boar, a venison brat, and a hickory-smoked sausage that combines the best of all worlds — venison, wild boar, jalapeño, and cheddar cheese. These are processed at a USDA-inspected plant using Broken Arrow's field-harvested meat and proprietary recipes.

Chris says he enjoys making sausage at home with deer that's been harvested at Broken Arrow, or with quail from its sister company, Diamond H Ranch. He also makes a signature sausage with Dorper lamb that Broken Arrow sources from a small group of family ranchers, all pasture and humanely raised (see page 199 for a rec-ipe). "I make fresh and smoked sausage, but it's so hot down here I haven't been able to figure out how to make a good curing chamber. It's on my wish list, though."

TRADE SECRETS

"Making sausage is a mix of science and art. Following recipes closely will make good sausage. Practice and experi-ence will make great sausage. Learn something from each batch you make."

— Chris Hughes

ALLIGATOR SAUSAGE

Jarred Zeringue (see profile, page 104) provided the recipe for this gator sausage, which he serves with creole mustard, pickled vegetables, and grilled French bread as part of his bar menu at Vacherie, where he is the owner and chef. The restaurant is located in the heart of the French Quarter of New Orleans but is named for the nearby town where Zeringue grew up that's located on the banks of the Mississippi River. Vacherie is one of several other establishments in Zeringue's portfolio, so you can be sure to find one that suits your fancy next time you're in the area.

1. Prepare the casing (see page 29).

2. Cut the alligator, pork, and fat into 1-inch cubes. Freeze the cubes for about 30 minutes to firm them up before grinding.

3. Grind the meat and fat through the medium disk of a meat grinder.

4. In a large bowl, combine the meat mixture with the water, onion, celery, bell pepper, parsley, scallion, salt, black pepper, and cayenne. Mix well with your hands until combined and the pork fat starts to coat the meat.

5. Stuff the mixture into the prepared casing, prick air pockets, and twist off into 6-inch lengths.

6. Place the sausage on a baking sheet fitted with a wire rack and refrigerate, uncovered, for 1 day.

7. Cut the links apart. Refrigerate, wrapped well in plastic, for 2 to 3 days, or freeze for up to 3 months; thaw overnight in the refrigerator before using.

8. Cook as desired (see methods on pages 38 to 40) to an internal temperature of 160°F (71°C).

Ingredients

- 6 feet medium hog casing
- 2 pounds boneless alligator meat
- 2 pounds boneless lean pork shoulder
- 8 ounces pork fat
- ½ cup ice water
- ¼ cup finely chopped onion
- ¼ cup finely chopped celery
- ¼ cup finely chopped red bell pepper (ribs and seeds removed)
- ¼ cup chopped fresh flat-leaf parsley
- ¼ cup thinly sliced scallion
- 2 tablespoons kosher salt
- ½ teaspoon freshly ground black pepper
- ¼ teaspoon cayenne pepper

PARTRIDGE AND WILD RICE SAUSAGE

MAKES 2 POUNDS

- 3 feet small hog or sheep casing
- 1½ pounds boneless partridge meat
- ½ pound chicken fat
- 1 cup apple cider
- 1 cup cooked wild rice
- ½ cup golden raisins
- 2 teaspoons kosher salt
- ½ teaspoon freshly ground black pepper
- ¼ teaspoon ground allspice
- ¼ teaspoon ground cinnamon
- 1 tablespoon fresh lemon juice

The boiled-down cider in the sausage mixture lends sweetness and helps bind the ingredients. Cinnamon and allspice round out the fall flavors, and plump raisins lend pleasant chewiness (and more sweetness). You can easily substitute the partridge meat with duck or dark-meat chicken. Serve as part of a harvest feast, with roasted pumpkin or butternut squash and sautéed kale or escarole.

1. Prepare the casing (see page 29).

2. Cut the partridge meat and the fat into 1-inch cubes. Freeze the cubes for about 30 minutes to firm them up before grinding.

3. Meanwhile, in a small saucepan, simmer the cider until it is reduced to about 3 tablespoons of syrupy liquid, 15 to 20 minutes. Remove from the heat and set aside.

4. Grind the meat and fat through the fine disk of a meat grinder.

5. In a large bowl, combine the ground meat and fat, reduced cider, rice, raisins, salt, pepper, allspice, cinnamon, and lemon juice. Mix well, using your hands.

6. Stuff the mixture into the prepared casing, prick air pockets, and twist off into 3-inch links.

7. Place the sausage on a baking sheet fitted with a wire rack and refrigerate, uncovered, for a few hours or preferably overnight.

8. Cut the links apart. Refrigerate, wrapped well in plastic, for 2 to 3 days, or freeze for up to 3 months; thaw overnight in the refrigerator before using.

9. Cook as desired (see methods on pages 38 to 40) to an internal temperature of 165°F (74°C); panfrying is recommended.

RABBIT SAUSAGE WITH FRESH HERBS

Although this recipe is specifically tailored to wild cottontail rabbits, you can substitute domestic rabbit, found in butcher shops, gourmet grocers, and some supermarkets. The domestic meat will be milder, so you may wish to reduce the amounts of the seasonings accordingly (start with less, then cook off a small portion and add more as needed).

1. Prepare the casing (see page 29).

2. Cut the meat and fat into 1-inch cubes. Freeze the cubes for about 30 minutes to firm them up before grinding.

3. In a large bowl, combine the meat, fat, salt, black pepper, white pepper, thyme, chives, celery, parsley, ginger, and wine. Mix well, using your hands.

4. Grind the mixture through the fine disk of a meat grinder.

5. Stuff the mixture into the prepared casing, prick air pockets, and twist off into 3-inch links.

6. Place the sausage on a baking sheet fitted with a wire rack and refrigerate, uncovered, for a few hours or preferably overnight.

7. Cut the links apart. Refrigerate, wrapped well in plastic, for 2 to 3 days, or freeze for up to 3 months; thaw overnight in the refrigerator before using.

8. Cook as desired (see methods on pages 38 to 40) to an internal temperature of 160°F (71°C).

Ingredients:

- 3 feet small hog or sheep casing
- 2½ pounds boneless rabbit meat
- ½ pound chicken fat or pork fat
- 1 tablespoon kosher salt
- ½ teaspoon freshly ground black pepper
- ½ teaspoon freshly ground white pepper
- ½ teaspoon dried thyme
- 2 tablespoons chopped fresh chives
- 2 tablespoons finely minced celery or lovage
- 2 tablespoons chopped fresh flat-leaf parsley
- 1 teaspoon grated fresh ginger
- 2 tablespoons dry white wine

WILD HARE: HANDLE WITH CARE

A wild cottontail usually weighs in at 2 to 4 pounds. Because wild rabbits are subject to tularemia, a highly contagious infectious disease, always wear rubber gloves when skinning or handling the animal. Wash your hands before and after handling the raw meat, and be sure to wash in hot, soapy water all utensils and surfaces that come into contact with the meat.

5 feet small hog or sheep casing

3½ pounds boneless wild turkey meat

½ pound turkey fat or chicken fat

1 tablespoon plus 2 teaspoons kosher salt

1 teaspoon celery seeds

1 teaspoon freshly ground white pepper

½ teaspoon ground bay leaf

¼ teaspoon ground cloves

2 tablespoons chopped fresh flat-leaf parsley

2 teaspoons grated lemon zest

¼ cup bourbon, preferably Wild Turkey

FRESH

WILD TURKEY SAUSAGE

This recipe is geared to the use of wild turkey, not the domestic bird. The difference in the flavor of the two meats is as great as the difference in the birds' native intelligence — in other words, wild turkey is vastly superior on both counts. Bourbon (of the Wild Turkey variety) is also essential.

1. Prepare the casing (see page 29).

2. Cut the turkey and fat into 1-inch cubes. Freeze the cubes for about 30 minutes to firm them up before grinding.

3. In a large bowl, combine the turkey, fat, salt, celery seeds, pepper, bay leaf, cloves, parsley, lemon zest, and bourbon. Mix well, using your hands.

4. Grind the mixture through the fine disk of a meat grinder.

5. Stuff the mixture into the prepared casing, prick air pockets, and twist off into 3-inch links.

6. Place the sausage on a baking sheet fitted with a wire rack and refrigerate, uncovered, for a few hours or preferably overnight.

7. Cut the links apart. Refrigerate, wrapped well in plastic, for 2 to 3 days, or freeze for up to 3 months; thaw overnight in the refrigerator before using.

8. Cook as desired (see methods on pages 38 to 40) to an internal temperature of 165°F (74°C).

HERBED GAME SAUSAGE

Got game? Then make this sausage, which is entirely adaptable to most wild-caught meats. Mixed fresh herbs are the ticket to success in this recipe, so avoid any temptation to substitute dried. You can use any combination you like, but we especially like parsley, chives, tarragon, and chervil — more classically known as *fines herbes* in classic French cooking.

1. Prepare the casing (see page 29).

2. Cut the game meat, pork, and fat into 1-inch cubes. Freeze the cubes for about 30 minutes to firm them up before grinding.

3. Grind the game meat and pork through the coarse disk of a meat grinder. Grind the fat through the coarse disk separately.

4. In a large bowl, combine the ground meats, fat, salt, pepper, parsley, mixed herbs, garlic, and wine. Mix well, using your hands. Freeze the mixture for 30 minutes.

5. Grind the seasoned mixture through the fine disk of the grinder.

6. Stuff the mixture into the prepared casing, prick air pockets, and twist off into 5-inch links.

7. Place the sausage on a baking sheet fitted with a wire rack and refrigerate, uncovered, for a few hours or preferably overnight.

8. Cut the links apart. Refrigerate, wrapped well in plastic, for 2 to 3 days, or freeze for up to 3 months; thaw overnight in the refrigerator before using.

9. Cook as desired (see methods on pages 38 to 40) to an internal temperature of 160°F (71°C).

Ingredients:

- 5 feet medium hog casing
- 1½ pounds venison, moose, bison, or elk
- 1 pound pork shoulder
- ½ pound pork fat
- 1 tablespoon kosher salt
- 1 teaspoon freshly ground black pepper
- ¼ cup chopped fresh flat-leaf parsley
- 3 tablespoons chopped fresh herbs (such as basil, chives, cilantro, marjoram, sage, savory, tarragon, and thyme, as desired)
- 5 garlic cloves, minced
- ½ cup dry red wine

VENISON AND LAMB CEVAPI

- 1 pound boneless venison

- 1 pound boneless lamb shoulder

- 3 tablespoons finely chopped onion

- 1 tablespoon minced garlic

- 2 teaspoons kosher salt

- 2 teaspoons paprika

- 2 teaspoons freshly ground black pepper

- ¾ teaspoon baking soda

  Olive oil, for cooking

This recipe is adapted from one provided by Chris Hughes, founder of Broken Arrow Ranch in beautiful Ingram, Texas (see profile, page 192). He tells its origin best: "I spent almost two years working in Bosnia-Herzegovina, where I developed a strong addiction to cevapi, a grilled, caseless sausage served with fresh bread, creamy cheese, roasted pepper relish, and sliced onions. Since there are not many Balkan restaurants in small-town central Texas, I've had to make my own cevapi to satisfy the cravings." Chris strongly suggests serving it the traditional way, with the aforementioned accompaniments, all available online. *See photo, page 122.*

1. Cut the venison and lamb into 1-inch cubes. Freeze for 30 minutes to firm them up before grinding.

2. Grind the meat separately through the medium disk of a meat grinder.

3. In a large bowl, combine the venison and lamb, onion, garlic, salt, paprika, pepper, and baking soda. Mix well, using your hands, until the meat just starts to become a bit tacky. Freeze for 30 minutes.

4. Scoop out a small amount of sausage and roll into a link, 3 to 4 inches long and ¾ inch thick. Repeat with the remaining sausage.

5. Coat the links lightly with olive oil to help prevent sticking to the grill. Grill the cevapi over a medium-hot fire, turning occasionally, until an instant-read thermometer reaches an internal temperature of 160°F (71°C). Alternatively, preheat the oven to 375°F (245°C) and roast the sausage for 8 to 10 minutes, until cooked through.

NOTE: *While freshly ground meat is usually preferable, you can still achieve very good, and rapid, results when using preground meat from the local butcher shop (steer clear of the supermarket variety, though). The baking soda added to the meat mixture in this recipe makes for a slightly springy texture and ensures the cevapi will hold up well on the grill.*

VENISON SUMMER SAUSAGE

MAKES 3 POUNDS

This recipe has been adapted from *Dishing Up Minnesota: 150 Recipes from the Land of 10,000 Lakes* by Teresa Marrone. According to Marrone, this isn't a true summer sausage because it isn't fermented or finished by smoking or drying, but it does have the characteristic color and texture thanks to the addition of curing salt, which also improves the safety of the finished product. The author prefers to use the food processor for the first chopping, and then the meat grinder fitted with a small disk for the final processing, but you can use the grinder for both. She also suggests substituting bison or grass-fed beef for the venison and, for more traditional links, using about 4 feet of large hog casings (prepared as described on page 29); the cooking time will be significantly shorter (closer to 30 minutes) because the sausages won't be as thick.

- 1½ pounds boneless lean, well-trimmed venison meat
- ¾ pound boneless lean pork butt or shoulder
- 1 tablespoon Morton Tender Quick
- 2 teaspoons kosher salt
- 1 teaspoon sugar
- ½ teaspoon garlic powder
- ⅛ teaspoon freshly grated nutmeg
- 1 teaspoon mustard seeds
- 1 teaspoon coarsely ground black pepper

1. Cut the meats into 1-inch cubes. Freeze the cubes for about 30 minutes to firm them up before grinding.

2. Combine the venison and pork cubes with the Tender Quick, salt, sugar, garlic powder, and nutmeg in a large bowl. Mix well, using your hands.

3. Chop the mixture very coarsely in a food processor (or grind through the coarse disk of a meat grinder). Pack into a ceramic or glass bowl. Cover tightly with plastic wrap and refrigerate for 2 days.

4. Grind the mixture through the fine disk of the grinder (or chop finely in a food processor).

5. Place in a large bowl and add the mustard seeds and pepper. Mix well with your hands, kneading and squeezing for several minutes.

6. Divide the mixture into three equal portions. Roll each portion into a log about 6½ inches long and 2 inches thick, packing firmly to eliminate air pockets. For each log, cut a 1-foot length of cheesecloth and open it up so there are two layers (most cheesecloth is folded into four layers). Wrap each log tightly in the cheesecloth, twisting tightly at the ends and tying with butcher's twine.

7. Refrigerate the wrapped logs while you preheat the oven to 325°F (163°C).

8. Place the wrapped logs on a broiler pan or in a roasting pan fitted with a wire rack. Bake for about 1¼ hours, or until an internal temperature of 160°F. Let the logs cool, then wrap in aluminum foil and refrigerate for at least 12 hours before sampling; peel away the cheesecloth before slicing. The cooked sausage will keep in the refrigerator for about 1 week; or up to 3 months in the freezer; thaw overnight in the refrigerator.

2–3 feet small hog or
    sheep casing

2 pounds boneless wild
   duck meat

2 teaspoons kosher salt

1 teaspoon ground
   coriander

1 teaspoon paprika

½ teaspoon freshly
   ground white pepper

¼ teaspoon ground
   mace

2 tablespoons finely
   chopped fresh chives

2 garlic cloves, minced

1 large egg white

COOKED

WILD DUCK DOGS

Wild duck meat is the reason to make these dogs, as is the streamlined technique — all that's needed is to grind the seasoned duck meat finely, without having to emulsify it. The larger species of wild ducks will yield about 12 ounces of meat per bird, meaning you will need three ducks to make this recipe. The seasonings are pretty traditional for franks, only used here in generous measure to stand up to the intensity of the duck meat.

1. Prepare the casing (see page 29).

2. Cut the meat into 1-inch cubes.

3. In a large bowl, combine the duck, salt, coriander, paprika, pepper, mace, chives, garlic, and egg white. Mix well, using your hands. Freeze for 30 minutes.

4. Grind the mixture through the fine disk of a meat grinder. Freeze for 30 minutes.

5. Stuff the mixture into the prepared casing, prick air pockets, and twist off into 4-inch links. Tie off each link with butcher's twine. Do not separate the links.

6. Bring a large pot of water to a boil, add the links, and reduce the heat to maintain a water temperature of 180 to 190°F (83 to 88°C). Simmer the sausages for 30 minutes, or until an internal temperature of 165°F (74°C).

7. Drain the sausages and let cool. Dry them thoroughly and wrap well in plastic wrap. Refrigerate for up to 1 week, or freeze for up to 2 months; thaw in the refrigerator overnight.

8. The dogs are best panfried in a little oil and served warm.

SMOKED GAME SAUSAGE

Take your pick: Venison, elk, moose, or bison, alone or in combination, will all work well in making this robust sausage. Worcestershire sauce adds depth of flavor, red wine imparts a welcome hit of acidity, and walnuts and chopped onion lend textural contrast (and more flavor). It's delicious grilled or cut into chunks and added to a hearty soup or stew.

1. Prepare the casing (see page 29).

2. Cut the meat and fat into 1-inch cubes. Freeze the cubes for about 30 minutes to firm them up before grinding.

3. Grind the meat through the coarse disk of a meat grinder. Grind the fat through the fine disk.

4. In a large bowl, combine the meat, fat, pepper, salt, coriander, mustard seeds, onion, walnuts, garlic, wine, and Worcestershire. Mix well, using your hands.

5. Stuff the mixture into the prepared casing, prick air pockets, and twist off into 6-inch links.

6. Place the sausages on a baking sheet fitted with a wire rack and refrigerate, uncovered, for 24 hours.

7. Hot-smoke the sausages at 170 to 180°F (77 to 82°C) for about 4 hours, or until the sausages reach an internal temperature of 160°F (71°C).

8. Refrigerate, wrapped well in plastic, for 2 to 3 days, or freeze for up to 3 months; thaw overnight in the refrigerator before using.

Ingredients

- 4 feet medium hog casing
- 2 pounds boneless lean venison, moose, bison, or elk
- 1 pound pork fat
- 1 tablespoon freshly ground black pepper
- 1 tablespoon kosher salt
- 2 teaspoons ground coriander
- 1 teaspoon mustard seeds
- 1 small onion, finely chopped
- ⅓ cup finely chopped walnuts
- 3 garlic cloves, minced
- ½ cup dry red wine
- 1 tablespoon Worcestershire sauce

MAKING JERKY

To begin, what does *jerky* mean anyway? Our English word *jerky* comes from the Spanish *charqui*, which means "dried meat." Basically, it's the oldest meat preservation technique known to mankind — sun-drying strips of meat. Evaporating the water from the meat stops enzyme reactions that lead to spoilage. Ancient Egyptians and others used the technique for any animal too big to eat all at once (we're talking elephant, crocodile, water buffalo, and so on).

Meat dries most quickly when it's cut into thin strips; salt and seasonings add flavor and help with preservation. In our more safety-conscious age, people who decide to make their own jerky might consider the USDA recommendation that meat be precooked to 160°F (71°C) before drying.

You can find jerky made from all types of meat and with all types of seasonings. This one is made with naturally lean bison, which is as good a place to begin your own jerky journey as you can find, though venison, elk, moose, emu, alligator, turkey, or beef (the original!) are worthy substitutes.

Bison Jerky

MAKES ABOUT ½ POUND

1 pound boneless bison flank steak or other lean cut, trimmed of exterior fat and connective tissue

6 juniper berries

1 tablespoon plus 1 teaspoon kosher salt

1 teaspoon mustard seeds

2 tablespoons pure maple syrup

1. Cut the bison with the grain into strips that are ¼ inch thick, 3 to 4 inches long, and about ¾ inch wide. Some bison flank steaks are about the right thickness, so you can just cut them into appropriately sized strips; slice thicker cuts into ¼-inch-thick strips. Place the meat in a bowl.

2. Crush the juniper berries and salt together with a mortar and pestle (or a spice mill), pulverizing the juniper berries until they are fairly fine. Add the mustard seeds, and crush fairly fine. Stir in the maple syrup. Scrape the mixture into the bowl with the bison, stirring to coat. Cover the bowl and refrigerate overnight, stirring once or twice.

3. The next day, blot the bison pieces lightly with paper towels to dry off excess liquid; the crushed seeds should still be clinging to the meat. Pound lightly on both sides with a meat tenderizer. Arrange the meat in a single layer on dehydrator trays (or smoker racks, if smoking). Dehydrate (or smoke) at 145°F (63°C) until the meat is dry and dark, 4 to 6 hours; if you bend a piece, it should not ooze any moisture. (If smoking, replenish the wood once or twice, but don't keep a heavy smoke going throughout or the jerky will be too smoky.) If you're concerned about the safety of the meat, bake the finished jerky in a 275°F (135°C) oven for 10 minutes before cooling and storing.

4. Let the jerky cool completely, then wrap in plastic wrap and refrigerate for up to 1 month; for longer storage, wrap the jerky very well and freeze it for up to 6 months.

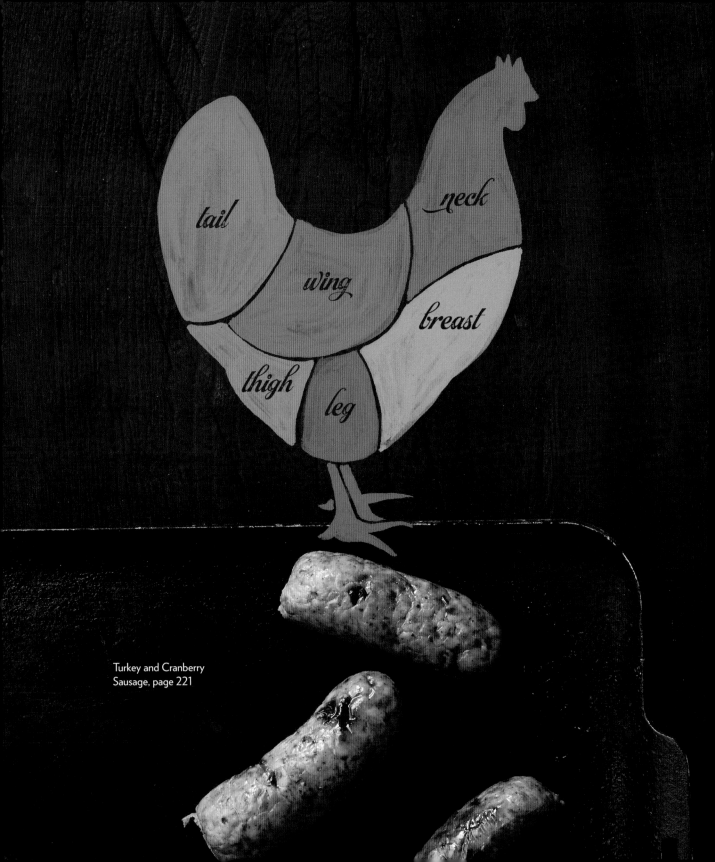

Turkey and Cranberry
Sausage, page 221

POULTRY SAUSAGES

A highbrow is the kind of person who looks at a sausage and thinks of Picasso.

— Sir Alan Patrick Herbert

Dr. Robert C. Baker of Cornell University has the distinction of being known as the inventor of the so-called chicken dog. In the 1970s, he and his colleagues in Cornell's College of Agriculture and Life Sciences realized that poultry could be used for virtually any sausage that is traditionally made with beef or pork, and with the extra advantages of being cheaper and lower in fat. Since then, commercial sausage makers, artisanal producers, and consumers have embraced poultry sausages with enthusiasm.

Sausage is one application in which the inherently mild taste of chicken and turkey is a real advantage, for it encourages all kinds of creativity on the part of the home sausage maker. All manner of seasonings can be added: fresh or dried herbs and spices, dried fruit, sharp cheeses, wines and liqueurs, nuts and even seeds, and grains, plus whatever else you might conjure up.

A larger proportion of dark meat to light meat seems to produce a more pleasing texture, and the judicious addition of some skin and fat will enhance the flavors without creating a greasy product. Enjoy your experiments with poultry sausage, and say a word of thanks to Dr. Baker.

COUNTRY CHICKEN SAUSAGE

This sausage relies on the traditional "country sausage" combination of herbs and spices — sage, thyme, ginger, and savory — plus the optional, but strongly suggested, kick of heat from cayenne. *See photo, page 58.*

1. Prepare the casing (see page 29).

2. Cut the chicken into 1-inch cubes and freeze for about 30 minutes to firm them up before grinding.

3. Grind the chicken through the fine disk of a meat grinder.

4. In a large bowl, combine the chicken, salt, black pepper, cayenne (if using), ginger, sage, summer savory, and thyme. Mix well, using your hands. Refrigerate for 30 minutes.

5. Grind the mixture through the fine disk of the meat grinder.

6. Stuff the mixture into the prepared casing, prick air pockets, and twist off into 2- to 3-inch links.

7. Place the sausage on a baking sheet fitted with a wire rack and refrigerate, uncovered, for a few hours or preferably overnight.

8. Cut the links apart. Refrigerate, wrapped well in plastic, for 2 to 3 days, or freeze for up to 3 months; thaw overnight in the refrigerator.

9. Cook as desired (see methods on pages 38 to 40) to an internal temperature of 165°F (74°C). Panfrying is recommended.

- 2 feet small hog or sheep casing
- 2 pounds boneless, skin-on chicken (mixture of dark and light meat)
- 2 teaspoons kosher salt
- 1 teaspoon freshly ground black pepper
- ½ teaspoon cayenne pepper (optional)
- ½ teaspoon ground ginger
- ½ teaspoon dried sage
- ½ teaspoon dried summer savory
- ½ teaspoon dried thyme

CHICKEN BRATWURST

3 feet small hog or sheep casing

3 pounds boneless, skin-on chicken (mixture of dark and light meat)

2 teaspoons kosher salt

1 teaspoon freshly ground white pepper

¾ teaspoon caraway seeds, crushed

¾ teaspoon dried marjoram

½ teaspoon ground allspice

For anyone preferring chicken over pork or beef, here's a way to have your brats and enjoy them, too. They feature the same classic spices (caraway seeds, marjoram, and allspice) and can be cooked (preferably grilled) and eaten the same way as the originals — namely, on a hard roll with brown mustard.

1. Prepare the casing (see page 29).

2. Cut the chicken into 1-inch cubes and freeze for about 30 minutes to firm them up before grinding.

3. Grind the chicken through the fine disk of a meat grinder.

4. In a large bowl, combine the chicken, salt, pepper, caraway seeds, marjoram, and allspice. Mix well, using your hands. Refrigerate for 30 minutes.

5. Grind the seasoned mixture through the fine disk of the meat grinder.

6. Stuff the mixture into the prepared casing, prick air pockets, and twist off into 3-inch links.

7. Place the sausage on a baking sheet fitted with a wire rack and refrigerate, uncovered, for a few hours or preferably overnight.

8. Cut the links apart. Refrigerate, wrapped well in plastic, for up to 3 days, or freeze for up to 2 months; thaw overnight in the refrigerator.

9. Cook as desired (see methods on pages 38 to 40) to an internal temperature of 165°F (74°C).

Meet the MAKERS

⚹

Brent Young and Ben Turley

The Meat Hook

Brooklyn, New York

PAYING IT FORWARD

"Our roster of sausages is around 80 right now, and we make about 14 different varieties at any given time. We always have a couple different Italian varieties, a couple Mexican chorizos, and everything else is on a rotating basis. We'll have a couple of 'classy' ones, a couple of 'trashy' ones, and then we do what makes people happy seasonally," says Brent Young, co-owner of The Meat Hook, along with Ben Turley. That's the way the sausages are categorized on their menu board and website: Classy (traditional) and Trashy (innovative). "Ben gets the credit for that idea."

That playful attitude pervades pretty much everything about this highly regarded butcher shop (where it seems a well-honed, slightly off-kilter sense of humor is right up there with knife skills as a job requirement). You'll find sausages with names like Franch Onion (phonetically spelled when spoken in French) and Long Dong Bud (described as *The classic. The ultimate. Pork, pepper Jack cheese, roasted jalapeños, and Texas Pete hot sauce. YOU'RE WELCOME*). "That was the first adventurous sausage we created, and it's still the universal favorite. It was 2009, the Pittsburgh Steelers were going to the Super Bowl, and I wanted to make a sausage that would be great for tailgating," says the Pittsburgh native. "I called my dad, whose name is Bud, to help me come up with what to call it. After a couple of other suggestions, he said, 'Why don't you call it Long Dong Bud!' and I basically laughed that off and proceeded to call it something else, like Pork and Pepper Jack sausage. No one bought it. The next day I decided what the heck, I'm going to call it Long Dong Bud, and it sold out within the hour. It might have just been the timing, but the name stuck." Soon after that, they created the Bacon Cheeseburger Sausage as a good way to use up an excess of beef (Ben gets the credit for this idea, too). "It became *the* thing we were known for very quickly. It's a huge part of our identity and ultimately we're proud of it, and it helps resonate that we enjoy our jobs and we like making fun, delicious stuff even if it's kind of goofy."

Not that he and his staff take all this in stride. "Sausage making is a skill and it's not easy to do. We take pride in our technique. It's not easy making blood sausage, and it's also not easy to take the flavors of coq au vin and make them into a sausage. You have to know how to actually execute this stuff." That they can produce such a broad range of sausages, including many familiar varieties like bratwurst, andouille, and merguez along with those that are less familiar, such as chipolata, boerewors, and lap cheong, speaks to their knowledge and know-how.

Another of The Meat Hook's signature sausages is their version of the stadium hot dog, which is also available at Rippers (where Young is a partner), on the boardwalk at Rockaway Beach. Because of the high demand, the hot dogs are processed, to their specifications, by a copacker. Plus, he says that hot dogs are by far the most difficult sausage to make, and there's just the economy of scale: The bigger, more industrial equipment you have, the easier it is to create that characteristic texture, at least the one that he was looking for. It took them five years of recipe development to land on the current one, made with grass-fed beef and pastured pork. "We tried it with 100 percent beef but it was too heavy, and it really did have too much flavor. The pork really lightens it up, so it's not such a kick in the gut."

He says that some customers (mostly Mets fans) think nothing of dropping over $100 on three dozen dogs for a tailgate party. And when anyone asks how to make hot dogs at home, he warns them off. "Do yourself a favor and don't try it. You're going to be really mad, it's going to get all over the place, and you're not going to be happy with the result. It's not about the skill level, it's about having the right equipment. A KitchenAid mixer just isn't strong enough to produce the proper emulsification, and the hot dogs will end up tasting sandy at best." He does, however, encourage people to try their hand at other types of sausages, and is more than happy to share tips and techniques with customers. Rule number one: Start with high-quality meat.

That was indeed the motivation behind The Meat Hook since day one: to change the way people buy meat and thereby create a market for the local farmers who were committed to adopting sustainable methods. After years working side by side at Brooklyn's Marlow & Daughters, Brent and Ben (and another co-owner who has since parted ways) came up with the idea of opening their own butcher shop, whose mission statement sums up their approach: *Everything we do is for the purpose of properly representing our farmers' hard work, deep knowledge and quality animals. We never forget that without them we're just another bunch of jerks selling pork chops.* To that end, they'll regularly put the "Closed" sign on the door and take the entire staff on a field trip. Spend the entire day at a farm, the idea being, and you'll never underestimate the value placed on the animals raised there or on the farmers who raised them — an idea that will translate to your customers, too.

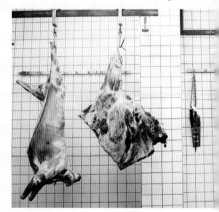

BUFFALO CHICKEN SAUSAGE

MAKES 5 POUNDS

Here's another popular item from The Meat Hook, and the recipe is courtesy of Brent Young, one of the owners (see profile, page 208). As they describe the sausage, "It's your favorite bar food without the Wet-Naps. Chicken and pork, Frank's hot sauce, blue cheese, and a hint of celery seed. Enough said, in our book." Grill, roast, or panfry, then serve with celery and more blue cheese. *See photo, page 188.*

1. Prepare the casing (see page 29).

2. Cut the chicken and pork into 1-inch cubes. Freeze for 30 minutes.

3. Grind the meats through the small disk of a meat grinder.

4. In a large bowl, combine the ground meat, hot sauce, salt, crushed red pepper, black pepper, garlic powder, onion powder, dry milk, and celery seeds. Mix very well with your hands until well combined and a little tacky.

5. Fold in the blue cheese gently, trying to keep it from breaking up too much.

6. Stuff the mixture into the prepared casing, prick air pockets, and twist off into 6-inch lengths.

7. Place the sausage on a baking sheet fitted with a wire rack and refrigerate, uncovered, for 1 day.

8. Cut the links apart. Refrigerate, wrapped well in plastic, for 2 to 3 days, or freeze for up to 3 months; thaw overnight in the refrigerator before using.

9. Cook as desired (see methods on pages 38 to 40) to an internal temperature of 165°F (74°C).

- 12 feet lamb casing
- 2½ pounds boneless, skin-on chicken thighs
- 2½ pounds boneless lean pork butt or shoulder
- ½ cup Frank's RedHot sauce
- 3 tablespoons kosher salt
- 1 tablespoon crushed red pepper
- 2 teaspoons freshly ground black pepper
- 1½ teaspoons garlic powder
- 1 teaspoon onion powder
- 3 tablespoons nonfat dry milk powder
- 1 teaspoon celery seeds
- ½ cup crumbled blue cheese

CHICKEN SAUSAGE WITH APPLES AND RIESLING

MAKES 2 POUNDS

- 2 feet small hog or sheep casing
- 2 pounds boneless, skin-on chicken thighs
- 2 teaspoons kosher salt
- 1 teaspoon ground ginger
- ½ teaspoon freshly ground black pepper
- 1 Granny Smith or other firm, tart apple, peeled and chopped
- 2 tablespoons minced onion
- ¼ cup dry Riesling

Tart apples provide contrasting flavor and bite to this fine-textured sausage that bears a subtle tang from the addition of wine. German Riesling works exceptionally well here — look for dry varieties with Trocken or Selection on the label — but you could also try it with Gewurztraminer, also on the drier side. Serve this sausage panfried, with roasted potatoes and homemade applesauce, or sliced and served atop crackers along with a creamy Brie or pungent blue cheese.

1. Prepare the casing (see page 29).

2. Cut the chicken into 1-inch cubes. Freeze for 30 minutes.

3. Grind the chicken and skin through the fine disk of a meat grinder.

4. In a large bowl, combine the ground chicken, salt, ginger, pepper, apple, onion, and wine. Mix well, using your hands.

5. Grind the seasoned mixture through the fine disk of the meat grinder.

6. Stuff the mixture into the prepared casing, prick air pockets, and twist off into 3-inch links.

7. Place the sausage on a baking sheet fitted with a wire rack and refrigerate, uncovered, for 1 day.

8. Cut the links apart. Refrigerate, wrapped well in plastic, for 2 to 3 days, or freeze for up to 3 months; thaw overnight in the refrigerator before using.

9. Cook as desired (see methods on pages 38 to 40) to an internal temperature of 165°F (74°C). Panfrying is recommended.

HERBES DE PROVENCE CHICKEN SAUSAGE

A gloriously juicy roasted chicken redolent with aromatic herbs was the influence for this delicious sausage, as was an extra supply of herbes de Provence — the famous blend of dried herbs that is the essence of southern French cooking. You can find the herbs in traditional terra-cotta crocks at most supermarkets these days and always at epicurean shops — or make your own blend; see below.

1. Prepare the casing (see page 29).

2. Cut the chicken into 1-inch cubes. Freeze for 30 minutes.

3. Grind the chicken and skin through the fine disk of a meat grinder.

4. In a large bowl, combine the chicken, herbes de Provence, salt, pepper, capers, oil, and lemon juice. Mix well, using your hands.

5. Grind the seasoned mixture through the fine disk of the meat grinder.

6. Stuff the mixture into the prepared casing, prick air pockets, and twist off into 3-inch links.

7. Place the sausage on a baking sheet fitted with a wire rack and refrigerate, uncovered, for a few hours or preferably overnight.

8. Cut the links apart. Refrigerate, wrapped well in plastic, for 2 to 3 days, or freeze for up to 3 months; thaw overnight in the refrigerator before using.

8. Cook as desired (see methods on pages 38 to 40) to an internal temperature of 165°F (74°C); panfrying is recommended.

Ingredients:

- 2 feet medium hog casing
- 2 pounds boneless, skin-on chicken (mixture of dark and light meat)
- 2 teaspoons herbes de Provence
- 1 teaspoon kosher salt
- ½ teaspoon freshly ground black pepper
- 1 tablespoon capers, rinsed and drained
- 1 tablespoon olive oil
- 2 teaspoons fresh lemon juice

HERBES DE PROVENCE

When making your own blend, you can tailor it to meet your own preferences (adding lavender, for instance) or to better suit a particular dish (adding rosemary for lamb dishes). Mix together (all dried) summer savory, marjoram, thyme, mint or basil, sage, and fennel seeds. Other additions: ground bay leaf, oregano, chervil, parsley, and orange zest. Experiment until you find the combination that you like best.

CHICKEN SAUSAGE WITH BROCCOLI RABE AND MOZZARELLA

MAKES 5 POUNDS

- 8 feet small hog casing
- 4 pounds boneless, skin-on chicken thighs
- 2 cups chopped broccoli rabe (about 1 pound)
- 1 tablespoon plus 1 teaspoon kosher salt
- 1 tablespoon minced garlic
- 2 teaspoons freshly ground black pepper
- ½ teaspoon crushed red pepper
- 7 ounces mozzarella cheese, cut into ¼-inch dice

Save the fresh mozzarella for a caprese salad — it's way too moist for this sausage. Use only the shrink-wrapped cheese you find in the supermarket instead. *See photo, page 214.*

1. Prepare the casing (see page 29).

2. Cut the chicken into 1-inch cubes. Freeze for 30 minutes.

3. While the meat is chilling, prepare the broccoli rabe: Bring 2 quarts water to a boil and add 1 teaspoon of the salt. Boil the broccoli rabe for 3 minutes, then drain. Let cool, and squeeze as much water as you can out of the broccoli rabe. Set aside.

4. Grind the chicken through the medium disk of a meat grinder.

5. In a large bowl, combine the chicken, the remaining 1 tablespoon salt, the garlic, black pepper, and crushed red pepper. Mix well, using your hands.

6. Fold in the cheese and broccoli rabe.

7. Stuff the mixture into the prepared casing, prick air pockets, and twist off into 6-inch lengths.

8. Place the sausage on a baking sheet fitted with a wire rack and refrigerate, uncovered, for 1 day.

9. Cut the links apart. Refrigerate, wrapped well in plastic, for 2 to 3 days, or freeze for up to 3 months; thaw overnight in the refrigerator before using.

10. Cook as desired (see methods on pages 38 to 40) to an internal temperature of 165°F (74°C).

Giardiniera, page 286

Chicken Sausage
with Broccoli Rabe
and Mozzarella,
page 213

Pickled Radishes,
page 286

Pickled Bell Peppers,
page 286

ITALIAN FLAVORS

Sicilian-Style Turkey
Sausage, page 224

Cotechino, page 63

Fruited Apricot
Mustard, page 293

Meet the
MAKERS

---◊---

Ben Siegel and Ted Prater

*Banger's Sausage
House & Beer Garden*

Austin, Texas

SAUSAGE MECCA

As Ben Siegel, the owner of Banger's Sausage House & Beer Garden, tells it, the concept for Banger's came about when he was attending the University of Texas at Austin. "There was a sausage cart called The Best Wurst down on Sixth Street, the main bar district. At two o'clock in the morning, when all the bars let out, that was the one place I wanted to go for snacks. The lines were longer there than anywhere else. At that time there were no brick-and-mortar places that focused on sausages. A lightbulb went off in my head."

He went on to pursue more traditional jobs, and anytime the idea popped back up in his mind, he'd just shrug it off. Fast-forward ten years: He was living in Los Angeles, where he grew up, and an upscale sausage restaurant called Wursteküche opened in the downtown area, and it soon became *the* hot spot. "That place validated my idea and was influential in convincing me to do what I wanted to do: Open a pure sausage and beer garden, where everything is made in-house, in Austin, where there was still nothing like it."

Besides, he knew there were plenty of Austinites who were fans of sausage, this being the land of barbecue — and beer. And while there were a lot of barbecue places and other restaurants that had sausage on the menu, there was nothing that had the same focus on sausage, or on pairing it with beer. "Which is ironic, since nothing goes better with sausage than beer."

Banger's opened its doors in July 2012, with about 150 seats, and has since grown to seat about 500 people. That's a lot of sausage, and Ben gives all the credit to Ted Prater, who has been working there since the beginning — and who, like Ben, had been wanting to move to Austin for over a decade and who was also trying to figure out what to do next in his career, which had involved working in some of the finer restaurants in Nashville. When Ted saw the posting for a chef at Banger's, he thought it was right up his alley. "I've been making sausage all my life. I come from a family of outdoorsmen in Tennessee. We would go hunting and fishing, and when you have venison, you make sausage. We'd try to outdo each other for bragging rights around the campfire. We knew you had to keep the meat cold but we didn't know why, we didn't know the science behind it. I've since done all I can over the past 15 years to learn about meat and charcuterie."

Now he oversees a crew that makes about 2,000 pounds of fresh sausage every week, seven days a week, to meet the needs of a restaurant that can serve as many as 2,000 people a day — and to wash all that down with beer (106 taps on the wall plus bottled options). Despite having to produce on such a large scale, Prater makes sausages in 20-pound batches. "They told me I could kick out 400 pounds per minute in my commercial grinder, but I still prefer to do only small batches. The cylinder of my hand-cranked piston stuffer can only hold 20 pounds, and I'm not going to get a bigger stuffer. My mixing bowl can only hold 20 pounds, and I'm not going to get a bigger mixer. I try to use only old-school methods rather than relying on fancy equipment. It takes time and effort and know-how, and that's something that Ben and I strongly agree on."

Which is not to say that he doesn't play around with tradition when it comes to the flavors of the sausages. Take, for instance, their shrimp and bacon sausage, served over white cheddar grits topped with barbecue sauce. The bulgogi dog puts all the usual components of the Korean dish in a sausage, which is grilled and served with house-made kimchi, dried shrimp, carrots, jalapeño, and lime and soy caramel sauce. "It's over the top but delicious." As for the namesake sausage, Prater describes it as more of an English-style banger, and he prefers to use more beef and less filler (such as bread crumbs and oats). But he serves it the typical way: with mashed potatoes, caramelized onions, and brown gravy. (See page 218 for another specialty that speaks to Prater's Southern roots.)

Despite all these unusual offerings, Prater says, "The bratwurst outsells everything. A lot of people like to stay with what they know, but it also gives new customers a way to gauge the quality of the restaurant, because it's a guarantee they've had bratwurst before and they can decide whether they think ours is up to par." Prater says their hot dog is also a favorite, for the same reasons. But that won't stop him from continuing to explore other, unique flavor combinations. For the past couple years, he's been experimenting with dry-cured meats and sausages for the new butcher shop that Siegel is opening next door. "My closet is filled with hams and salami from recipes I've been testing. We are going to have a curing room and I'll be able to teach classes on making charcuterie. We started off four years ago with 150 seats, and here we are four years later with 500 seats, so if people keep coming the way they have been, who knows where this will take us in a couple years. We want to keep all our options open."

FRIED CHICKEN SAUSAGE

This sausage, a regular feature on the menu at Banger's Sausage House & Beer Garden in Austin, Texas (see page 216), is just like it sounds — meaning downright delicious! Seasoned ground chicken thighs and skins are stuffed into chicken skins, rather than casings, and then the links are dipped in buttermilk, dredged in flour, and fried to a golden brown crisp. Chef Ted Prater happily shared this recipe, and suggests asking for chicken skins from a butcher or supermarket. Asian grocers are another good source. Or you can buy an extra chicken and remove the skin yourself, taking care to avoid any rips or tears; you can then cook the rest of the chicken as desired. Prater also encourages serving these sausages the way they were intended for the restaurant: accompanied by mashed potatoes, milk gravy, and buttermilk biscuits with honey butter.

1. Combine the salt, thyme, paprika, pepper, sage, granulated garlic, and onion powder in a large bowl. Add the chicken, turn to coat evenly and thoroughly, and refrigerate, covered, overnight.

2. Grind the seasoned meat mixture through the medium disk of a meat grinder.

3. Cut the chicken skin into 7- by 8-inch rectangles. One at a time, lay a piece of skin flat on the counter, and portion ½ cup of the chicken mixture onto the skin. Roll up the chicken mixture in the skin, tucking in the sides to enclose the filling. Secure with toothpicks (about 4 per sausage).

4. Soak the chicken links in the buttermilk to cover for 30 minutes. Remove from the buttermilk, letting any excess drip back into the bowl, then dust in flour to coat, tapping off any excess. Then dip in the buttermilk again, and dust with flour once more.

5. Heat several inches of oil in a Dutch oven or heavy pot to 350°F (177°C). Fry the links, a few at a time, until they float to the top of the oil and reach an internal temperature of 165°F (74°C).

6. Drain on paper towels and serve warm.

3 tablespoons kosher salt

2 tablespoons dried thyme

2 tablespoons paprika

1 tablespoon plus 1 teaspoon freshly ground black pepper

1 tablespoon dried rubbed sage

2 teaspoons granulated garlic

2 teaspoons onion powder

10 pounds boneless chicken thighs, cut into 1-inch pieces

Chicken skins, as needed to roll sausage

2–3 quarts buttermilk

4 cups flour, plus more as needed

Vegetable oil, for deep frying

3 feet medium hog casing

3 pounds boneless, skin-on chicken thighs

1 tablespoon kosher salt

1 teaspoon freshly ground black pepper

½ teaspoon crushed red pepper

2 tablespoons chopped fresh basil, preferably Thai or holy basil

2 tablespoons chopped fresh cilantro

1 tablespoon chopped fresh mint

2 teaspoons minced garlic

2 teaspoons grated fresh ginger

1 tablespoon Asian fish sauce

2 teaspoons Thai green curry paste

1 teaspoon grated lime zest

1 tablespoon fresh lime juice

FRESH

THAI CHICKEN SAUSAGE

All the fresh, vibrant flavors that are used in Thai cooking — garlic, ginger, chiles, cilantro, mint, and lime juice — make an excellent seasoning base for sausage, too. Fish sauce, another staple of Southeast Asia, is pungent, with a funky aroma, but tastes neither salty nor fishy when used appropriately (a little goes a long way); here, it gives these links that desirable umami quality. Slice them and serve as is with cocktails, or add them to stir-fries or pad thai. *See photo, page 85.*

1. Prepare the casing (see page 29).

2. Cut the chicken into 1-inch cubes. Freeze for 30 minutes.

3. Grind the chicken and skin through the coarse disk of a meat grinder.

4. In a large bowl, combine the ground chicken, salt, black pepper, crushed red pepper, basil, cilantro, mint, garlic, ginger, fish sauce, curry paste, lime zest, and lime juice. Freeze for 30 minutes.

5. Grind the seasoned mixture through the fine disk of the grinder. Freeze for 30 minutes.

6. Stuff the mixture into the prepared casing, prick air pockets, and twist off into 4- to 5-inch links.

7. Place the sausage on a baking sheet fitted with a wire rack and refrigerate, uncovered, for 1 day.

8. Cut the links apart. Refrigerate, wrapped well in plastic, for up to 3 days, or freeze for up to 2 months; thaw overnight in the refrigerator before using.

9. Cook as desired (see methods on pages 38 to 40) to an internal temperature of 165°F (74°C).

GOOSEWURST

Whether you use fresh or frozen goose meat, this may be one of the best wursts you've ever tasted, with the distinctive flavor of Drambuie (Scotch-based liqueur) playing off the blend of warming spices.

1. Prepare the casing (see page 29).

2. Cut the goose into 1-inch cubes. Freeze for 30 minutes.

3. Grind the goose through the coarse disk of a meat grinder.

4. In a large bowl, pour the Drambuie over the goose and mix well. Cover and refrigerate for at least 3 hours or overnight.

5. Add the salt, paprika, pepper, coriander, mace, cayenne, onion, and chives to the goose mixture. Mix well, using your hands. Freeze for 30 minutes.

6. Grind the seasoned mixture through the fine disk of the meat grinder.

7. Stuff the mixture into the prepared casing, prick air pockets, and twist off into 3- to 4-inch links.

8. Place the sausage on a baking sheet lined with a wire rack and refrigerate, uncovered, for at least a few hours or preferably overnight.

9. Cut the links apart. Refrigerate, wrapped well in plastic, for 2 to 3 days, or freeze for up to 3 months; thaw overnight in the refrigerator before using.

10. Cook as desired (see methods on pages 38 to 40) to an internal temperature of 165°F (74°C). Panfrying is recommended.

- 5 feet medium hog casing
- 5 pounds boneless, skin-on goose meat
- ½ cup Drambuie (Scotch-based liqueur)
- 2 tablespoons kosher salt
- 1 tablespoon paprika
- 2 teaspoons freshly ground white pepper
- 1 teaspoon ground coriander
- 1 teaspoon ground mace
- ½ teaspoon cayenne pepper
- ½ cup minced onion
- ¼ cup chopped fresh chives

2 feet medium hog casing

2 pounds boneless, skin-on turkey thighs

Zest from 1 orange

2 teaspoons kosher salt

1 teaspoon dried marjoram

1 teaspoon dried sage

½ teaspoon freshly ground black pepper

½ cup dried cranberries, chopped

¼ cup roasted chestnuts, chopped (see Note)

NOTE: *Fresh chestnuts are generally available in grocery stores from October through January. You can find peeled roasted chestnuts in vacuum-packs or jars throughout the year.*

FRESH

TURKEY AND CRANBERRY SAUSAGE

How to capture Thanksgiving flavors in a sausage: Season turkey with dried sage and Grand Marnier before grinding, then fold in some dried cranberries and chopped roasted chestnuts before stuffing. The links are best roasted and served with — what else? — creamy mashed potatoes (regular or sweet) and any other of your favorite holiday fixings. *See photo, page 189.*

1. Prepare the casing (see page 29).

2. Cut the turkey into 1-inch cubes. Freeze for 30 minutes.

3. Grind the turkey through the fine disk of a meat grinder.

4. In a large bowl, combine the ground turkey, orange zest, salt, marjoram, sage, and pepper. Mix well, using your hands. Freeze for 30 minutes.

5. Grind the seasoned mixture through the fine disk of the grinder. Fold in the cranberries and chestnuts. Freeze for 30 minutes.

6. Stuff the mixture into the prepared casing, prick air pockets, and twist off into 3-inch links.

7. Place the sausage on a baking sheet fitted with a wire rack and refrigerate, uncovered, for a few hours or preferably overnight.

8. Cut the links apart. Refrigerate, wrapped well in plastic, for 2 to 3 days, or freeze for up to 3 months; thaw overnight in the refrigerator before using.

9. Cook as desired (see methods on pages 38 to 40) to an internal temperature of 165°F (74°C).

ROASTED CHESTNUTS

Your kitchen will be heady with the aroma of street-cart vendors when you roast chestnuts, and they can be added to all manner of holiday dishes, from stuffings to desserts. Here, they add welcome chewiness and nutty sweetness to the turkey sausages. To roast them: Cut an *X* in the flat side of each chestnut with a paring knife and place the chestnuts, *X* side down, in a single layer on a rimmed baking sheet. Roast in a 350°F (177°C) oven until darkened and they split open, about 30 minutes. Remove from the oven and, when cool enough to handle, peel off and discard the shells and papery skins.

TURKEY SAUSAGE WITH ROASTED CHILES AND CILANTRO

MAKES 2 POUNDS

Perfect in a fajita, terrific in a black bean stew, and absolutely decadent with huevos rancheros — these sausages can add new life to your familiar Tex-Mex favorites. Roasting the chiles (see below) is an extra step that intensifies the flavors, while red wine vinegar and lime juice combine to give them a welcome tang. *See photo, page 248.*

1. Prepare the casing (see page 29).

2. Cut the turkey into 1-inch cubes. Freeze for 30 minutes.

3. Grind the turkey through the coarse disk of a meat grinder.

4. In a large bowl, combine the ground turkey, chili powder, salt, cumin, pepper, fresh chiles, cilantro, garlic, vinegar, and lime juice. Freeze for 1 hour.

5. Grind the seasoned mixture through the coarse disk of the grinder. Freeze for 30 minutes.

6. Stuff the mixture into the prepared casings, prick air pockets, and twist off into 4- to 5-inch links.

7. Place the sausage on a baking sheet fitted with a wire rack and refrigerate, uncovered, for a few hours or preferably overnight.

8. Cut the links apart. Refrigerate, wrapped well in plastic, for up to 3 days, or freeze for up to 2 months; thaw overnight in the refrigerator.

9. Cook as desired (see methods on pages 38 to 40) to an internal temperature of 165°F (74°C).

ROASTED CHILES

Roasting chile peppers deepens their flavors and makes them wonderfully tender. You can use this method for any type of fresh chile as well as for bell peppers: Place the chiles directly over a gas burner set on high until blackened and blistered all over, turning with tongs. Place in a bowl, cover with plastic wrap, and let steam for about 10 minutes, then rub off the papery skins with paper towels. Use a paring knife to remove any stubborn spots. The stem will usually slip right out, or you can slice it off and then scrape out the seeds. Roasted chiles can be refrigerated, covered, for up to 3 days before using. To store a bit longer, place them in a jar or other tightly sealed container with extra-virgin olive oil to cover.

Ingredients:

- 2 feet medium hog casing
- 2 pounds boneless, skin-on turkey thighs
- 2 tablespoons chili powder
- 2 teaspoons kosher salt
- 1½ teaspoons ground cumin
- ¾ teaspoon freshly ground black pepper
- 1 Anaheim chile, roasted, seeded, and chopped (see below)
- 1 jalapeño chile, roasted, seeded, and chopped (see below)
- ¼ cup chopped fresh cilantro
- 2 teaspoons minced garlic
- 2 tablespoons red wine vinegar
- 1 tablespoon fresh lime juice

MAKES 5 POUNDS

- 5 feet small hog or sheep casing
- 5 pounds boneless chicken (mixture of dark and light meat)
- 1 cup nonfat dry milk powder
- 1 tablespoon plus 2 teaspoons kosher salt
- 1 tablespoon onion powder
- 1 tablespoon paprika
- 1 tablespoon sugar
- 2 teaspoons ground coriander
- 1 teaspoon garlic powder
- 1 teaspoon ground mace
- 1 teaspoon dried marjoram
- 1 teaspoon finely ground mustard seeds
- ¼ teaspoon ascorbic acid
- Curing salt (use supplier's recommended quantity for 5 pounds of meat; see Note)

CHICKEN DOGS

Because the meat mixture is ground three times, the texture of this sausage is extra-fine, almost like a hot dog, but without having to go to the trouble of making an emulsification (see page 156). Make sure your grinder plates are properly sharp before you begin, or the meat mixture will be smeared, resulting in a greasy link. Keeping the meat mixture cold at all times is also essential, so be sure to freeze it before each grinding.

1. Prepare the casing (see page 29).

2. Cut the chicken into 1-inch cubes. Freeze for 30 minutes.

3. In a large bowl, combine the chicken, dry milk, salt, onion powder, paprika, sugar, coriander, garlic powder, mace, marjoram, mustard seeds, ascorbic acid, and curing salt. Mix well, using your hands.

4. Grind the mixture through the coarse disk of a meat grinder. Freeze for 30 minutes.

5. Grind the mixture through the fine disk of the grinder. Freeze for 30 minutes.

6. Grind the mixture through the fine disk a second time.

7. Stuff the mixture into the prepared casing, prick air pockets, and twist off into 5-inch links.

8. Arrange the sausage on a baking sheet fitted with a wire rack and refrigerate, uncovered, for a few hours or preferably overnight.

9. Cold-smoke at 110 to 120°F (44 to 49°C) for 2 hours.

10. Bring a large pot of water to a boil. Add the links and reduce the heat to maintain a water temperature of 180 to 190°F (83 to 88°C). Simmer the links for 30 minutes, or until an instant-read thermometer shows an internal temperature of 165°F (74°C).

11. Drain the links and let them cool. Pat the links dry, wrap in plastic wrap, and refrigerate for 3 days to cure before heating and eating. Use within 1 week, or freeze for up to 2 months; thaw in the refrigerator overnight.

NOTE: *We recommend that you use a commercial premixed cure in any recipe for cured sausage. Premixed cures replace the saltpeter in older recipes.*

SICILIAN-STYLE TURKEY SAUSAGE

MAKES 5 POUNDS

Sicilian sausages rely principally upon fennel seeds for their distinctive taste. Here, the hot smoking cooks the sausage and adds its own inimitable flavor. *See photo, page 215.*

1. Prepare the casing (see page 29).

2. Cut the turkey into 1-inch cubes. Freeze for 30 minutes.

3. Grind the turkey through the fine disk of a meat grinder.

4. In a large bowl, combine the turkey, salt, crushed and whole fennel seeds, black pepper, crushed red pepper (if using), and garlic. Freeze for 30 minutes.

5. Grind the mixture through the fine disk of the grinder.

6. Stuff the mixture into the prepared casing, prick air pockets, and twist off into 3-inch links.

7. Place the sausage on a baking sheet fitted with a wire rack and refrigerate, uncovered, for 1 day.

8. Hot-smoke the sausages at 180 to 190°F (73 to 88°C) for 2 to 4 hours, until an instant-read thermometer shows an internal temperature of 165°F (74°C).

9. Cut the links apart. Refrigerate, wrapped in plastic, for up to 1 week, or freeze for up to 2 months; thaw in the refrigerator overnight.

5 feet small hog or sheep casing

5 pounds boneless, skin-on turkey (mixture of dark and light meat)

2 tablespoons kosher salt

2 teaspoons crushed fennel seeds

2 teaspoons whole fennel seeds

2 teaspoons freshly ground black pepper

2 teaspoons crushed red pepper (optional)

2 garlic cloves, minced

5 feet small hog or sheep casings

5 pounds boneless, skin-on turkey

2 tablespoons paprika

1 tablespoon plus 2 teaspoons kosher salt

2 teaspoons cayenne pepper

2 teaspoons ground cumin

2 teaspoons dried oregano

2 teaspoons freshly ground black pepper

1 teaspoon ground coriander

1 teaspoon ground ginger

½ teaspoon celery seeds

½ teaspoon ground cinnamon

Curing salt (use supplier's recommended quantity for 5 pounds of meat; see Note)

½ cup dry red wine

TURKEY CHORIZO

Just because you don't eat pork — or simply want to switch things up a bit now and then — doesn't mean you have to forgo Spanish-style chorizo. Turkey makes a delicious alternative to pork in this recipe. As in all chorizo-style sausages, hot red pepper carries the day. The curing process intensifies the flavors of the spices.

1. Prepare the casing (see page 29).

2. Cut the turkey into 1-inch cubes. Freeze for 30 minutes.

3. Grind the turkey through the fine disk of a meat grinder.

4. In a large bowl, combine the turkey, paprika, salt, cayenne, cumin, oregano, black pepper, coriander, ginger, celery seeds, cinnamon, curing salt, and wine. Mix well, using your hands.

5. Spread the mixture in a large, shallow pan, cover loosely with waxed paper, and cure it in the refrigerator for 24 hours.

6. Grind the seasoned mixture through the fine disk of the grinder.

7. Stuff the mixture into the prepared casing, prick air pockets, and twist off into 3-inch links. Do not separate the links.

8. Hang the sausage to dry in a cold place for 6 to 8 weeks. Test the sausage after 6 weeks by cutting off one link and slicing through it. If the texture is firm enough to suit your taste, the remaining sausage can be cut down and wrapped tightly in plastic for storage in the refrigerator for several months.

NOTE: *We recommend that you use a commercial premixed cure in any recipe for cured sausage. Premixed cures replace the saltpeter in older recipes.*

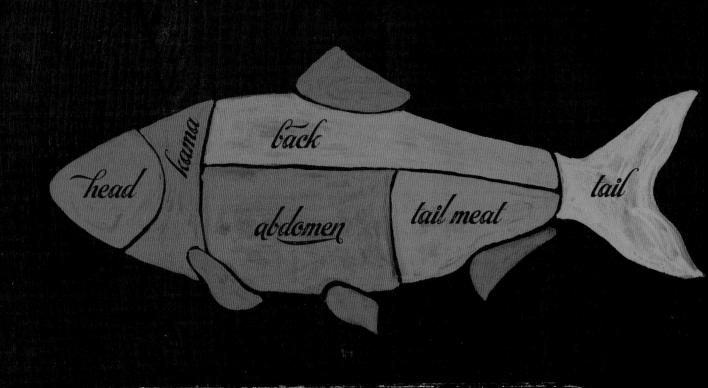

head · kama · back · abdomen · tail meat · tail

Louisiana Crayfish
Sausage, page 229

SEAFOOD SAUSAGES

But there is something about meat or fowl or fish that has been ground . . . then properly seasoned and correctly cooked, that appeals to me immediately. . . .

— Colman Andrews

Fish, shrimp, lobster, clams — these are undoubtedly not the first ingredients that come to mind when thinking about making sausage, but seafood can be a wonderful alternative to pork, beef, and other meats as the main component.

Most of the recipes in this chapter can be made quickly using a food processor to mince the ingredients — no grinder required. You don't even have to stuff the seasoned mixtures into casing: A simple wrap of buttered parchment paper or even plastic wrap will suffice for most, or the mixture can be shaped into patties for panfrying.

Seafood is also rich in protein, low in saturated fat, and rich in heart-healthy omega-3 fatty acids. Even crustaceans (shrimp and lobster) and shellfish (clams and oysters) are lower in cholesterol than chicken. All have fewer calories, making them an excellent bang for the buck.

Finally, fish and seafood can be a more economical choice for the home cook. You can make delicious sausage with relatively inexpensive fish, such as pollack, haddock, and grouper — really whatever is available at your local fish market (and fishmongers are great sources of recommendations). Plus, there's no butchering to be done, and virtually no waste — save the bones and shells for making broth, which can then be used to make soups and risottos and pasta sauces.

The real reason to make these seafood sausages, however, is to enjoy their distinctive taste and texture — and that's true for pescatarians and carnivores alike.

Crayfish Sausage and
Grits, page 273

- 4 feet small hog or vegetarian casing
- 1 pound picked-over crayfish tail meat, well chilled
- 1 pound firm white fish fillets such as pollock, haddock, or grouper, cubed and well chilled
- 2 teaspoons kosher salt
- ½ teaspoon cayenne pepper
- ½ teaspoon freshly ground black pepper
- 1 tablespoon minced celery
- 1 tablespoon minced green bell pepper
- 1 tablespoon minced shallot
- 2 tablespoons chili sauce
- 1 large egg or 2 large egg whites
- 1 tablespoon fresh lemon juice

FRESH

LOUISIANA CRAYFISH SAUSAGE

The succulent freshwater crustacean of the Louisiana bayou country is more tender than lobster, more delicate than shrimp. It finds its way into chili, mousse, the spicy stew called étouffée, and other dishes — as well as these seasonal sausages. Enjoy them with a glass of cold Dixie beer. *See photo, page 226.*

1. Prepare the casing (see page 29).

2. In a food processor, pulse the crayfish and white fish a few times, just until broken up.

3. Add the salt, cayenne, black pepper, celery, bell pepper, shallot, chili sauce, egg, and lemon juice. Process just until everything is well blended.

4. Stuff the mixture into the prepared casing, prick air pockets, and twist off into 3- to 4-inch links. Cut the links apart. Refrigerate, wrapped well in plastic, for up to 1 day.

5. Cook as desired (see below).

COOKING SEAFOOD SAUSAGES

Because of their leaner, more delicate texture and flavor, poaching is one of the best ways to cook seafood sausages, either fully or as a first step before grilling or broiling. They'll take 10 to 12 minutes to cook all the way through, or 5 to 7 minutes to partially cook. Then, when ready to serve, you can grill or broil the sausages just long enough to heat through and crisp the skins, about 5 minutes.

Panfrying is another good option, and you don't need to poach them first. Simply panfry in butter or oil for about 10 minutes, until golden brown and cooked through. Whichever method you use, be careful not to overcook seafood sausages.

CHESAPEAKE BAY SAUSAGE

MAKES 2 POUNDS

If you've ever come across the yellow tin of Old Bay Seasoning that is ubiquitous at restaurants along the Chesapeake Bay, you'll appreciate how well it goes with seafood (and everything else, according to the company's catchphrase; see below). We've replicated the flavor of that proprietary blend of 18 herbs and spices with our own seasoning combination to make sausage with remarkable flavor.

1. Prepare the casing (see page 29).

2. In a food processor, pulse the fish a few times, just until broken up.

3. Add the kosher salt, celery salt, mustard seeds, paprika, bay leaf, cardamom, cloves, ginger, mace, pepper, egg, and lemon juice. Process just until everything is well blended.

4. Stuff the mixture into the prepared casing, prick air pockets, and twist off into 3- to 4-inch links. Cut the links apart. Refrigerate, wrapped well in plastic, for up to 1 day.

5. Cook as desired (see page 229); panfrying is recommended.

- 4 feet small hog or vegetarian casing
- 2 pounds firm white fish fillets such as pollock, haddock, or grouper, cubed and well chilled
- 1½ teaspoons kosher salt
- 1½ teaspoons celery salt
- 1¼ teaspoons crushed mustard seeds
- 1 teaspoon paprika
- ¼ teaspoon ground bay leaf
- ¼ teaspoon ground cardamom
- ¼ teaspoon ground cloves
- ¼ teaspoon ground ginger
- ¼ teaspoon ground mace
- ¼ teaspoon freshly ground black pepper
- 1 large egg or 2 large egg whites, lightly beaten
- 1 teaspoon fresh lemon juice

OLD BAY, OLD FAVORITE

You can replace the herbs and spices in the Chesapeake Bay Sausage with 2 teaspoons of Old Bay Seasoning, the beloved herb-and-spice blend that was created in 1939 in Baltimore by a German-Jewish refugee named Gustav Brunn. Having opened a shop near the city's fish market, Brunn developed a new recipe to showcase the famous Chesapeake Bay crabs using a secret blend of more than a dozen herbs and spices. (This much we know: It contains bay leaf, celery salt, cloves, ginger, mustard, and red pepper.) Old Bay is very popular in the Mid-Atlantic area (and elsewhere) and is used on everything from seafood to hamburgers to even french fries and potato chips. You can find it in any supermarket and fish market.

- 4 feet small hog or vegetarian casings

- 1 pound salmon, cubed and well chilled

- 1 pound firm white fish fillets such as pollock, haddock, or grouper, cubed and well chilled

- 1 tablespoon chopped fresh dill

- 2 teaspoons chopped fresh chives

- 2 teaspoons chopped fresh flat-leaf parsley

- 1½ teaspoons kosher salt

- ½ teaspoon freshly ground black pepper

- 1 large egg or 2 large egg whites, lightly beaten

- 2 tablespoons sour cream

FRESH

SALMON AND DILL SAUSAGE

From the don't-overthink-it school of sausage making: Stick to flavor pairings that have stood the test of time and you're sure to have a hit on your hands. A case in point is this salmon-and-dill combo, a classic if ever there were one. Fresh chives and parsley, also commonly used with salmon, round out the flavors; sour cream adds welcome tang and keeps the sausage ultra moist.

1. Prepare the casing (see page 29).

2. In a food processor, pulse the salmon and white fish a few times, just until broken up.

3. Add the dill, chives, parsley, salt, pepper, egg, and sour cream. Process just until everything is well blended.

4. Stuff the mixture into the prepared casing, prick air pockets, and twist off into 3- to 4-inch links. Cut the links apart. Refrigerate, wrapped well in plastic, for up to 1 day.

5. Cook as desired (see page 229); poaching is recommended.

ODOR CONTROL

Some people avoid cooking with fish because of the lingering odors, but that doesn't have to be the case. If you rinse your hands under cold water before you handle fish, they won't smell as fishy. After working with fish, try rubbing your hands, knife, and cutting board with lemon wedges and/or salt.

SMOKED SALMON SAUSAGE

The smoky flavor and rosy color of smoked salmon predominates in this recipe, even though a milder fish is the main ingredient and no real smoking is involved. The links would be superb for brunch, with soft scrambled eggs, red onion, and bagels with a schmear of cream cheese. Or cut them into slices and serve with a tangy dip (see below), which would also go well with the brunch spread, as it happens (dill has an affinity for eggs, too).

1. Prepare the casing (see page 29).

2. In a food processor, pulse the white fish and smoked salmon a few times, just until broken up.

3. Add the parsley, onion, lemon juice, egg whites, salt, and cayenne, and process just until well blended. Freeze for 30 minutes.

4. Stuff the mixture into the prepared casing, prick air pockets, and twist off into 3- or 4-inch links. Cut the links apart. Refrigerate, wrapped well in plastic, for up to 1 day.

5. Cook as desired (see page 229); poaching is recommended.

SOUR CREAM DIP

Smoked salmon is often partnered with sour cream, dill, and capers — and here they all come together in a super-simple dip.

To make it: Stir together ¼ cup sour cream, ¼ cup plain yogurt, 1 tablespoon chopped fresh dill, and 1 teaspoon rinsed and drained capers. Season to taste with salt. Refrigerate, covered, until ready to serve, preferably within a few hours. Makes about ½ cup.

- 4 feet small hog or vegetarian casing
- 1½ pounds skinned, firm white fish fillets, such as pollack, haddock, or grouper, cubed and well chilled
- ¼ pound smoked salmon, chopped and well chilled
- 2 tablespoons chopped fresh flat-leaf parsley
- 1 tablespoon minced onion
- 1 teaspoon fresh lemon juice
- 2 large egg whites, lightly beaten
- 1½ teaspoons kosher salt
- Pinch of cayenne pepper

GINGER-SCALLION SEAFOOD SAUSAGE

MAKES 2 POUNDS

- 2 fcct medium hog or vegetarian casing
- ½ pound large shrimp, peeled and deveined, cubed and well chilled
- ½ pound bay scallops, well chilled
- 1 pound firm white fish, such as pollack, haddock, or grouper, cubed and well chilled
- ¼ cup minced scallions
- ¼ cup minced fresh cilantro
- 1 tablespoon fresh lemon juice
- 1 tablespoon grated fresh ginger
- 2 teaspoons minced garlic
- 2 teaspoons kosher salt
- 2 teaspoons freshly ground white pepper
- 2 teaspoons Asian fish sauce

Cilantro, scallions, fresh ginger, and Asian fish sauce give these shellfish sausages the telltale flavors of Vietnamese cooking. They are delicious sliced and served with nuoc chom, the essential condiment that is a vibrant blend of sweet, sour, salty, and bitter (recipe below). You could also add the sausages to pho, the heavenly Vietnamese soup (see recipe, page 350). *See photo, page 84.*

1. Prepare the casing (see page 29).

2. In a food processor, pulse the shrimp, scallops, and white fish until chunky.

3. Add the scallions, cilantro, lemon juice, ginger, garlic, salt, pepper, and fish sauce, and pulse just until the mixture comes together. If it is too stiff, add a spoonful of water to loosen. Freeze for 30 minutes.

4. Stuff the mixture into the prepared casing, prick air pockets, and twist off into 3-inch lengths. Cut the links apart.

5. Refrigerate, wrapped well in plastic, for up to 1 day.

6. Cook as desired (see page 229). Poaching is recommended.

NUOC CHOM

The name is actually a generic term for "dipping sauces," for which you will find many different variations on the same theme. Feel free to play around with the quantities to find your preferred taste; you can also replace half the lime juice with unseasoned rice vinegar for a less assertive sauce.

To make it: Stir together 6 tablespoons fresh lime juice (from 2 to 3 limes), ¼ cup sugar, 1 cup water, and ¼ cup Asian fish sauce. Stir in 2 minced garlic cloves, 2 or 3 thinly sliced Thai chiles, and Asian chili sauce to taste. Makes about 1½ cups.

LAKE TROUT AND CHIVE SAUSAGE

MAKES 2 POUNDS

A mere tablespoon of long-grain rice goes a long way to achieving the desired texture in these lean links, as does puréeing some of the trout and grinding the remaining. You may use any slightly flaky, delicately flavored fish you have available. Try it with arctic char, which is related to lake trout and salmon and has the best characteristics of both.

5 feet medium hog or vegetarian casing

1 tablespoon unsalted butter

1 tablespoon long-grain rice

⅓ cup fish broth or chicken stock

2 pounds trout fillets, skinned and boned

¼ cup crushed ice

2 teaspoons kosher salt

Pinch of cayenne pepper

2 tablespoons minced fresh chives

1. Prepare the casing (see page 29).

2. Heat the butter in a small skillet over medium-high heat. Add the rice and sauté for 1 minute. Add the fish broth and reduce the heat. Cover and simmer until the rice is tender, about 20 minutes. Remove from the heat and let cool.

3. Cut the trout into 2-inch pieces. Freeze for 30 minutes.

4. Add ¾ pound of the trout to a blender with the crushed ice, salt, and cayenne. Purée until smooth, then refrigerate.

5. Grind the remaining trout through the fine disk of the meat grinder.

6. In a large bowl, combine the ground fish with the fish purée and the chives.

7. Stuff the mixture into the prepared casing, prick air pockets, and twist off into 4-inch lengths.

8. Cut the links apart. Refrigerate, wrapped well in plastic, for 2 to 3 days, or freeze for up to 3 months; thaw overnight in the refrigerator before using.

9. Cook as desired (see page 229).

4 feet small hog or vegetarian casing

2 pounds shucked oysters, drained, liquor reserved

2 cups fresh bread crumbs

1½ teaspoons kosher salt

1 teaspoon freshly ground black pepper

1 large egg or 2 large egg whites, lightly beaten

¼ cup chopped fresh flat-leaf parsley

1 garlic clove, minced

FRESH

OYSTER SAUSAGE

Fresh oysters are preferred here, although canned may be substituted in a pinch. When buying fresh oysters, make sure they are still alive (they should open when tapped on the shell); they should also be wet and have a mild, not strong, smell.

You could do worse than to panfry the links to use in po'boys, in place of fried oysters, with the essential toppings: remoulade (see recipe, page 238) and hot pepper sauce. Or slice and serve with your favorite oysters-on-the-half-shell accompaniments, such as mignonette, grated fresh horseradish, and cocktail sauce — or a simple spritz of lemon juice.

1. Prepare the casing (see page 29).

2. In a food processor, pulse the oysters a few times, just until coarsely chopped.

3. Add the bread crumbs, salt, pepper, egg, parsley, and garlic; process until well blended. Add enough of the reserved oyster liquor to make a stiff paste.

4. Stuff the mixture into the prepared casing, prick air pockets, and twist off into 3- to 4-inch links. Cut the links apart. Refrigerate, wrapped well in plastic, for up to 1 day.

5. Cook as desired (see page 229).

HOW TO SHUCK OYSTERS

First hold an oyster under cold running water and scrub with a stiff brush, then insert an oyster knife (or other sturdy small knife) between the shells at the narrow end and slide the blade back and forth to separate the muscle from the upper shell. Twist to remove the top shell, then cut the muscle from the bottom shell.

FISH AND LOBSTER SAUSAGE

There's no need to break the bank to make these luscious links. A half-pound of lobster meat (the typical amount of meat from an average one-pound lobster) is all that's needed for each two-pound batch, but its rich flavor and meaty texture permeates the entire sausage.

1. Prepare the casing (see page 29).

2. In a food processor, pulse the fish a few times, just until broken up.

3. Add the salt, paprika, coriander, mustard seeds, pepper, lemon juice, and egg. Process just until well blended.

4. In a large bowl, mix together the fish mixture and the lobster.

5. Stuff the mixture into the prepared casing, prick air pockets, and twist off into 3- to 4-inch links. Cut the links apart. Refrigerate, wrapped well in plastic, for up to 1 day.

6. Cook as desired (see page 229). Poaching is recommended.

SEAFOOD STOCK

It would be a shame to let those shells go to waste, as they can yield a delicious dividend: a flavorful seafood stock.

To make it: Freeze shrimp and lobster shells after you've removed their meat. When you've amassed a bunch, simmer them in a mixture of water, bottled clam juice, chopped celery-heart leaves, onion skins, and whole peppercorns. Add other spices as desired. Let the liquid reduce to concentrate its flavor, then strain the stock, let cool, and refrigerate for up to 2 days or freeze for up to 2 months in a covered container. Use for making fish soups and stews, risotto, paella, and seafood sauces.

4 feet small hog or vegetarian casing

1½ pounds firm white fish fillets such as pollock, haddock, or grouper, cubed and well chilled

2 teaspoons kosher salt

1 teaspoon paprika

½ teaspoon ground coriander

½ teaspoon ground mustard seeds

½ teaspoon freshly ground white pepper

1 teaspoon fresh lemon juice

1 large egg or 2 large egg whites, lightly beaten

½ pound lobster meat (from 1 lobster), coarsely chopped

MIXED SEAFOOD SAUSAGE WITH SHRIMP, CRABMEAT, AND SCALLOPS

MAKES 2 POUNDS

- 4 feet small hog or vegetarian casing
- 1 pound firm white fish fillets, such as pollack, haddock, or grouper, cubed and well chilled
- 2 teaspoons kosher salt
- ½ teaspoon ground allspice
- 1 tablespoon minced garlic
- 2 teaspoons minced onion
- 2 large eggs or 4 large egg whites, lightly beaten
- 1 teaspoon unsalted butter, melted and cooled
- 1 teaspoon fresh lemon juice
- 1 teaspoon extra-virgin olive oil
- ½ pound small shrimp, peeled and deveined
- ¼ pound crabmeat, picked over
- ¼ pound bay scallops or quartered sea scallops

We used restraint in seasoning these sausages, to allow the delicate flavor of the shellfish — shrimp, crabmeat, and scallops — to shine through. Just a bit of minced garlic and onion, and a drizzle each of lemon juice, melted butter, and extra-virgin olive oil, are all that's needed to bind the seafood mixture (which also includes firm-fleshed fish) and provide extra richness. The links would be wonderful poached, then sliced or diced and tossed with pasta or stirred into risotto. Or serve like a summer shrimp boil, with corn on the cob and potatoes.

1. Prepare the casing (see page 29).

2. In a food processor, pulse the fish a few times, just until broken up.

3. Add the salt, allspice, garlic, onion, eggs, butter, lemon juice, and oil. Process until well blended.

4. Add the shrimp, crab, and scallops. Process just until blended. Do not overprocess.

5. Stuff the mixture into the prepared casing, prick air pockets, and twist off into 3- to 4-inch links. Cut the links apart. Refrigerate, wrapped well in plastic, for up to 1 day.

6. Cook as desired (see page 229).

CLAM DOGS

A Down East (that would be Maine) favorite is the clam roll — fried clams heaped into a toasted hot dog bun (New England–style, split at the top) and dolloped with tartar sauce (see recipe below). You might want to eat these clam dogs in the same manner.

1. Prepare the casing (see page 29).

2. In a food processor, pulse the fish a few times, just until broken up.

3. Add the parsley, salt, pepper, basil, onion, garlic, and egg. Process until well blended.

4. In a large bowl, mix together the fish mixture and the clams until well blended.

5. Stuff the mixture into the prepared casing, prick air pockets, and twist off into 4- to 5-inch links. Cut the links apart. Refrigerate, wrapped well in plastic, for up to 1 day.

6. Cook as desired (see page 229); grilling is recommended.

- 4 feet small hog or vegetarian casing
- 1½ pounds firm white fish fillets, such as cod or haddock, cubed and well chilled
- 2 tablespoons chopped fresh flat-leaf parsley
- 2 teaspoons kosher salt
- ½ teaspoon freshly ground black pepper
- ½ teaspoon dried basil
- 2 teaspoons minced onion
- 2 garlic cloves, minced
- 1 large egg or 2 large egg whites, lightly beaten
- 2 cans (6 ounces each) chopped clam meat, drained

TARTAR SAUCE

Some people insist that unless it contains chopped hard-cooked eggs, it just isn't tartar sauce, and if you fall in that camp then by all means fold in one or two at the end. Otherwise, this recipe is as classic as it comes.

To make it: Combine 1 cup mayonnaise, 2 tablespoons each finely chopped cornichons and capers (drained), 1 tablespoon each finely chopped chives and fresh flat-leaf parsley, 2 teaspoons each Dijon mustard and fresh lemon juice (more to taste), ½ teaspoon finely chopped fresh tarragon, and salt and cayenne (optional) to taste. Whisk to combine, then cover and refrigerate for at least 1 hour and up to 2 days before serving. Makes about 1¼ cups.

To make a classic remoulade sauce, simply add 2 teaspoons anchovy paste or finely chopped anchovy fillets to tartar sauce.

For a Louisiana-style version, skip the anchovy and add 2 teaspoons whole-grain mustard, 2 minced garlic cloves, 1 teaspoon Worcestershire sauce, and hot pepper sauce to taste (meaning a lot).

SQUID SAUSAGE WITH TOMATOES AND PARSLEY

MAKES 1 POUND

- 1 pound cleaned squid (tubes and tentacles)
- 1 teaspoon kosher salt
- 1 teaspoon freshly ground black pepper
- ½ teaspoon dried oregano
- 1 cup chopped fresh flat-leaf parsley
- 1 garlic clove, minced
- 1 cup fresh bread crumbs
- ¼ cup plus 1 tablespoon olive oil
- ½ cup white wine
- 2 cups crushed tomatoes (canned)

These "sausages" are stuffed into squid bodies rather than casings. They're inspired by stuffed calamari, a traditional part of the Italian-American Feast of the Seven Fishes on Christmas Eve. Most often, squid is sold cleaned and separated into the body (or tubes) and the tentacles. If whole squid — or "dirty squid"—is available, clean it following the directions below. Save the ink from the squid to make risotto, for serving alongside.

1. Rinse the squid tubes in cold water and lay onto paper towels to drain.

2. Chop the tentacles coarsely in a food processor. Add the salt, pepper, oregano, parsley, and garlic. Pulse two times to combine.

3. Add the bread crumbs and 1 tablespoon of the oil. Pulse two times to combine.

4. Fill the bodies of the squid between one-half and three-quarters full. (Do not overfill, or they will burst during cooking.) Thread the openings shut with toothpicks.

5. Heat the remaining ¼ cup oil in a large skillet over medium heat and add the stuffed squid. Panfry until lightly browned, turning often to prevent them from sticking, about 10 minutes.

6. Add the wine and cook until it has nearby evaporated, about 3 minutes. Add the crushed tomatoes and simmer, stirring occasionally, for about 20 minutes. Serve warm (remind your guests to watch out for the toothpicks).

HOW TO CLEAN SQUID

Buying whole squid and cleaning them yourself will save you money, and you'll also be rewarded with the ink sac, which you can use to make risotto or fresh pasta noodles. First, however, you must know how to prepare them: Remove the head from the mantle (body) of the squid by cutting just below the eyes. Cut the tentacles from the head and reserve; discard the rest of the head (or reserve for making broth). Push gently to remove the internal organs and ink sac from the mantle, then rinse inside the mantle under cold running water. Pull out the pen, or feather-shaped cartilage, and discard. Peel off the fins by hand, then peel off the skin, from the mantle's opening to the tip. Use a paring knife for any stubborn spots.

CONFETTI SEAFOOD SAUSAGE

MAKES 3 POUNDS

This recipe is adapted from one that appeared in *Fresh Fish: A Fearless Guide to Grilling, Shucking, Searing, Poaching & Roasting Seafood* by Jennifer Trainer Thompson. Besides giving this sausage its rich, sweet flavor, the pieces of lobster in the otherwise smooth stuffing provide contrasting texture as well as specks of red color, aided by flecks of copious fresh herbs — chives, tarragon, thyme, and cilantro. Hence the sausage's name. Any firm, flaky white-fleshed fish will do in place of the cod; others to try include halibut or haddock. For best results, the sausages are partially cooked in boiling water before grilling.

1. Prepare the casing (see page 29).

2. In a food processor, pulse together the cod, apple, egg whites, chives, shrimp paste, tarragon, cilantro, thyme, black pepper, salt, and cayenne five or six times. Scrape down the sides with a spatula. Add the cream and purée until thoroughly combined, 20 to 30 seconds.

3. Add the lobster and mushrooms, pulse two or three times (only until coarsely chopped), and transfer to a large container.

4. Stuff the mixture into the prepared casing, prick air pockets, and twist off into 6-inch lengths.

5. Heat a grill to medium-hot and bring a large pot of water to a boil. Reduce the heat to medium, add the sausages, and simmer until sausages are slightly soft, 6 to 8 minutes.

6. Remove the sausage from the water and cut at the tie-off points.

7. Transfer the links to the grill and cook until browned and cooked through (they will be firm to the touch), 10 to 12 minutes. Serve immediately.

Ingredients:

- 12 feet small hog or vegetarian casing
- 1 pound cod, cubed and well chilled
- 1 small apple, peeled and coarsely chopped (about 1 cup)
- 4 large egg whites
- ¼ cup chopped fresh chives
- ¼ cup Shrimp Paste (see recipe opposite, or substitute ½ pound peeled and deveined shrimp)
- ¼ cup chopped fresh tarragon
- ¼ cup chopped fresh cilantro
- 2 tablespoons fresh thyme leaves
- 2 teaspoons freshly ground black pepper
- ½ teaspoon flaky sea salt, such as Maldon
- ¼ teaspoon cayenne pepper
- 1 cup heavy cream
- ½ pound lobster meat (from 1 lobster; or substitute crab), finely chopped
- 1 cup finely chopped white mushrooms

SHRIMP PASTE

Jennifer Trainer Thompson also suggests using this shrimp paste to make a tasty hors d'oeuvre, by spreading it on crostini or on pieces of pumpernickel cocktail bread, as well as a dip for crudités. She also likes to mix 2 to 3 tablespoons into rice as it cooks, and to swirl some into chowders or fish soups just before serving. Galangal, also called Thai ginger, has a kick of heat that does not require the addition of hot peppers. If you cannot find galangal, substitute an equal amount of fresh ginger and add a little pure ground chile powder or crushed red pepper.

To make it: Peel a 2-inch piece of galangal, cut into chunks, and process in a food processor until very finely chopped, about 1 minute. Scrape down the sides of the bowl and add 6 ounces cooked shrimp (peeled and deveined); process until well combined, about 30 seconds. Add ¼ cup fresh tangerine or orange juice and 1 tablespoon low-sodium soy sauce and process until smooth. If not using immediately, transfer to a covered container and refrigerate for up to 4 days. Makes about 1¼ cups.

beet

Beet and Walnut
Sausage, page 253

VEGETARIAN SAUSAGES

Sir Andrew Aguecheek: "I am a great eater of beef,
and I believe that does harm to my wit."
Sir Toby Belch's reply: "No question."

— William Shakespeare, *Twelfth Night*

Sir Andrew was apparently ahead of his time. Today, many people have decided, for health or ethical or other reasons, to cut back on their consumption of meat. Just consider the Meatless Monday movement, started by another Sir — as in Sir Paul McCartney — as well as Vegan Before Six and other initiatives that have resonated with Americans. This chapter will provide you with ample opportunity to embark on such a path without feeling the least bit deprived. Or perhaps you just want to expand your repertoire to please the vegans in your life.

The dedicated carnivore may regard the term *vegetarian sausage* as an oxymoron. But sausage is more than just a foodstuff, it's a state of mind — what else offers the devourer a portable meal-in-one with beloved flavor combinations and unparalleled chew factor? Those qualities remain no matter the actual ingredients inside the casing. Besides, the same herbs and spices that go into traditional sausages can also be used with the whole grains, beans and legumes, and fresh vegetables that are the mainstays of meat-free sausage.

What's more, the offerings in this chapter will be far superior to the processed varieties — with their lengthy list of unrecognizable ingredients — found in the freezer case at supermarkets. They'll be a revelation to devoted meat-eaters and vegetarians alike.

Same Formula, Different Foundation

When making any type of sausage, including the ones in this chapter, you are essentially following the same method: foundation (traditionally meat) plus seasonings and other add-ins (for flavor and texture), with binders and thickeners as needed or desired.

Foundation ingredients. Cooked beans or legumes; cooked barley, rice, or other grains; and tofu or tempeh. These foundation ingredients are usually grated, mashed, or ground and replace the ground meat in a sausage. Nuts are sometimes added, and are a good source of protein — and flavor.

Vegetables and fruits. Fresh carrots, beets, mushrooms, celery, onion, squash, and apples, dried apricots, dates, and so on. They are usually minced, chopped, or grated, and sometimes sautéed before they are added to the sausage mixture.

Moistening and binding ingredients. Usually eggs or egg whites act as a binder. Butter, oil, brandy, wine, or reduced cider might be used to moisten the mixture; tomato paste moistens and imparts umami.

Thickeners. Bread crumbs, flour, rolled oats, wheat germ, or other ingredients thicken and bulk up the mixture so it can be shaped. Bread crumbs are probably the most versatile. (See opposite page.)

Seasonings. Parsley, sage, rosemary, thyme, salt, pepper, and the whole cast of characters used in traditional meat sausages are used in vegetarian sausages. Please feel free to personalize our recipes to suit your own tastes. As with meat sausages, seasonings provide the greatest opportunity for experimentation.

Casings. Vegetarian casings are more delicate than natural casings, so you'll need to handle them with care when stuffing, twisting, and cooking. You can also use edible cellulose casings with great results.

Besides casings, there are other ways to give your sausage mixture shape that will hold up during cooking: Employ natural casings in the form of cabbage or chard leaves (see page 250) or other sturdy wrappers such as corn husks or the outer layer of leeks (see page 260). Parchment paper, plastic wrap, and aluminum foil can be employed as nonedible casings. Unless otherwise specified in a recipe, you should use about ½ cup of mixture and shape it into a sausage that's about ¾ inch thick and 3 to 4 inches long.

Many of the recipes that follow are shaped into patties (or logs or balls, as you wish), and you could do this as well for any of the other vegetarian sausages that are stuffed into casings (and vice versa). A good rule is to use about ½ cup of the sausage mixture for each patty, which should be about ½ inch thick and 3 inches in diameter.

To stuff any of the vegetarian sausage mixtures in this chapter, prepare small or medium vegetarian casings. Freeze the seasoned sausage mixture for 30 minutes, then stuff into the prepared casing, prick air pockets, and twist off into 3- to 4-inch links.

Cooking Vegetarian Sausages

Because they are made from precooked or quick-cooking ingredients, vegetarian sausages tend to take little time to reach the desired doneness. Depending on the size, shape, makeup, and casings (if any) of the sausage, panfrying, grilling, and poaching are all viable options, but we've provided recommendations for each recipe in this chapter. The exception is that you should not poach sausages that are stuffed into vegetarian casings, which will dissolve during cooking. Vegetarian sausages can be refrigerated, wrapped well in plastic, for up to 3 days, or frozen for up to 3 months; you can typically cook them without thawing.

- 1½ cups cooked white or navy beans, or 1 can (15 ounces), rinsed and drained
- 1 cup cooked brown lentils
- ¼ cup toasted wheat germ
- 1 teaspoon kosher salt
- ½ teaspoon dried marjoram
- ½ teaspoon freshly ground black pepper
- ½ teaspoon dried sage
- ½ teaspoon dried thyme
- ½ cup minced celery
- 2 teaspoons chopped fresh flat-leaf parsley
- 2 tablespoons unsalted butter, melted, or 2 tablespoons vegetable oil, plus more for baking sheet
- ¼ cup fresh bread crumbs, or more as needed
- ⅔ cup milk
- 1 large egg, lightly beaten
- 1 cup cornmeal

VEGETARIAN BREAKFAST SAUSAGE

Any number of different "vegetarian breakfast links" are sold in the freezer case of the supermarket, right next to the convenient pork sausage "brown 'n' serves." What makes them breakfast sausages? The seasonings are largely the same — as is the coarser, crumblier texture. Here, we forgo any wrapper and give the sausage shapes a golden cornmeal crust instead. *See photo, page 59.*

1. Mash the beans and lentils in a large bowl. Stir in the wheat germ, salt, marjoram, pepper, sage, thyme, celery, parsley, and butter, and knead together until well blended. Add the bread crumbs as needed to make a mixture that holds its shape.

2. With wet hands, roll small handfuls of mixture into sausage shapes, about ¾ inch in diameter and 3 to 4 inches long. Place on a baking sheet fitted with a wire rack and refrigerate for at least 1 hour, or overnight.

3. Preheat the oven to 450°F (230°C) and grease a roasting pan.

4. In a shallow bowl, combine the milk and egg. Sprinkle the cornmeal in a pie plate. Dip each sausage into the egg mixture, then roll in the cornmeal. Arrange in the pan and roast for about 10 minutes, or until golden brown, turning frequently. Serve hot.

HOMEMADE BREAD CRUMBS

It's worth making your own bread crumbs and keeping them on hand for sausage and all manner of other preparations. Most any type of bread will do, but we are especially fond of rustic loaves and crusty baguettes, or whole-grain varieties.

To make them: For fresh bread crumbs, tear the bread into pieces and then pulse in a food processor until the desired texture is reached. For dried bread crumbs, let the bread pieces sit out overnight, or dry them slightly in an oven set to low heat, before pulsing. Stored in a tightly sealed container in the freezer, bread crumbs will keep for months.

BLACK BEAN AND SMOKED CORN SAUSAGE

Grilling the sweet corn over a charcoal fire gives the sausage patties a smoky edge. Black beans, fresh cilantro, cumin, and salsa lend more Southwestern flavors. The sausages are terrific in a tortilla, topped with guacamole, queso fresco, and more salsa. You could also save the corn husks for wrapping the sausage mixture; blanch them in boiling water until softened, then drain and dry thoroughly. Follow the technique in the recipe on page 250 for the chard leaves. *See photo, page 248.*

1. Heat a charcoal grill to medium-hot. Grill the corn, turning frequently, until the kernels are lightly browned on all sides, about 5 minutes. Set aside.

2. Heat the oil in a skillet on top of the grill, then sauté the onion, garlic, and cumin until the onion is transparent, about 5 minutes. Remove the pan from the grill.

3. Using a sharp knife, slice the kernels from the cobs, working over a shallow bowl to capture the corn milk.

4. In a food processor, combine the corn and its liquid, onion mixture, beans, bread crumbs, salt, pepper, cilantro, egg, and the 2 tablespoons salsa. Pulse until the ingredients are blended and the mixture holds together.

5. With wet hands, form about ½ cup of the mixture into ½-inch-thick patties or 3-inch-long sausage shapes. Place on a baking sheet fitted with a wire rack and refrigerate for at least 1 hour, or until firm.

6. To cook, panfry or grill over medium heat until browned on all sides, about 10 minutes. Serve with more salsa and guacamole, if desired.

2 ears corn, husked

1 tablespoon olive oil

¼ cup minced onion

2 garlic cloves, minced

1 teaspoon ground cumin

1½ cups cooked black beans, or 1 can (15 ounces), rinsed and drained

½ cup dried bread crumbs

½ teaspoon kosher salt

½ teaspoon freshly ground black pepper

2 tablespoons chopped fresh cilantro

1 large egg, lightly beaten

2 tablespoons salsa, plus more for serving

Guacamole, for serving (optional; see opposite)

CLASSIC GUACAMOLE

It hardly merits a recipe, except if you've never ventured past the basic version, which eschews anything other than lime juice, jalapeño, and salt to heighten the creaminess of avocados. Taking that as a starting point, we've incorporated a few other ingredients, all authentic and equally traditional to the original spirit of everyone's favorite Mexican dip.

To make it: Halve 4 ripe avocados lengthwise, remove the pits, and scoop the flesh into a bowl. Add 3 or 4 minced garlic cloves, ¼ cup finely chopped red onion or scallion, 2 finely chopped jalapeño or serrano chiles (seeds removed for less heat), ½ cup finely chopped cilantro, and ¼ cup fresh lime juice. Mash to combine, leaving some chunks if desired. Season with salt, then add more lime juice as needed. Sometimes we like to mix in a teaspoon of ground cumin, for depth, or to crumble queso fresco over the top, along with more chopped cilantro, before serving. Makes about 2 cups.

Black Bean and Smoked
Corn Sausage, page 246

Fresh Chorizo,
page 67

Turkey Sausage with
Roasted Chiles and
Cilantro, page 222

LATIN FLAVORS

Chimichurri,
page 296

ITALIAN BULGUR AND BEAN SAUSAGE ROLLS

MAKES ABOUT 6
SAUSAGE ROLLS

Cabbage leaves stuffed with hearty, meaty fillings are a long-standing tradition in Germany and Eastern Europe. Here, that idea takes on Italian accents, with milder-tasting Swiss chard being used for the wrapper and a meat-free filling given heft with red kidney beans and bulgur and seasoned with fresh herbs, capers, and tomato paste. Serve with pasta and a robust marinara sauce (see opposite for a recipe).

1. In a medium bowl, pour the boiling water over the bulgur. Cover and let steam for about 10 minutes.

2. In a food processor, combine the beans, parsley, basil, oregano, capers, and garlic. Process until fairly smooth.

3. Drain any unabsorbed water from the bulgur. Add the bulgur, bread crumbs, salt, pepper, egg, oil, and tomato paste to the bean mixture and pulse just until combined. Refrigerate for about 30 minutes.

4. Place the chard leaves in a steamer basket and set over a pan of simmering water. Cover and steam until the leaves are pliable, 30 to 60 seconds.

5. Form about ½ cup of the sausage mixture at a time into a sausage shape and roll it up in a chard leaf, as if making an egg roll, tucking in the sides of the leaf as you go. Secure each roll with a toothpick. Repeat until all the filling and leaves are used.

6. To cook, in a large high-sided skillet, heat enough broth to cover the rolls to a simmer. Add the rolls and simmer for about 15 minutes, or until heated through. Lift out the rolls with a slotted spoon and serve hot. (Alternate method: Heat enough marinara sauce, opposite, to cover the rolls in the skillet, cover the pan, and simmer until heated through, about 15 minutes.)

1⅓ cups boiling water

⅔ cup uncooked bulgur wheat

1½ cups cooked red kidney beans, or 1 can (15 ounces), rinsed and drained

¼ cup chopped fresh flat-leaf parsley

2 tablespoons chopped fresh basil

1 teaspoon chopped fresh oregano

1 teaspoon small capers, rinsed and drained

1 teaspoon minced garlic

½ cup dried bread crumbs

½ teaspoon kosher salt

½ teaspoon freshly ground black pepper

1 large egg, lightly beaten

1 tablespoon olive oil

1 tablespoon tomato paste

6–8 fresh Swiss chard leaves

Vegetable broth or water, for cooking

MARINARA SAUCE

MAKES 4 CUPS

1 tablespoon olive oil

3 garlic cloves, crushed

1–2 teaspoons crushed red pepper

¼ cup dry red wine, such as Chianti or Montepulciano

4 cups peeled and chopped ripe plum tomatoes (see below), or 1 can (28 ounces) whole peeled tomatoes in juice, cut into pieces

Kosher salt

2 tablespoons chopped fresh basil

Homemade marinara sauce will taste much better than anything from a jar — and doesn't take much more effort to prepare. Plus, you can tailor the seasonings to your taste, adding fresh mint or oregano along with basil, and upping the amount of crushed red pepper and garlic (or omitting either or both) according to your whim. You can also make a big batch and keep the sauce on hand for whenever you need it. Just double the recipe and freeze in pint-size airtight containers.

For this sauce, use either dead-ripe plum tomatoes from your garden or farm stand, peeled and chopped, or best-quality canned whole peeled tomatoes packed in juice. If using the latter, here's a neat trick: Use kitchen shears to cut up the tomatoes right in the can. You can also pop them into a food processor and pulse one or two times, just to break them up. Or do as any respectable *nonna* would do and crush the tomatoes with your hands as you add them to the pot. This all-purpose sauce tastes wonderful over pasta, in lasagna, dolloped onto pizza — basically anywhere you want summery flavor.

1. In a heavy pot, heat the oil over low heat. Add the garlic and sauté just long enough to take the edge off the raw taste, about 1 minute. Add the crushed red pepper and cook, stirring, for 30 seconds.

2. Add the wine and cook, stirring, until almost evaporated. Add the tomatoes, season with salt, and simmer, uncovered, for about 30 minutes, stirring frequently and breaking up the tomatoes with the back of the spoon.

3. Remove from the heat and add the basil. If not using immediately, let cool completely before storing in airtight containers. Sauce can be refrigerated for up to 3 days, or frozen for up to 3 months.

HOW TO PEEL FRESH TOMATOES

When using fresh tomatoes in cooking, such as in a sauce or soup where a velvety consistency is the goal, you may want to first peel the tomatoes. It's an easy step that's worth the two or three minutes' effort.

Bring a pot of water to a boil and have ready a bowl of ice water. Rinse the tomatoes, remove any stems, and cut an X in the other side with a sharp paring knife. Blanch the tomatoes, working in batches if needed, until the skins start to look wrinkled, 30 to 60 seconds. Remove with a slotted spoon and immerse in the ice-water bath; once cool, peel off the skins with your fingers, using a paring knife for stubborn spots. Halve the tomatoes and squeeze or scoop out the seeds, as desired (more important for fleshy tomatoes, not so much for plum tomatoes).

PORTOBELLO SAUSAGE

You'd be hard-pressed to find a vegetarian restaurant that doesn't have at least one portobello dish on the menu, since the large cepes offer satisfying chewiness. They're also incredibly versatile and a boon for the home cook, who can stuff the caps and bake or broil to a golden top, or slice the mushrooms and broil, roast, sauté, stir-fry, or even coat in tempura batter. Or, as in this recipe, you can use them in making sausages that have chickpeas as the base. They're just right for a fall harvest feast, with creamy polenta (or crisped and fried polenta squares); roasted or grilled red, yellow, and orange bell peppers; and steamed greens from the garden.

- ¾ **pound portobello mushrooms, cut into chunks**
- ½ **cup chopped onion**
- 2 **garlic cloves, chopped**
- 1 **tablespoon olive oil**
- 1½ **cups cooked chickpeas, or 1 can (15 ounces), rinsed and drained**
- ¼ **cup chopped fresh flat-leaf parsley**
- 1 **teaspoon kosher salt**
- ½ **teaspoon freshly ground black pepper**
- 1 **tablespoon fresh lemon juice**
- ½ **cup fresh bread crumbs**
- 2 **tablespoons grated Parmesan cheese**

1. Preheat the oven to 450°F (230°C).

2. In a medium baking dish, toss the mushrooms, onion, and garlic with the oil until well coated. Roast for 8 to 10 minutes, until the mushrooms release their juices and start to brown.

3. Spoon the mixture, including juices, into a food processor and add the chickpeas, parsley, salt, pepper, and lemon juice. Process briefly, until the ingredients are combined but the mixture is still chunky.

4. Transfer the mixture to a large bowl, add the bread crumbs and Parmesan, and mix well.

5. With wet hands, form about ½ cup of the mixture into ½-inch-thick patties or 3-inch-long sausage shapes. Place on a baking sheet fitted with a wire rack and refrigerate for at least 1 hour, or until firm.

6. Panfry over medium heat or grill the sausages until crisp and heated through, about 10 minutes. Serve hot.

- ½ cup dried chickpeas
- 2 medium beets (about ¾ pound), peeled and coarsely chopped
- ½ onion, coarsely chopped
- ½ cup walnuts, toasted
- 2 garlic cloves, peeled
- 2 tablespoons chopped fresh dill
- 1½ teaspoons kosher salt

FRESH

BEET AND WALNUT SAUSAGE

These sausages combine the earthy flavor of beets with rich walnuts and nutty chickpeas, with dill adding a fresh note. If you prefer not to wait a day to make them, you can use one 15-ounce can of chickpeas, rinsed and drained, instead of starting with dried, and skip step 1. Enjoy these sausages tucked into warm pita bread with yogurt, cucumbers, and pickled onions.

1. Put the chickpeas in a small bowl and add enough water to cover by 1 inch. Let the chickpeas soak overnight.

2. Drain the chickpeas, rinse and drain again. Put the chickpeas into a food processor with the beets, onion, walnuts, garlic, dill, and salt. Process until you have an evenly blended, just slightly grainy mixture.

3. With wet hands, form about ½ cup of the mixture into ½-inch-thick patties or 3-inch-long sausage shapes. Place on a baking sheet fitted with a wire rack and refrigerate for at least 1 hour, or until firm.

4. To cook, panfry over medium heat until browned on all sides, about 10 minutes. Serve hot.

SPICY TEMPEH SAUSAGE

Tempeh is made from fermented cooked soybeans and sometimes grains, pressed into a firm cake. It can be crumbled or grated to use as a meat substitute, and it's famous for absorbing the flavors of whatever it is combined with. Find tempeh in the refrigerated section of most natural food stores. This is a good sausage to make in bulk (see page 34) and then crumble and use in breakfast scrambles and also huevos rancheros (in lieu of eggs).

1. Bring a pot of water with a steamer basket to a boil. Steam the tempeh, covered, for about 10 minutes. Let cool, then grate with the coarse side of a box grater.

2. In large bowl, combine the tempeh, rice, chili sauce, chile peppers, oil, egg white, chili powder, salt, garlic, cumin, and hot sauce. Mix well, using your hands.

3. With wet hands, form about ½ cup of the mixture into ½-inch-thick patties or 3-inch-long sausage shapes. Place on a baking sheet fitted with a wire rack and refrigerate for at least 1 hour, or until firm.

4. To cook, panfry over medium heat until browned on all sides, about 10 minutes. Serve hot.

½ pound tempeh

½ cup cooked brown rice

¼ cup chili sauce

2 Anaheim or poblano chiles, seeded and minced

2 tablespoons vegetable oil, plus more for panfrying

1 large egg white

1 teaspoon chili powder

¾ teaspoon kosher salt

½ teaspoon minced garlic

½ teaspoon ground cumin

Dash of hot pepper sauce, such as Tabasco

⚶

Aubry and Kale Walch

The Herbivorous Butcher

Minneapolis, Minnesota

VEGANS ON A (DELICIOUS) MISSION

Why would a vegan brother-and-sister duo decide to open a "butcher shop" that specializes in sausage? "We grew up in Guam, where eating with your family is the most important thing you do. Nothing grows there, so you eat meat at every meal. There was this Hawaiian version of Portuguese sausage that came in a large green can, which was just so good," Aubry recalls.

Even though they had given up eating meat — Aubry more than 18 years ago, Kale within the past 5 years — they didn't want to give up the satisfaction that comes from biting into a juicy link. Finding commercial varieties of vegan sausage lacking, Aubry embarked on a quest to make a better sausage herself and came across a basic seitan recipe. "I made my own vegan broth and tested with different types of beans and also beans in different forms — whole, mashed, puréed — before landing on the right texture."

The first sausage she ever made successfully was an Italian sausage with sun-dried tomatoes and herbs, and it's still the most popular item on the menu today. "I remember when finally, finally, finally! I got it right and I gave it to my dad to taste for me." Even though she will still tweak that recipe now and then, she usually prefers to just leave that alone and use it as a springboard for developing other flavor profiles.

Although Aubry is not willing to share her recipe (and who can blame her, after perfecting it over years of testing), she happily shares her standard method: "Almost everything starts out as a dough made by mixing together the wet and dry ingredients like you would for any dough." Then, depending on the desired outcome, she either bakes, boils, or steams the sausages. "This step is what took us so long with the first sausage, since it turned out we had been doing it one way when it was the wrong one." Now they usually either roll them up in aluminum foil and bake them, or wrap them in cheesecloth and boil them. If the dough is particularly tough (meaning potentially too dry), she might boil them first, then finish by steaming. To give sausages that desirable snap, she will boil them and then bake them. "It all depends on the desired texture, and all of our sausages are done differently. Our doughs are all different, too. The chorizo and Portuguese sausages in particular are firmer and more crumbly than others, in the best way."

What started as a stall at the local farmers' market in Minneapolis is now a full-fledged storefront, thanks to a successful Kickstarter campaign where they exceeded their goal. Aubry and Kale also attribute their success to their many loyal customers. "They are just such a big part of every day, especially those who were there when we first set out our 6-foot table at the farmers' market. They convinced us to open up a brick-and-mortar shop, so they are 100 percent a big part of our business."

Aubry currently makes about 45 batches of sausages per week, each 21 to 24 pounds (or about 80 sausages per batch). "I'm definitely the resident sausage maker here. I can roll them better than anyone else, and I also handle the business," she says. "My brother is the face of the shop, plus as a recent convert to veganism he still has meat memories, so he is the official taste tester." Most of the sausage goes to walk-in clientele, with about 30 percent shipped across the country. "When we opened the business we didn't necessarily think it would be this popular, so we restricted the amount that could be shipped, but that's going to change soon."

In addition to a regular rotation of sausages — breakfast sausages, Mexican chorizo, Sriracha brats, and pepperoni — they make "meat-free meats" like smoky house-barbecued ribs, porchetta, and rib-eye steak, plus bacon, jerky, deli meats, and even "cheese-free cheese." They are also always trying out new sausages for one reason or another. "Sometimes Kale will say he's craving a certain sausage for lunch, so we'll give it a go. We tried a boudin and it's not perfect yet, but we'll get there." They're also developing a kimchi sausage based on popular demand.

As devout vegans, they are thrilled when their products inspire others to reconsider their food choices. "Initially I chose to give up meat for ethical reasons, and then as I got older the environmental impact became more important. Some customers stop by and they say right away, 'Listen, I'm a meat-eater, but I heard about you guys.' They'll usually just take one or two sausages home, but then soon they're back and taking home more and then all of a sudden they are completely vegetarian. I can see that eating this sausage is making a big difference. It's the little things that I notice each day, and then it becomes this big thing that they want to share with their family and friends. I've learned that you can't rush change."

TRADE SECRETS

"With sausage, there are so many places you can go with it. For sweetness, you can use apple juice instead of maple syrup. For Italian sausage, you could include tomato juice. For an iron-y (read: meaty) flavor, there's tomato paste. You can make a whole meal in a sausage, and that's why I love it so. You can do chicken and rice and broccoli in a sausage. You don't even need to put anything with that."

— Aubry Walch

PIMENTÓN LENTIL PATTIES

MAKES ABOUT 16 PATTIES

Lentils have a meaty texture, and when paired with Spanish pimentón, they take on a subtle smokiness that's just right for using as the foundation of these sausages, which cook up to a delightful crisp exterior thanks to dredging in chickpea flour before panfrying. Serve with labneh or Greek yogurt, sprinkled with more pimentón and parsley. *See photo, page 123.*

1. Heat the oil in a 3-quart saucepan over medium-high heat until hot but not smoking, then add the onion and garlic, and cook until the onion is softened and fragrant. Add the pimentón and cook, stirring, for 1 minute.

2. Add the lentils and salt to the pan with enough water to cover and bring to a boil. Reduce the heat to a simmer and cook uncovered until the lentils are completely tender, 30 to 40 minutes, adding water as needed to keep the lentils covered.

3. When the lentils are tender, drain off any remaining liquid and stir in the mayonnaise, vinegar, parsley, lemon zest, black pepper, coriander, and cayenne until well combined, mashing some of the lentils with the back of the spoon.

4. Taste the mixture and add additional salt and/or vinegar if needed. The mixture should be well seasoned with just a little hit of vinegar.

5. Put the chickpea flour on a plate. Form ¼ cup of the lentil mixture at a time into ½-inch-thick patties, dusting with chickpea flour as needed to keep from sticking.

6. To cook, panfry over medium heat until browned on both sides, about 10 minutes. Serve hot.

Ingredients:

- 1 tablespoon olive oil, plus more for cooking patties
- ½ cup minced onion
- 1 teaspoon minced garlic
- 2 teaspoons pimentón (smoked paprika)
- 1½ cups brown lentils
- 2 teaspoons kosher salt, plus more as needed
- 2 tablespoons mayonnaise
- 1 tablespoon sherry vinegar, plus more as needed
- 1 tablespoon minced fresh flat-leaf parsley
- 2 teaspoons grated lemon zest
- ½ teaspoon freshly ground black pepper
- ½ teaspoon ground coriander
- Pinch of cayenne pepper
- 1 cup chickpea flour

SMOKY APPLE AND TOFU SAUSAGE

- 2 teaspoons vegetable oil, plus more for baking sheet
- ½ cup chopped onion
- 2 tart apples, peeled, grated, and tossed with 1 teaspoon fresh lemon juice
- 1 pound extra-firm tofu, drained (see box below)
- 1 cup fresh bread crumbs
- 2 teaspoons Lapsang souchong tea, ground in a spice mill if coarse
- 1 teaspoon minced fresh sage
- 1 teaspoon kosher salt
- ¾ teaspoon freshly ground black pepper
- ½ teaspoon ground allspice
- ½ teaspoon freshly grated nutmeg
- ½ teaspoon ground ginger
- 2 large egg whites, lightly beaten

This savory low-fat mixture makes a delicious breakfast or lunch offering. A smoky note is introduced with Lapsang souchong tea, available at natural food stores and some supermarkets. If you prefer, you can substitute ¼ teaspoon liquid smoke for the tea.

1. Heat the oil in a medium skillet over medium-high heat. Add the onion and sauté until softened, about 3 minutes. Add the grated apples and sauté for 3 minutes longer. Transfer the mixture to a large bowl to cool.

2. Crumble or grate the tofu into the bowl; toss to combine with the onion mixture. Add the bread crumbs, tea, sage, salt, pepper, allspice, nutmeg, ginger, and egg whites. Mix together well.

3. With wet hands, form about ½ cup of the mixture into ½-inch-thick patties or 3-inch-long sausage shapes. Place on a baking sheet fitted with a wire rack and refrigerate for at least 1 hour, or until firm.

4. To cook, preheat the oven to 450°F (230°C). Grease a baking sheet.

5. Bake the sausages for 10 to 15 minutes, until the outsides are golden brown and the insides are cooked through and firm.

HOW TO DRAIN TOFU

To remove most of the excess moisture from tofu, cut the block into three horizontal slabs. Put folded paper towels between the layers and rest a heavy pan on top of the tofu for about 15 minutes.

VEGETARIAN KISHKE IN LEEK CASINGS

Kishke is a Jewish-American sausage made from flour, matzo meal, fat, onions, and ground beef. This tasty version replaces the meat with a flavorful mix of carrots, celery, and leeks. And instead of beef casing, the outer leaves of the leeks form an elegant wrapper for these sausages. Serve with a mix of roasted vegetables for an outstanding vegetarian dinner.

1. Trim off the root end of the leeks, leaving about 6 inches of the green part intact. Blanch the leeks in simmering water for 2 minutes, then plunge them into cold water. Carefully push out the centers of the leeks, leaving the outer skins whole. Coarsely chop the white portions of the centers.

2. Combine the leek centers, celery, and carrot in a food processor and pulse until finely chopped. Add the oil, matzo meal, wheat germ, salt, pepper, and egg, and pulse until the mixture holds together.

3. Preheat the oven to 350°F (175°C). Grease a baking sheet.

4. Carefully fill the leek skins with the filling mixture. (If any of the mixture is left over, you can roll it in squares of aluminum foil.)

5. Arrange the leek sausages on the baking sheet, cover with foil, and bake for 25 minutes.

6. If desired, tie each of the baked sausages with two chive stems and decorate with chive blossoms.

6 leeks (1–1½ inches in diameter)

2 celery stalks, chopped

1 large carrot, grated

3 tablespoons vegetable or olive oil, plus more for baking sheet

1 cup matzo meal or fresh bread crumbs

¼ cup raw wheat germ

1 teaspoon kosher salt

½ teaspoon freshly ground pepper

1 large egg

12 long chive stems with blossoms (optional)

FIG AND WALNUT SALAMI

MAKES ONE 14-OUNCE SALAMI

- 6 ounces dried yellow or brown figs, preferably Kalamata, stemmed
- 2 ounces pitted dates
- ½ cup dry red wine
- 2 tablespoons balsamic vinegar
- Pinch of kosher salt
- ¾ cup walnuts, toasted and very coarsely chopped
- 1 teaspoon grated orange zest
- ½ teaspoon fennel seeds, toasted
- Generous pinch of freshly ground black pepper
- Vegetable oil, for greasing parchment paper

Salame di fichi is a treat offered in Italy during the winter months. You will find many variations on the same theme, and this one was inspired by one from Boston chef Jody Adams, who posted it on her blog, *The Garum Factory*. This sweet salami pairs beautifully with sherry and blue cheese, but wouldn't be out of place on a breakfast buffet alongside smoked sausages and biscuits. You can substitute blanched almonds or shelled pistachios for the walnuts. The addition of dates in this recipe helps bind the ingredients together, and the brief drying time helps intensify the flavors.

1. Cut the figs and dates into bite-size pieces.

2. Put the fruit into a saucepan large enough to hold it. Pour in the wine and vinegar and add the salt. Bring to a simmer and cook, stirring occasionally, until the liquid has just about simmered away. Remove from the heat.

3. Stir in the walnuts, zest, fennel seeds, and pepper until well combined. Let the mixture cool until it is not too hot to touch, 10 to 15 minutes.

4. Very lightly oil an 8- by 12-inch sheet of parchment paper. Scrape the fig mixture onto the sheet and form into a log 1½ to 2 inches in diameter and about 6 inches long along the long edge of the parchment. Roll the parchment over the log and, using your hands and fairly firm pressure, form into an even cylinder. Twist the ends to tighten the parchment around the salami and tie them with butcher's string to secure.

5. If desired, hang in a cool place for up to 1 week to dry, but the salami may be eaten right away.

PART 3
COOKING WITH SAUSAGE

Pita Crisps, page 302

Shakshuka with Sausage,
page 272

SAUSAGE FOR
BREAKFAST
OR BRUNCH

What? Sunday morning in an English family and no sausages?
God bless my soul, what's the world coming to, eh?

— Dorothy L. Sayers

The English aren't the only ones who consider sausage to be a welcome (if not essential) sight at the Sunday breakfast table. You'll find preparations across many cultures that put the salty, savory taste of sausage to excellent use, and not just astride a tall stack or "two over easy." Though there is something about the classic pairing; opinions as to which came first — the sausage or the egg — depends on who you ask. Here you'll find recipes for sausages paired with eggs in quiche, frittata, and more. But sausage plays well with other morning-meal staples, including biscuits and gravy, grits, and bacon, as well as other unexpected ingredients like quinoa (in a breakfast bowl) and kale or chard (in a strata). Some are easy enough for busy weekdays, but all are ideal for sharing with family and friends over a leisurely weekend brunch, when you can show off your sausage-making skills to delicious effect.

COUNTRY-STYLE EGGS WITH SAUSAGE

Here's a simple recipe for two that pairs the versatile egg with equally adaptable sausage. It's easily doubled, tripled, or more — for however many hungry guests you need to feed.

Serves 2

- 2 teaspoons olive oil or vegetable oil
- ½ pound Country-Style Breakfast Sausage (page 56), removed from casing
- 4 large eggs, well beaten
- 2 tablespoons grated Parmesan cheese
- 2 tablespoons milk
- 2 teaspoons chopped fresh chives
  Kosher salt and freshly ground black pepper

1. Heat the oil in a medium skillet over medium-high heat. Add the sausage and sauté, crumbling it with the back of a spoon, until it is browned and cooked through, about 10 minutes. Drain off almost all the grease; set aside the skillet.

2. In a mixing bowl, combine the eggs, Parmesan, milk, chives, and salt and pepper to taste. Mix well.

3. Return the skillet with the sausage to the burner over medium heat. Pour in the egg mixture and cook, stirring constantly, until the eggs are set, about 2 minutes. Serve immediately.

BISCUITS WITH SAUSAGE GRAVY

Using the drippings from cooking sausage to make gravy has a strong tradition in the South, where the economics of feeding a large family demanded stretching every morsel. Also known as sawmill gravy (most often used with chicken-fried steak) or milk or cream gravy, sausage gravy is just one of many variations on the same idea, and it has long been served with featherlight biscuits for a hearty breakfast.

This culinary signpost can be found up and down the southeastern states, where it is a must on any restaurant breakfast menu — or home cook's breakfast table. Hot sauce is considered an essential component of the dish; the more the better, some say.

Makes 12 biscuits and about 12 cups gravy

BISCUITS
- 3 cups all-purpose flour, plus more for dusting
- 4 teaspoons baking powder
- 1 tablespoon sugar (use less sugar if you prefer)
- 1 teaspoon salt
- 1 teaspoon baking soda
- ¾ cup (6 ounces) cold unsalted butter, cut into ¼-inch cubes
- 1 cup cold buttermilk, plus more as needed

SAUSAGE GRAVY
- 1 pound breakfast sausage (any recipe will do), removed from casing
- ¼ cup finely chopped onion
- ⅓ cup all-purpose flour
  Kosher salt
- 4 cups milk
- 1 tablespoon minced fresh sage
  A few dashes of hot pepper sauce, such as Tabasco
  Freshly ground black pepper

Make the biscuits

1. Preheat the oven to 450°F (230°C). Line a baking sheet with parchment paper.

2. In a medium bowl, mix the flour, baking powder, sugar, salt, and baking soda together.

3. Using a pastry cutter, two knives, or your fingertips, cut the butter into the flour mixture until the mixture is pebbly and crumbly.

4. Using a flexible spatula, fold in the buttermilk until it is just absorbed into the dough. If the dough seems very dry, add a little more buttermilk. The dough should not be thoroughly moist but should be slightly tacky.

5. Use a ¼-cup measure to scoop out dough and drop each mound onto the baking sheet at least 2 inches apart.

6. After all the dough has been portioned onto the sheet, use lightly floured hands to pat each biscuit down just a little.

7. Bake for 15 to 18 minutes, until light golden brown, rotating the baking sheet halfway through.

Make the gravy

1. Heat a 12-inch cast-iron or other heavy skillet over medium-high heat.

2. Add the sausage and onion to the pan and cook, crumbling the sausage with the the back of a spoon, until the sausage is cooked through and the onions are softened.

3. Add the flour to the pan, season with salt, and stir to coat the meat with the flour.

4. Slowly pour in the milk, stirring constantly. Bring to a simmer and stir occasionally until the gravy is thickened, about 15 minutes.

5. Stir in the sage and hot sauce, and season with salt and pepper to taste.

To serve

Split the biscuits in half and top with the gravy.

SAUSAGE AND PEPPER QUICHE

The best excuse for using up a few eggs is to make a quiche — and it's quest-worthy to boot. This version is a take on old-school quiche Lorraine and is best with highly spiced sausages like Merguez (page 132), Boerewors (page 89), or Luganega (page 66), but feel free to substitute what you have on hand. The seasoning is reminiscent of chermoula, a fragrant North African herbed sauce, which of course would be more than suitable served alongside.

Serves 6 to 8

PASTRY

  1 cup sifted all-purpose flour, plus more
     for dusting
  ¼ teaspoon kosher salt
  ⅓ cup chilled vegetable shortening, cut into
     marble-size pieces
    About ¼ cup cold water

FILLING

  1 teaspoon vegetable oil
  ½ pound fresh sausage, removed from casing
  1 small onion, chopped
  2 garlic cloves, minced
  ½ cup roasted red bell peppers, chopped
  ¼ cup minced fresh cilantro
  2 tablespoons minced fresh flat-leaf parsley
    Zest of 1 lemon
  3 large eggs
  ½ teaspoon ground cumin
    Pinch of cayenne pepper
    Kosher salt and freshly ground black pepper
  2 cups half-and-half
  ¼ cup grated Romano cheese

Make the pastry

1. Sift the flour together with the salt into a medium bowl. Using a pastry blender or two knives, mix the shortening into the flour mixture.

2. Sprinkle the water onto the flour mixture, a little at a time, mixing continuously. Form the dough into a ball.

3. On a lightly floured surface, roll out the pastry into a 10-inch-diameter circle. Fold the pastry onto itself, then gently drape the dough over a 9-inch pie plate. Unfold the dough, fit the pastry into the plate, then crimp the edges with your fingers or a fork.

Make the filling

1. Preheat the oven to 400°F (200°C). Heat the oil in a large skillet over medium-high heat. Crumble the sausage into the skillet and cook, breaking it up with the back of a spoon, until the sausage is browned, about 10 minutes. Remove the sausage with a slotted spoon and set aside.

2. Add the onion and garlic to the sausage drippings in the skillet and sauté over medium heat until softened, about 10 minutes. Remove from the heat and stir in the roasted peppers, cilantro, parsley, and lemon zest. Set aside.

3. In a medium bowl, beat the eggs, then beat in the cumin, cayenne, and salt and black pepper to taste. Stir in the half-and-half.

To assemble

1. Sprinkle the sausage in the pastry shell. Layer the onion mixture evenly over the sausage. Sprinkle with the cheese and pour in the egg mixture.

2. Bake for 30 minutes, or until firm and lightly browned. Let the quiche cool for about 10 minutes before slicing and serving.

SAUSAGE AND CHARD FRITTATA

Leave it to the Italians to put a rustic spin on the French-style omelet. It's practically foolproof, and it tastes equally good warm from the oven, at room temperature, or even chilled the next day, assuming you have any leftovers — meaning it's a great make-ahead option for brunch with friends. You can swap out the chard for other hearty greens, such as Tuscan kale or escarole. Just a little sausage goes a long way to season the filling; serve more links alongside, if you like.

Serves 6 to 8

- 1 bunch Swiss chard
- 12 large eggs
- ½ cup milk
- ¾ cup grated cheddar cheese (about 3 ounces)
  Kosher salt and freshly ground black pepper
- 2 tablespoons olive oil
- ½ medium onion, chopped
- ½ fresh sausage link, such as Hot Italian Sausage (page 64), removed from casing
- ½ teaspoon crushed red pepper

1. Remove the stems from the Swiss chard. Cut the stems into ½-inch-thick pieces, and the leaves into ½-inch-wide ribbons.

2. Preheat the broiler. In a large bowl, whisk the eggs and milk together. Mix in half of the cheese, and season with salt and pepper.

3. Heat the oil in a large cast-iron skillet. Add the Swiss chard stems, onion, and sausage, and cook, breaking up the sausage with the back of a spoon, until the sausage is cooked through, about 10 minutes. Add the Swiss chard leaves and crushed red pepper, and cook until the chard is wilted, about 3 minutes.

4. Pour the egg mixture into the skillet and reduce the heat to low. Cook until the eggs appear set around the edges of the skillet. Sprinkle the remaining cheese over the top of the frittata.

5. Put the skillet under the broiler, and cook until the frittata is nearly cooked and the cheese is golden, 5 to 8 minutes. The frittata will continue to cook after it is removed from the broiler.

6. Let the frittata rest for at least 5 minutes before serving.

TOAD-IN-THE-HOLE

Like many a popular British dish (bubble and squeak comes to mind), toad-in-the-hole has a wonderfully eccentric name for what is really the uber-comfort food — or "a homely but savoury dish," as described by Mrs. Beeton. Somehow that name has been translated in the United States to mean a piece of toast with an egg cooked in a hole cut out of the center — and nary a sausage in sight.

But the original version, whose origins are a bit murky, is more like a type of Yorkshire pudding: Whole sausages are covered with an eggy custard and the whole thing is baked until puffed and golden brown on top. It's often served with onion gravy, though we find it completely satisfying on its own.

Serves 4

- ½ cup all-purpose flour
- ½ teaspoon dry mustard
- ½ teaspoon kosher salt
- 1 large egg
- 1¼ cups milk
- ½ teaspoon chopped fresh thyme
- 1 pound sausages, such as bangers or breakfast sausages
- 1 tablespoon vegetable oil

recipe continued on next page

1. Preheat the oven to 425°F (220°C).

2. In a medium bowl, combine the flour, dry mustard, and salt. Whisk in the egg and milk to produce a creamy batter, then whisk in the thyme.

3. Place the sausages in a 9-inch square baking dish. Add the oil and roll to coat the sausages. Bake the sausages for about 15 minutes.

4. Pour the batter over the sausages, return the pan to the oven, and bake until the batter has risen and is well browned with crispy edges. Serve hot.

SKILLET STRATA WITH GREENS AND SAUSAGE

Strata is the busy host's secret weapon: Make it the day before your guests arrive and then just pop it in the oven the next morning. You can also bake it straight away. Depending on what's available, kale, Swiss chard, and collards will all work well in this recipe, as would spinach. Breakfast sausage is a natural choice, but feel free to substitute with Italian Sausage (hot or sweet, page 64), Luganega (page 66), Fresh Chorizo (page 67), or any other fresh options you have on hand. *See photo, opposite.*

Serves 6 to 8

2 tablespoons olive oil
½ pound sausage, removed from casing
1 large onion, chopped
1 bunch kale or other hearty greens, stems and leaves cut ½ inch thick
  Kosher salt and freshly ground black pepper
  Pinch of freshly grated nutmeg
3 cups milk
8 large eggs
2 cups grated Gruyère cheese
1 cup grated Parmesan cheese
2 teaspoons dry mustard

8 cups cubed bread, such as brioche, challah, or rustic loaf (about ½ pound)

1. Heat a large cast-iron or oven-safe nonstick skillet over medium-high heat. Add the oil, sausage, and onion. Cook, stirring occasionally, until the sausage is cooked through and the onion is tender. Add the kale stems and leaves to the pan, and cook until wilted. Remove from the heat and season with salt, pepper, and the nutmeg.

2. In a large bowl, whisk together the milk, eggs, 1½ cups of the Gruyère, ½ cup of the Parmesan, and the dry mustard.

3. Add the bread and the sausage mixture to the bowl and stir gently to combine.

4. Transfer the contents of the bowl to a 3- to 4-quart casserole. Top with the remaining cheeses and cover with plastic wrap. Store in the refrigerator overnight.

5. Remove the strata from the refrigerator 30 to 45 minutes before baking. Preheat the oven to 350°F (175°C).

6. Bake the strata for 45 minutes to 1 hour, until puffy and golden. When a knife is poked into the center, there should be no liquid present. Let the strata rest for 5 to 10 minutes before serving.

Skillet Strata with
Greens and Sausage

SAUSAGE PIPÉRADE

A pipérade is a Basque dish of tomatoes and peppers. Eggs can be added to make an open-faced omelet. This recipe is a variation on that theme.

Serves 4 to 6

> 2 tablespoons plus 1 teaspoon olive oil or vegetable oil
> 1 pound Italian Hot or Sweet Italian Sausage (page 64), removed from casing
> 1 small onion, finely chopped
> 4 garlic cloves, finely chopped
> ½ cup chopped red bell pepper
> ½ cup chopped orange bell pepper
> ½ cup chopped yellow bell pepper
> 3–4 ripe tomatoes, chopped
> 1 tablespoon minced fresh basil
> 1 tablespoon minced fresh flat-leaf parsley
> 2 teaspoons minced fresh oregano
> 1 teaspoon minced fresh thyme
> Kosher salt and freshly ground black pepper
> 6 large eggs

1. Heat 1 teaspoon of the oil in a medium skillet over medium-high heat. Add the sausage and sauté, crumbling it with the back of a spoon, until it is cooked through and browned, about 10 minutes. Remove the sausage with a slotted spoon. Discard the drippings.

2. In the same skillet, heat the remaining 2 tablespoons oil over medium-high heat. Add the onion and garlic, and sauté until the onion is translucent, about 5 minutes. Add the bell peppers and sauté for 5 minutes. Add the tomatoes and sauté, shaking the pan, until the tomato juices have mostly evaporated.

3. Add the sausage, basil, parsley, oregano, thyme, and salt and pepper to taste. Stir gently to combine.

4. Crack an egg into a cup or ramekin. With a large spoon, make a small well in the simmering sauce, and slide in the egg. Repeat with the remaining eggs. Cover the skillet and cook over medium-low heat until the eggs are cooked to your preference, about 4 minutes for eggs with runny yolks and 7 minutes for firm yolks. Serve immediately.

SHAKSHUKA WITH SAUSAGE

This one-skillet meal-in-one, a specialty of North Africa, happens to be the breakfast of choice in Israel — but it also makes an excellent (and easy) supper for a busy weeknight. Merguez (page 132) is our favorite choice for this dish, matching the seasonings tit for tat. Serve with Pita Crisps (page 302). *See photo, page 264.*

Serves 4 to 6

> 3 tablespoons extra-virgin olive oil
> ½ pound fresh sausage, removed from casing
> 1 large onion, halved and thinly sliced
> 1 large red bell pepper, seeded and thinly sliced
> 3 garlic cloves, thinly sliced
> 1 teaspoon ground cumin
> 1 teaspoon paprika
> ⅛ teaspoon cayenne pepper
> 2 pounds fresh tomatoes, cored and halved, or 1 can (28 ounces) whole peeled plum tomatoes
> Kosher salt and freshly ground black pepper
> 6 large eggs
> Chopped fresh cilantro, for serving
> Hot pepper sauce, such as Tabasco, for serving

1. Heat the oil in a large skillet over medium-high heat. Add the sausage, onion, and bell pepper. Cook, crumbling the sausage with the back of a spoon, until the sausage is browned and the vegetables are very soft and caramelized, about

8 minutes. Add the garlic, cumin, paprika, and cayenne, and cook for 1 minute.

2. If using fresh tomatoes, hold them with the skin side against your palm. Grate on the large holes of a box grater, leaving the skin behind. If using canned tomatoes, cut or crush them.

3. Pour the tomatoes into the skillet and season with salt and pepper. Simmer until the tomatoes have thickened, about 10 minutes.

4. Crack an egg into a cup or ramekin. With a large spoon, make a small well in the simmering tomatoes, and slide in the egg. Repeat with the remaining eggs. Season with more salt and pepper. Cover the skillet and cook until the eggs are just set, 5 to 7 minutes. Sprinkle with cilantro and serve with hot sauce.

CRAYFISH SAUSAGE AND GRITS

Think of this low country–style breakfast as an update on the traditional shrimp and grits. Other seafood sausages will do just fine, but none will taste better than the Louisiana Crayfish Sausage on page 229. *See photo, page 228.*

Serves 4 to 6

- 1 cup stone-ground grits
- ¾ cup grated cheddar cheese
- ¼ cup grated Parmesan cheese
- 3 tablespoons unsalted butter
  Kosher salt
- 2 tablespoons vegetable oil
- 2 slices bacon, chopped
- 1 pound Louisiana Crayfish Sausage (page 229)
- 2 garlic cloves, minced
- ½ cup chicken broth
- 1 tablespoon fresh lemon juice
- ½ teaspoon hot pepper sauce, such as Tabasco
- 6 scallions, thinly sliced
- 2 tablespoons minced fresh flat-leaf parsley
  Lemon wedges, for garnish

1. In a medium saucepan, bring 4 cups water to a boil. Reduce the heat to low and whisk in the grits. Cook, whisking or stirring, until the grits are cooked, about 30 minutes. Add more water as needed if the grits start to get dry. Stir in the cheddar, Parmesan, and 1 tablespoon of the butter. Season with salt to taste. Set aside.

2. Heat a 10-inch skillet over medium heat and add the bacon. Cook until crisp, then remove the bacon from the pan, crumble it, and set aside. Leave the bacon fat in the skillet.

3. Lower the heat to medium-low and cook the crayfish sausage in the skillet until cooked through and the casing is crispy. Remove from the skillet. If desired, cut the sausage into bite-size pieces.

4. Add the garlic to the skillet and cook over medium heat. Cook, stirring, until fragrant, about 1 minute. Add the chicken broth and scrape up any browned bits from the bottom of the skillet. Increase the heat to medium-high and reduce the broth by about half.

5. Add the sausage, lemon juice, and hot sauce to the skillet. Add the remaining 2 tablespoons butter and swirl to melt the butter into the sauce.

6. To serve, pour the grits onto a serving platter, top with the sausage mixture, and garnish with the bacon, scallions, parsley, and lemon wedges.

Savory Quinoa
Breakfast Bowl with
Vegetarian Sausage

SAVORY QUINOA BREAKFAST BOWL WITH VEGETARIAN SAUSAGE

Breakfast bowls are a great way to start off any day. With the right balance of grains, protein, and vegetables, they offer a convenient and well-balanced meal that can be adapted in countless ways. This version calls for crumbling in some meatless sausage and an optional soft-boiled egg. You could go with many of the sausages in chapter 9, but we especially like to use Vegetarian Breakfast Sausage (page 245), shown opposite, Black Bean and Smoked Corn Sausage (page 246), or Pimentón Lentil Patties (page 258). *See photo, opposite.*

Serves 6

DRESSING
- 1 shallot, minced
- ¼ cup fresh lime juice
- ¼ cup extra-virgin olive oil
- Kosher salt and freshly ground black pepper

QUINOA AND SAUSAGE
- 2 tablespoons extra-virgin olive oil
- ½ onion, finely chopped
- 2 small garlic cloves, minced
- 1 cup quinoa, preferably red
- 1 teaspoon ground cumin
- Pinch of crushed red pepper
- Kosher salt and freshly ground black pepper
- 1 pint grape tomatoes, halved
- 2 tablespoons finely chopped fresh cilantro
- ½ pound vegetarian sausage

BOWL
- 1 ripe but firm avocado, pitted, peeled, and sliced or diced
- 6 large eggs, soft-boiled or poached (optional)
- Cilantro leaves
- Lime wedges

Make the dressing

In a small bowl, soak the shallot in the lime juice for 10 minutes. Gradually whisk in the oil, then season with salt and pepper to taste.

Make the quinoa and sausage

1. Heat 1 tablespoon of the oil in a medium saucepan over medium heat. Add the onion and garlic, and cook, stirring, until the onion is tender, about 2 minutes. Stir in the quinoa and cook, stirring, until toasted, about 1 minute. Stir in the cumin and crushed red pepper, and cook, stirring, for 30 seconds. Season with salt and black pepper to taste.

2. Add 2 cups water to the pan, stir once to combine, then cover and reduce the heat to low. Simmer until the quinoa is tender and most of the liquid has evaporated, about 15 minutes. Turn off the heat and let stand for 10 minutes, covered. Fluff the quinoa with a fork.

3. Meanwhile, heat the remaining 1 tablespoon oil in a medium skillet. Cook the sausage, breaking it up with a spoon, until browned and heated through.

4. While still warm, transfer the quinoa to a large bowl and toss with the tomatoes, cilantro, sausage, and some of the dressing.

Assemble the bowl

Divide the quinoa mixture among serving bowls and top with the avocado, eggs (if desired), cilantro leaves, and lime wedges for squeezing.

Pigs in a Blanket, page 290

SAUSAGE STARTERS

A document from the Spanish vice-regent in Palermo on January 30, 1415, shows that lamb, pork, or sausages were bought thirteen days out of the month and macaroni only once a month.

— Clifford A. Wright

Sausage and good times go together like backyard barbecues and beer, Fourth of July and fireworks. Not only are they fun to make, sausages are great fun to share — and just right for entertaining. Throwing a cocktail party? Sausages are your go-to salty snacks. Having friends over for dinner? Sausages are great for waking up the palate for the meal to come. Next to the cracker, sausage is among the most versatile ingredient in nibbles and noshes, capable of being served on its own — set out some toothpicks and assorted mustards, for dipping — or as part of a charcuterie board (see page 148), with cheese, pickled anything, fruity chutneys, and other savory or sweet sidekicks.

Sausage also imparts incredible flavor (similar to bacon) to a host of appetizers, from Scotch eggs and egg rolls to delectable spreads and everyone's favorite: pigs in a blanket. No matter how you serve them, sausages are sure to get the party started.

SCOTCH QUAIL EGGS

This recipe was developed by Cathy Barrow (see profile on page 130), using her own Merguez sausage. It makes the daintiest Scotch eggs you've ever seen — and because they are as rich and filling as ever, the smaller stature is a welcome change of pace. If you can't find quail eggs, use four of the smallest hen's eggs you can find, boil them for 4 minutes in step 1, and increase the frying time to 8 to 10 minutes in step 7.

According to Barrow, "Peeling quail eggs can be a nightmare, and there was no guidance on the entire Internet on how long to cook a quail egg, so I had to keep cooking them and opening them up to see if it was right. I went through dozens of quail eggs before I got to that place, but I figured it out. You want it to be slightly past soft-boiled, which makes it challenging to peel. Get into your Zen meditation and peel those tiny blue shells off the eggs very carefully." *See photo, opposite.*

Makes 12 eggs

> 12 quail eggs
> ½ cup all-purpose flour
> 2 large eggs, well beaten
> 1 cup fine fresh bread crumbs
>   Kosher salt and freshly ground black pepper
> 1 pound fresh sausage, such as Merguez (page 132), removed from casing
> 2 cups grapeseed oil
>   Mustard and pickles, for serving

1. Fill a medium bowl with cold water and a few ice cubes. Bring a small saucepan of water to a boil. Immerse the quail eggs for precisely 2 minutes. Drain and plunge them into the ice water.

2. Gently peel the eggs under a stream of running water. They will not be cooked through, so be careful not to break into the yolks. Rinse the peeled eggs and place them on a paper towel–lined plate. Chill them well before proceeding.

3. Place the flour on a plate, and place the beaten eggs in a bowl. Place the bread crumbs on another plate. Season the flour, egg, and bread crumbs with salt and pepper.

4. Divide the sausage into 12 equal parts. Form ¼-inch-thick oblong flat patties with the sausage.

5. One at a time, roll the quail eggs in the flour, then wrap each egg with the sausage. Gently coax the sausage around each egg until it is evenly covered and smooth. Place the sausage-covered eggs on a sheet pan as they are completed.

6. Bread the eggs by rolling each in flour, then dipping into the egg to coat. Then roll the egg in the bread crumbs to coat evenly. Return to the sheet pan and chill the breaded eggs for at least 1 hour.

7. Heat the oil in a straight-sided 1- to 2-quart saucepan to 375°F (190°C). Gently lower the eggs into the hot oil and fry for 6 minutes. Drain on paper towels. Serve with mustard and pickles.

SPANISH TORTILLA

In Spain, this dish is called tortilla espagnola, and it is a mainstay at tapas bars and on restaurant menus across the country. In the United States, it is often called Spanish omelet to avoid confusion (there are no actual tortillas involved), and that's a pretty good approximation of what it is.

First, thin slices of potatoes are cooked in copious amounts of olive oil, then are drained before cooking with the egg mixture. Save that extra oil to make Aioli (page 281) for dolloping

recipe continued on page 281

Scotch Quail Eggs

Spanish Tortilla

Aioli,
page 281

Romesco,
page 296

on the tortilla wedges along with Romesco (page 296), another Spanish specialty. Dried chorizo is added to this version; you could also use fresh chorizo, but you will need to cook that off before combining it with the onions and garlic. *See photo, opposite.*

Serves 12 or more as a tapa or 4 as a main dish

- 2 cups olive oil
- 4 medium Yukon Gold potatoes, peeled and sliced ⅛ inch thick
-   Kosher salt and freshly ground black pepper
- 2 onions, diced
- 4 garlic cloves, chopped
- ½ pound Spanish-Style Chorizo (page 110), crumbled
- 6 large eggs

1. In a 10-inch cast-iron or other nonstick skillet, heat the oil over medium-high heat.

2. Toss the potatoes with 1½ teaspoons salt. Set a colander in a large bowl.

3. When the oil is very hot (a potato slice will immediately start sizzling when added to it), very carefully add the potatoes to the oil. Cook, turning the potatoes occasionally with a spatula or spoon, until tender enough to be pierced with a paring knife. Lift the potatoes with a slotted spatula or spoon into the colander and let drain.

4. Carefully pour off all but about 3 tablespoons oil from the pan. Add the onions, garlic, and sausage to the pan and cook over medium heat until the onions are tender and starting to brown.

5. In a large bowl, beat the eggs, and season with salt and pepper. Add the potatoes and the onion mixture, turning carefully to evenly combine.

6. Heat a large skillet over medium-high heat. When very hot, add the egg mixture, gently smoothing it flat. Cook for 2 minutes, then reduce the heat to medium-low. Cook until the eggs are set at the edges and the center is beginning to set, and the bottom is golden brown. Shake the pan to loosen the tortilla from the bottom.

7. Lay a plate large enough to cover the skillet over the skillet. Using pot holders to hold the plate firmly against the skillet, flip the tortilla over and onto the plate. Return the pan to the stove and gently slide the tortilla back into the skillet. Neaten the edges of the tortilla with a spatula if needed.

8. Cook over medium-low heat until the tortilla is cooked through, about 5 minutes.

9. Unmold the tortilla onto a serving platter and let rest for 10 minutes before cutting into wedges or 1-inch squares.

AIOLI

Think of this as Spain's answer to mayonnaise — it's just as versatile and goes well with all manner of meat, fish, and vegetables.

To make it: Mince 2 peeled garlic cloves with a generous pinch of kosher salt to a paste (a mortar and pestle is traditional, but you can do this on a cutting board with the flat side of a chef's knife).

In a bowl, whisk together 2 egg yolks and 1 tablespoon each fresh lemon juice and water until combined. Whisking constantly, add ¼ cup extra-virgin olive oil in a slow, steady stream until emulsified. Whisk in the garlic paste and season with salt to taste. Aioli can be refrigerated, covered, for up to 2 days.

MINI CORN DOGS

This recipe was adapted from *Dishing Up Minnesota: 150 Recipes from the Land of 10,000 Lakes* by Teresa Marrone. Deep-fried hot dogs on a stick are found all over the grounds of the Minnesota State Fair, and each vendor has its own style. Some, such as Poncho Dogs, use a batter that contains a lot of cornmeal; others, such as Pronto Pup, use a smoother, floury batter that doesn't have the grainy texture associated with cornmeal. The recipe here is for a cornmeal-style batter.

At home, it's lots easier to deep-fry mini dogs, because home fryers typically aren't tall enough to hold a standard-length hot dog. You can make your own mini sausages using lamb casings; otherwise, cut full-size sausages into smaller pieces. Good sausage options include any smoked sausage, Fresh Chorizo (page 67), or Loukaniko (page 152).

For serving, you could stick with good old-fashioned golden mustard, or try any of the variations on page 293. You will also need 32 (or more) thin bamboo skewers, 8 to 10 inches long. *See photo, opposite.*

Makes about 32

    Vegetable oil, for frying
1 cup yellow cornmeal (not coarse ground)
¾ cup all-purpose flour
3 tablespoons sugar
1½ teaspoons baking powder
¾ teaspoon kosher salt
¼ teaspoon baking soda
¼ teaspoon onion powder (optional)
⅛ teaspoon freshly ground black pepper
1 large egg
1½ cups milk, plus more as needed
1 pound mini sausages (see headnote), cooked
    Mustard, for serving

1. Heat the oil in a deep fryer to 360°F (180°C). Alternatively, pour enough oil into a deep stockpot to reach a depth of 4 inches and heat the oil over medium heat to 360°F (180°C). Preheat the oven to 250°F (120°C). Line a plate with paper towels.

2. In a large bowl, whisk together the cornmeal, flour, sugar, baking powder, salt, baking soda, onion powder (if desired), and pepper. In a small bowl, beat the egg and 1 cup of the milk. Whisk the milk mixture into the cornmeal mixture. Add the remaining ½ cup milk as needed; the batter should be thick enough to cling to a spoon.

3. Spear a sausage lengthwise on the sharp end of each skewer. Dip the skewered sausage into the batter, rolling it around to coat completely, then hold it above the batter and twirl the skewer so the batter coats the sausage evenly, with no drips.

4. Fry the corn dogs two or three at a time, turning them to brown evenly; they will float, and you may find it easier to hold on to the ends of the skewers and push the tips down to submerge them completely in the oil. Cook until richly browned, 3 to 4 minutes. Drain on the paper towel–lined plate while you fry the next batch; as you finish a batch, transfer the previous batch to a baking sheet and keep warm in the oven while you continue frying. The batter may thicken as you continue; thin it with a little more milk as necessary.

5. When all the corn dogs are fried, transfer to a serving plate. Serve with mustard.

NOTE: *You will probably have some batter left after you fry all the corn dogs. You can cut dill pickles into ¾-inch chunks, dip them in the batter, and fry to serve alongside the corn dogs.*

Mini Corn Dogs

SAUSAGE-STUFFED EGG ROLLS

Commercially prepared egg roll skins will work very well in this recipe. Plan ahead to allow refrigeration time for the egg rolls to firm up before they are fried. Serve these with Chinese (hot) mustard or your favorite dipping sauce.

Makes 10 to 12 egg rolls

> 1 tablespoon olive oil or vegetable oil
> ½ pound Country-Style Breakfast Sausage (page 56), removed from casing
> 1 cup finely shredded cabbage
> 1 cup finely chopped celery
> 1 cup finely chopped scallions
> ½ cup finely shredded carrots
> ½ cup finely chopped mushrooms
> ¼ cup finely chopped green bell pepper
> 1 tablespoon grated fresh ginger
> ½ teaspoon freshly ground white pepper
> 2 teaspoons soy sauce
> 1 large whole egg
> 1 large egg white, lightly beaten
> 10–12 egg roll wrappers
> Vegetable oil, for frying

1. Heat the olive oil in a large skillet over medium-high heat. Add the sausage and sauté, crumbling it with the back of a spoon as it cooks, until it is lightly browned, about 10 minutes. Drain off most of the fat.

2. Add the cabbage, celery, scallions, carrots, mushrooms, bell pepper, ginger, white pepper, soy sauce, and egg to the skillet and sauté for 2 minutes, stirring. Remove the mixture from the heat; let cool.

3. To assemble the egg rolls, spoon about ¼ cup of the filling onto the center of an egg roll skin and fold two sides over to meet in the center. Brush the two sides and the two open ends with the egg white, then roll up the skin, beginning at one end, to enclose the center. Carefully place the egg rolls, seam side down, on a baking sheet. Refrigerate for 2 hours.

4. Heat 3 inches of vegetable oil to 375°F (190°C) in a pot. Deep-fry the egg rolls until they are crisp and golden, about 5 minutes. Drain on paper towels. Serve hot.

MORTADELLA MOUSSE

Here's a dead-simple recipe that yields crowd-pleasing results. Essentially a lightened purée of mortadella, it tastes wonderful on toasted or grilled bread to nibble along with cocktails, or as part of a charcuterie board. While this recipe calls for mortadella, there is no reason a high-quality bologna or smooth-textured smoked sausage couldn't be used instead. *See photo, opposite.*

Makes about 2½ cups

> ¾ pound Mortadella (page 90), cut into ½-inch cubes
> 4 ounces ricotta cheese
> ¼ cup grated Parmesan cheese
> Pinch of freshly grated nutmeg
> ¼ cup heavy cream
> Kosher salt and freshly ground black pepper
> Finely chopped toasted pistachios, for garnish

1. Place the mortadella in a food processor and pulse until finely chopped.

2. Add the ricotta, Parmesan, and nutmeg, and pulse to combine. With the motor running, quickly pour in the heavy cream. Continue to blend until smooth.

3. Taste and add salt and pepper as desired. Scrape into a small crock or bowl and sprinkle with the pistachios.

4. Serve cold or at room temperature.

Mortadella Mousse

ESSENTIAL SIDEKICKS
PICKLES AND SUCH

Giardiniera

If you've only ever had the store-bought variety of this Italian-American relish, you're in for a treat. Making it is just a matter of prepping a bunch of vegetables and aromatics, then letting the vinegar do its magic. Don't be tempted to serve it right away; you'll need to wait a few days to allow the giardiniera to reach its full flavor potential.

Makes 2 quarts

- 1 medium head cauliflower, cut into small florets
- 2 celery stalks, sliced into ¼-inch pieces
- 2 carrots, sliced into ¼-inch rounds
- 2 red, orange, or yellow bell peppers, cut into strips
- 2 jalapeño or serrano chiles, cut into thin rings
- 4 garlic cloves, thinly sliced
- ½ cup kosher salt
- ½ cup green olives, pitted and chopped
- 1 tablespoon dried oregano, preferably Greek oregano
- 1 teaspoon crushed red pepper
- 1 teaspoon whole black peppercorns
- ½ teaspoon celery seeds
- 1 cup white vinegar, plus more as needed
- 1 cup extra-virgin olive oil, plus more as needed

1. In a large bowl, mix the cauliflower, celery, carrots, bell peppers, chiles, garlic, and salt together until well combined. Add enough water to cover, then cover the bowl and leave at room temperature to brine overnight.

2. Drain and rinse the vegetables. Add the olives, oregano, crushed red pepper, peppercorns, and celery seeds, and mix to combine. Pack the vegetables into two quart-size Mason jars.

3. Pour ½ cup vinegar into each jar. Pour ½ cup oil into each jar. If there is not enough liquid to cover the vegetables, add more vinegar and oil in equal amounts to cover the vegetables. Cap each jar and shake well to combine the marinade.

4. Refrigerate for at least 3 or 4 days before serving. The giardiniera will keep in the refrigerator for several months.

Pickled Anything

Consider this method as a blueprint for creating your own pickled (or marinated) vegetables — no canning required. Change up the herbs and spices depending on the vegetable and your own preferences. Some of our favorite flavor combinations are listed on page 288.

Makes 1 quart

- 2 cups white wine vinegar, red wine vinegar, or apple cider vinegar
- 2 cups water
- 1 tablespoon kosher salt
- 3 cups thinly sliced vegetables (trimmed as necessary)
- 1 tablespoon spices, such as peppercorns, coriander seeds, fennel seeds, mustard seeds, celery seeds, juniper berries, crushed red pepper, and so on.

recipe continued on page 288

Pickled Pearl Onions

Pickled Radishes

Pickled Bell Peppers

Giardiniera

Other seasonings as desired, such as peeled
garlic cloves, strips of orange or lemon zest,
sliced chile peppers, and so on.

Extra-virgin olive oil, for storing

In a large saucepan, bring the vinegar, water, and salt
to a boil. Place the vegetables, spices, and any other
seasonings in a quart-size Mason jar. Pour the vinegar
mixture into the jar. Top with oil (leaving about 1 inch
of headspace above). Screw on the jar lid, and when
cool enough to handle, shake the contents well to
combine. Refrigerate for at least 2 days before serving. The pickles will keep in the refrigerator for several
months.

SUGGESTED COMBINATIONS

➤ Roasted red peppers: garlic, strips of lemon zest,
crushed red pepper, dried oregano

➤ Carrots: mustard seeds, green chiles, sliced onion,
garlic

➤ Pearl onions: coriander seeds, strips of orange
zest, peppercorns, cinnamon stick

➤ Beets: dill seeds, fennel seeds, coriander seeds,
strips of orange zest

Hungarian Celeriac Kraut

This recipe is excerpted from *Fermented Vegetables*
by Kristen K. Shockey and Christopher Shockey.

Makes about 2 quarts

  2 pounds celeriac, shredded

5–6 wax peppers (use Hungarian for heat, banana
for sweet), thinly sliced

  1 generous tablespoon caraway seeds

  1 teaspoon ground paprika

1–1½ teaspoons unrefined sea salt, or to taste

1. In a large bowl, combine the celeriac with the
peppers, caraway seeds, paprika, and 1 teaspoon
of the salt, and massage well, then taste. It should
taste slightly salty without being overwhelming.
Add more salt, if necessary, until it's to your liking.
The celeriac will become limp and liquid may
begin to pool. The dry quality of these roots
means that sometimes the brine isn't obvious until
the celeriac is pressed into the vessel.

2. Transfer a few handfuls at a time to a 2-quart jar
or a 1-gallon crock. Press down on each portion
with a tamper to remove air pockets. You should
see some brine on top when you press. Pack the
vessel, leaving 4 inches of headspace for a crock,
or 2 to 3 inches for a jar. Top with a piece of plastic wrap or grape leaves, if you have some. For a
crock, follow this with a plate that fits the opening
of the container and covers as much of the vegetables as possible; then weight down with a sealed
water-filled jar. For a jar, use a sealed water-filled
jar or ziplock bag.

3. Set aside your vessel on a baking sheet to ferment, somewhere nearby, out of direct sunlight,
and cool, for 5 to 10 days. Check daily to make
sure the celeriac is submerged, pressing down as
needed to bring the brine back to the surface.
You may see scum on top; it's generally harmless.

4. You can start to test the ferment on day 5. You'll
know it's ready when it's pleasingly sour and
pickle-y tasting, without the strong acidity of
vinegar, and the flavors have mingled.

5. Tamp down to make sure the kraut is submerged,
screw on the lid, and store in the refrigerator for
about 10 months.

Kimchi

This recipe is also excerpted from *Fermented Vegetables*. The authors describe this as their basic kimchi recipe, which they make in the fall in one 3-gallon batch to (almost) last through the entire winter — and also one that makes an excellent springboard for experimentation. The cabbage can be sliced in a variety of ways after brining: chopped, cut into bite-size pieces (the most common option in the United States), quartered, halved, or left whole, as you wish.

Makes about 1 gallon

    1 gallon unchlorinated water
    1 cup unrefined sea salt
    2 large heads napa cabbage
    ½ cup crushed red pepper or salt-free gochugaru
    ½ cup shredded daikon radish
    ¼ cup shredded carrot
    3 scallions, white and green parts, sliced
  ½–1 head garlic, cloves separated and minced
    1 tablespoon minced fresh ginger

1. In a crock or a large bowl, combine the water and salt and stir to dissolve.

2. Remove the coarse outer leaves of the cabbages; rinse a few of the unblemished ones and set aside. Rinse the cabbages in cold water, trim off the stalk end, and cut in half. Submerge the cabbage halves and the reserved outer leaves in the brine. Use a plate as a weight to keep the cabbages submerged. Set aside, at room temperature, for 6 to 8 hours.

3. Drain the cabbage for 15 minutes, reserving about 1 cup of the soaking liquid. Set the separated outer leaves aside.

4. Meanwhile, in a large bowl, combine the crushed red pepper, daikon, carrot, scallions, garlic, and ginger, and blend thoroughly.

5. Chop the brined cabbage into bite-size pieces, or larger if you prefer, and add them to the bowl. Massage the mixture thoroughly, then taste for salt. Usually the brined cabbage will provide enough salt, but if it's not to your liking, sprinkle in a small amount, massage, and taste again.

6. Transfer the vegetables, a few handfuls at a time, to a crock or jar, pressing with your hands as you go. Add reserved brine as needed to submerge the vegetables and leave about 4 inches of headspace for a crock, or 2 to 3 inches for a jar. Cover with the brined leaves. For a crock, top with a plate and weight down with a sealed water-filled jar. For a jar, you can use a sealed water-filled jar or ziplock bag.

7. Set aside on a baking sheet to ferment, somewhere nearby, out of direct sunlight, and cool, for 7 to 14 days. Check your ferment daily to make sure the vegetables are submerged. You may see scum on top; it's generally harmless.

8. You can start to test the kimchi after 1 week. It will taste mild at this point, like a half-sour pickle. The cabbage will have a translucent quality and the brine will be an orange-red color. Kimchi is often quite effervescent; it's normal whether it's bubbly or not.

9. When it's ready, spoon the kimchi into smaller jars, making sure the veggies are submerged; screw on the jar lids, and store in the refrigerator for about 9 months.

QUESO FLAMEADO

Also known as queso fundido, this melty delight is cozy comfort food. It's traditionally made with chorizo, as here (you can find versions using shrimp or fajita meat instead). We added a roasted poblano chile for smoky flavor.

When possible, look for Mexican melting cheeses such as asadero, panela, or queso de Oaxaca (known as the "mozzarella of Mexico"), available at Latin food markets and many supermarkets (where Mexican cheese is sometimes labeled as queso blanco, which just means "white cheese").

Serves 8 to 10

- 1 poblano or pasilla chile (or substitute a small green bell pepper)
- 2 links Fresh Chorizo (page 67), removed from casing
- 3 cups shredded Mexican cheese (see headnote)
- 1 cup shredded Monterey Jack cheese
  Corn or flour tortillas, warmed, for serving

1. Heat the poblano directly over a gas burner until blackened and blistered, turning with tongs. Place in a bowl, cover with plastic wrap, and let steam for 10 minutes. Rub off the skins with a paper towel, then remove the stems, ribs, and seeds and cut the chile into fine strips.

2. Preheat the oven to 350°F. Oil an 8-inch cast-iron skillet or broiler-proof baking pan.

3. Cook the chorizo in a skillet over medium heat, breaking it up with a spoon, until browned.

4. Place both cheeses in the prepared cast-iron skillet. Top with the chile strips and crumbled chorizo. Bake until bubbling, about 10 minutes.

5. Turn on the broiler and cook until the top is golden in spots. Serve hot, spooned into tortillas.

PIGS IN A BLANKET

Trade in packaged mini sausages for your own fresh variety (and the dough that pops out of a can for frozen puff pastry) and the popular party food is instantly more guest-worthy — and not just for Super Bowl Sunday. Including a little mustard inside each one also helps. *See photo, page 276.*

Makes about 4 dozen

- 2 pounds fresh sausage, cooked
- 1 large egg
  All-purpose flour, for dusting
- 1 pound frozen puff pastry (2 sheets), thawed
- ½ cup Dijon Mustard (page 293), plus more for serving
- ½ cup sesame seeds

1. Preheat the oven to 450°F (230°C).

2. Quarter the sausages lengthwise and cut into 1½- to 2-inch pieces.

3. Make an egg wash by beating the egg with 2 tablespoons water.

4. On a lightly floured surface, roll a puff pastry sheet into a 14- by 10-inch rectangle. Cut lengthwise into 1½-inch-wide strips. Cut each strip into four pieces.

5. Brush each piece with a little Dijon mustard. Lay a piece of sausage across the strip and roll up. Place 2 inches apart on a baking sheet. Repeat with the second sheet of puff pastry.

6. Brush each bundle with egg wash and sprinkle with the sesame seeds. Chill for 15 minutes.

7. Bake for 20 minutes, or until the pastry is puffed and evenly browned. Serve hot, with mustard for dipping.

MINI RICOTTA AND SAUSAGE CALZONES

Calzones are an Italian invention that can be a main dish, a hearty snack, or a delicious appetizer. As appetizers, the smaller-scale versions can be quite filling, so they are best served when a relatively light meal is to follow.

The dough for calzones is a standard Neapolitan pizza dough. You can substitute a ready-made pizza dough from the supermarket or a local pizzeria instead of preparing your own.

Makes 8

DOUGH

- 1 package (¼ ounce) active dry yeast
- ¾ cup plus ½ cup warm water (105 to 115°F; 41 to 46°C)
- 1 teaspoon kosher salt
- 3 cups all-purpose flour, plus more for dusting
- 1 tablespoon extra-virgin olive oil, plus more for greasing and brushing

FILLING

- 2 tablespoons plus 1 teaspoon olive oil
- 1 pound Sweet Italian Sausage (page 64), removed from casing
- 1 small onion, diced
- ½ pound Pepperoni (page 170), Genoa Salami (page 161), or other dry sausage, or a combination, cut into ¼-inch cubes
- 1 tablespoon chopped fresh flat-leaf parsley
  Assorted add-ins, such as chopped pitted black olives, mashed anchovies, capers, and/or crushed red pepper (all optional)
  Kosher salt and freshly ground black pepper
- 2 cups shredded mozzarella cheese

Make the dough

1. Mix the packet of yeast into ¾ cup of the warm water. Let stand for 15 minutes, or until bubbles form.

2. In a large bowl, stir the salt into the flour. Gradually add the yeast mixture and the oil. Turn the dough out onto a floured surface. Gradually add the remaining ½ cup warm water, while kneading the dough for about 10 minutes. It should be smooth and elastic.

3. Transfer the dough to a lightly greased bowl, turning to coat. Cover the bowl and let rise in a warm place until the dough doubles in bulk, about 1 hour.

Make the filling

1. Heat 1 teaspoon of the oil in a large skillet over medium-high heat. Add the sausage and sauté, crumbling it with the back of a spoon, until it is cooked through and browned, about 10 minutes. Transfer the sausage with a slotted spoon to a large bowl.

2. Add the onion to the pan with the sausage drippings and sauté until it is translucent, about 5 minutes. Drain the onion and add it to the sausage. Stir in the pepperoni, parsley, any optional filling ingredients you are using, and season with salt and pepper to taste.

Make the calzones

1. Preheat the oven to 450°F (230°C). Lightly oil a baking sheet.

2. To assemble the calzones, on a floured work surface, divide the dough into eight equal portions. Roll out each portion to form a circle about ⅛ inch thick. Brush the middle of each circle with some of the remaining oil, leaving a 1-inch

recipe continued on next page

border. Divide the sausage mixture equally among the circles of dough, placing it to one side of center. Sprinkle an equal amount of mozzarella on each. Carefully fold over each piece of dough to form a semicircle and press to seal the edges.

3. Brush the tops of the calzones with oil and arrange them, 1 inch apart, on the baking sheet.

4. Bake the calzones for 25 minutes, or until golden brown and the filling is bubbling. Serve piping hot.

NOTE: *To make calzones for a main course, make four calzones instead of eight.*

MINI SAUSAGE TOSTADAS

Tostados are often overshadowed by tacos as the all-time favored Mexican food, but they are just as appealing and versatile when it comes to the toppings. They are also equally portable when made with tortillas that are cut into small rounds and baked in muffin tins for edible cups. We used a spicy turkey sausage for the filling, but Fresh Chorizo (page 67) would also work well here.

Makes 20

- 10 corn or flour tortillas (6-inch size)
- 2 teaspoons olive oil or vegetable oil, plus more for greasing
- 1 pound Turkey Sausage with Chiles and Cilantro (page 222), removed from casing
- ½ pound Monterey Jack or cheddar cheese, shredded
- ½ cup chopped pitted black olives
- ¼ cup chopped jalapeño chiles (ribs and seeds removed for less heat)

1. Using a 3-inch biscuit cutter, cut two circles out of each tortilla.

2. Preheat the oven to 425°F (220°C). Grease 20 cups of two standard 12-cup muffin tins.

3. Heat 1 teaspoon of the oil in a large skillet over medium-high heat. Working in batches, briefly soften the tortilla circles in the hot skillet, flipping once to ensure they have a thin layer of oil on both sides. Press the tortilla circles into the cavities of the muffin tins to make cups.

4. In the same skillet over medium-high heat, heat the remaining 1 teaspoon oil. Add the sausage and sauté, crumbling it with the back of a spoon, until it is cooked through and lightly browned, about 10 minutes. Drain well on paper towels.

5. Sprinkle an equal amount of sausage, cheese, olives, and jalapeños into each tortilla cup. Bake the tostadas for 1 to 2 minutes, just until the cheese melts. Serve warm.

MORE ESSENTIAL SIDEKICKS
MUSTARD, MOSTARDA, CHUTNEY, AND SUCH

It's hard to imagine sausage without its constant companion — and mustard is easy enough to whisk together while you are waiting for the sausage meat to chill. Mustards also provide ample opportunity for improvisation. Then there are the nonmustardy mostardas (fruited mustards that are not really mustard), courtesy of northern Italy, and chutneys from points far east. Plus pickled this and that from everywhere, since every culture has its own fermented favorites. Here are a few recipes to get you started. Keep in mind that most of these condiments benefit from a few days of mellowing after they're made, and are best used within a month. Store in a covered container in the refrigerator and stir until smooth before serving, at room temperature for fullest flavor.

Dijon or Grainy Mustard

Combine ½ cup each dry white wine and white wine vinegar or apple cider vinegar with ¼ cup each brown and yellow mustard seeds in a small nonreactive bowl. Add ½ teaspoon kosher salt, or to taste. Cover and let sit at room temperature for 2 to 3 days. Purée in a blender to break up the mustard seeds. Stop puréeing when you reach the desired consistency: very smooth for Dijon and chunky for grainy mustard. Makes about 1½ cups.

Honey Mustard

Follow the recipe for Dijon mustard, blending until the mustard seeds are broken up. Add ½ cup honey and purée until the mustard is very smooth. Makes about 2 cups.

Spicy Horseradish Mustard

Follow the recipe for Dijon mustard, replacing the wine with ½ cup dark beer, such as Guinness Stout. Blend until the mustard seeds are broken up. Add ½ cup drained prepared horseradish and purée until the mustard is very smooth. Makes about 2 cups.

Peppercorn Mustard

Follow the recipe for grainy mustard, blending until the mustard seeds are broken up and the mustard is chunky. Stir in 1 to 2 tablespoons cracked black peppercorns and 1 teaspoon chopped fresh tarragon. Makes about 1½ cups.

Fruited Apricot Mustard

A cousin to Italian *mostarda*, fruited mustards incorporate fresh, cooked, or dried fruit.

To make it: Combine ½ cup each dry white wine and apple cider vinegar, ½ cup brown mustard seeds, and ½ teaspoon kosher salt (or to taste) in a small nonreactive bowl. Cover and let sit at room temperature for 2 to 3 days. Purée in a blender to break up the mustard seeds. Add ½ cup chopped dried apricots and purée until the mustard is very smooth. Stir in 2 teaspoons chopped fresh thyme. Makes about 2 cups.

Strawberry Mostarda

Traditional recipes for *mostarda di frutta*, a northern Italian condiment, involve a days-long process of macerating fruit in sugar, boiling off the syrup, and repeating the whole thing again and again until you have a

recipe continued on page 295

Fruited Apricot Mustard, page 293

Tomato Chutney

Dijon Mustard, page 293

Strawberry Mostarda

Grainy Mustard, page 293

Shallot Jam

sweet-sour-spicy relish (akin to chutney) that's ideal for serving with charcuterie. This version accomplishes the same feat in a much faster time. Strawberries are called for here, but blueberries, peaches, or nectarines are also exceptional options.

To make it: Rinse, hull, and slice 1 quart strawberries. Toss together with 3 cups sugar and ¼ teaspoon kosher salt, then cover and let macerate overnight in the refrigerator. Drain off the accumulated juice through a sieve into a skillet. Boil until reduced to a syrup. Stir in the berries, 1 tablespoon yellow mustard seeds, ¼ cup apple cider vinegar, and 1 teaspoon mustard powder, and boil until reduced to a jammy consistency. Let cool before serving. The mostarda can be refrigerated in an airtight container for up to 1 month. Makes about 2 cups.

Shallot Jam

A slightly sweeter and mellower version of onion jam, this deeply flavorful spread will earn a permanent spot on any charcuterie board (or sandwich).

To make it: In a saucepan, heat 2 tablespoons vegetable oil over medium-high heat. Add 1 pound shallots, peeled and thinly sliced, ½ teaspoon kosher salt, and ¼ teaspoon freshly ground black pepper to the pan, and cook until the shallots are softened, stirring often. Add ½ cup unsweetened apple cider, ¼ cup brown sugar, 3 tablespoons apple cider vinegar, and ½ cup dried currants. Reduce the heat to medium-low and cook, stirring often to prevent the jam from sticking to the pan, until the shallots are deeply caramelized and any liquid left in the pan is thick and glossy. If the pan dries out before the shallots are done, add a few tablespoons of water to loosen. Let cool before serving or storing. The jam can be refrigerated in a tightly sealed jar for up to 1 month. Makes about 1½ cups.

Tomato Chutney

Here's to remembering that tomato is a fruit, and like other fruits makes an astounding basis for Indian-style chutneys. If you don't mind a bit of skins in your chutney, skip the blanching and peeling part and just chop the tomatoes. This recipe calls for *asafetida*, a pungent spice that's a staple in Indian cooking; though optional, it lends its own brand of flavor to the condiment.

To make it: Score an *X* in the bottom of 8 ripe beefsteak or other slicing tomatoes, then blanch in a pot of boiling water for 30 seconds. Remove with a slotted spoon and rinse under cool running water. Peel off the skins, then cut the tomatoes in half, remove the seeds, and thinly slice.

In a saucepan, combine the tomatoes with 2 tablespoons water and kosher salt and freshly ground black pepper to taste. Cook over medium heat until the tomatoes are soft and the liquid has evaporated. Remove from the heat and mash until smooth.

Heat 2 tablespoons vegetable oil in a small skillet. Add ½ teaspoon each cumin seeds, yellow mustard seeds, and fennel seeds, and cook, stirring, until they begin to sputter. Add 1 tablespoon each ground cumin and cayenne pepper along with ¼ teaspoon asafetida powder (optional), and cook, stirring, for 1 minute. Remove from the heat, stir in the tomato mixture, and let cool before serving or storing. The chutney can be refrigerated in a tightly sealed jar for up to 3 weeks. Makes about 2 cups.

SAUSAGE, SAUCED

Grilled sausage is great on its own, but even better when paired with a delicious sauce. After all, you've gone to the trouble of making the sausage by hand, so any sauce will be a breeze in comparison.

Chimichurri

This bright, garlicky Argentinian herb sauce was created with one thing in mind: grilled meats, making it a slam-dunk choice for serving alongside grilled fresh beef, lamb, or pork sausages, either cut into slices for an easy appetizer or in *choripàn* — a sandwich of grilled chorizo slathered with chimichurri on a toasted baguette.

To make it: In a blender or food processor, combine 4 peeled garlic cloves, 2 cups packed fresh flat-leaf parsley, ½ cup packed fresh cilantro (leaves and tender stems), 2 tablespoons fresh oregano, ¼ teaspoon crushed red pepper, and kosher salt and freshly ground black pepper to taste. Blend or process until the herbs are finely chopped. Drizzle in ¼ cup red wine vinegar, then 1 cup extra-virgin olive oil, while blending or processing until smooth. Serve immediately, or, better yet, refrigerate in an airtight container for at least 2 days (and up to 1 week). Whisk until smooth again before serving. Makes about 1½ cups. See photo, page 249.

Romesco

Spain's contribution to the sauce arsenal is made with roasted red pepper, almonds, tomato, and garlic, sometimes with bread (as here), other times not. Sherry vinegar is essential.

To make it: Toast ½ cup blanched almonds and a thick slice of crusty bread, cut into cubes, on a baking sheet in a 375°F (190°C) oven until fragrant and the bread is starting to brown, about 10 minutes. Let cool.

Meanwhile, roast 2 red bell peppers over a gas burner, turning with tongs, until blackened and blistered. Place in a bowl, cover with plastic wrap, and let stand for 10 minutes. Then rub off the skins with paper towels, remove the ribs and seeds from the peppers, and chop the peppers.

In a food processor, pulse the bread, almonds, peppers, and 3 peeled garlic cloves until finely chopped, then add ¼ cup extra-virgin olive oil, 2 tablespoons sherry vinegar, and 1 tablespoon pimentón, and process until smooth. Season with kosher salt to taste. The romesco will keep, covered, in the refrigerator for up to 3 days. Serve at room temperature. Makes about 1½ cups. See photo, page 280.

Muhammara

Muhammara is a Middle Eastern condiment also made with roasted red bell pepper and nuts, in this case walnuts. Pomegranate molasses adds sweetness; look for it at gourmet grocers or online.

To make it: Toast ¾ cup walnuts on a baking sheet in a 375°F (190°C) oven until fragrant, about 20 minutes. Let cool; set aside 2 tablespoons for garnish.

In a food processor, combine the remaining walnuts with 1 chopped roasted red bell pepper (see Romesco, above), peeled and with seeds and ribs removed, ½ cup chopped scallions, 2 teaspoons fresh lemon juice, 1 teaspoon ground cumin, and 1 teaspoon Aleppo pepper. Pulse until finely chopped. Add 2 teaspoons pomegranate molasses and ¼ cup extra-virgin olive oil and purée until smooth. Add ¼ cup fresh bread crumbs and pulse to combine, adding more bread crumbs as needed to achieve the desired consistency. Season with kosher salt and more Aleppo pepper to taste.

The sauce can be refrigerated, covered, for up to 3 days. Serve at room temperature, drizzled with more olive oil and pomegranate molasses and topped with the reserved toasted walnuts. Makes about 2 cups.

Salsa Verde

Not to be confused with the Mexican sauce made with tomatillos, this classic Italian salsa verde is a zesty, zippy relish that capitalizes on the bright, fresh flavor of herbs and the briny taste of capers and anchovies. When possible, buy salt-packed capers and anchovy fillets, which after rinsing will be less salty than those packed in brine (or oil, for anchovies).

To make it: In a small bowl, soak ¼ cup finely chopped shallots in 2 tablespoons red or white wine vinegar for about 10 minutes. In a food processor, pulse 5 anchovy fillets, 3 tablespoons capers, and 1 teaspoon Dijon mustard together until combined. Add the shallot mixture, 3 cups chopped fresh flat-leaf parsley, ¼ cup chopped fresh tarragon, and 1 teaspoon grated lemon zest, and pulse to combine. Add 1 cup extra-virgin olive oil and process until smooth. Season with salt to taste. The salsa verde can be refrigerated, covered, for up to 1 day. Serve at room temperature. Makes about 2 cups.

Thai-Style Chile Sauce (Jaew)

Although not as well known in the United States as the ubiquitous sweet chili sauce served in Thai restaurants, *jaew* is equally popular in Thailand, where it is served up with grilled skewered pork and chicken. It is decidedly smokier, with a pronounced salty-sour flavor provided by fish sauce. You can find toasted rice powder and galangal at Asian food markets and many natural food stores.

To make it: In a bowl, whisk together ¼ cup thinly sliced shallots, ¼ cup finely chopped fresh cilantro, ⅓ cup Asian fish sauce, the zest and juice of 1 lime, 2 teaspoons grated palm sugar or brown sugar, 1 tablespoon toasted rice powder, 1 tablespoon crushed red pepper, and ½ teaspoon ground galangal. Taste and adjust with more fish sauce, lime juice, and/or sugar as desired. The sauce can be refrigerated, covered, for up to 3 days. Serve at room temperature. Makes about 1 cup.

POTATO AND SAUSAGE BALLS

Surprise: In these Swedish-inspired bites, the sausage is hidden inside — and is the reason to make them. They go great with chilled vodka or aquavit.

Makes about 4 dozen

- 1 pound Country-Style Breakfast Sausage (page 56), removed from casing
- 2–3 cups slightly stiff mashed potatoes
- 2 large eggs, lightly beaten
- 2 tablespoons water
- 1 cup dried bread crumbs
- 1 teaspoon freshly ground white pepper
- ½ teaspoon dried basil
- ½ teaspoon dried oregano
- Pinch of cayenne pepper
- 2 tablespoons grated Parmesan cheese
- 1 tablespoon chopped fresh flat-leaf parsley
- Vegetable oil, for frying

1. Shape the sausage meat into 1-inch balls. Enclose each ball in mashed potatoes to make a ball about 2 inches in diameter.

2. In a small shallow bowl, whisk together the eggs and water.

3. In a second shallow bowl, combine the bread crumbs, white pepper, basil, oregano, cayenne, Parmesan, and parsley.

4. Dip each sausage ball into the egg mixture, then gently roll it in the seasoned bread crumbs. Arrange the balls on a baking sheet. Refrigerate for 1 hour to allow the crumb coating to set.

5. Heat several inches of oil in a deep saucepan to 375°F (190°C). Fry the sausage balls, a few at a time and turning occasionally, until they are golden, about 3 minutes. Drain on paper towels. Serve hot.

SPICY SAUSAGE AND CHEESE BALLS

Betty Crocker (or was it Bisquick?) may well have been the first to popularize these tasty bites, but many a host has tweaked the original recipe over the generations since. That's an easy proposition, since you can switch up the flavor with different types of sausage, cheese, and seasonings, and it's hard to mess up.

Most recipes, like this one, call for breakfast sausage and cheddar, but you could also try Fresh Chorizo (page 67) paired with manchego or Hot Italian Sausage (page 64) with mozzarella and Parmesan. Double the recipe for a bigger batch; these little morsels go quickly.

Makes about 4 dozen

- 1¼ cups all-purpose flour
- 1½ teaspoons baking powder
- 1 teaspoon kosher salt
- ½ teaspoon freshly ground black pepper
- ½ teaspoon cayenne pepper
- 2 cups grated cheddar cheese (or substitute pepper Jack cheese and omit jalapeño)
- 1 pound Country-Style Breakfast Sausage (page 56), removed from casing
- ½ cup finely chopped onion
- 2 tablespoons finely chopped jalapeño chile (ribs and seeds removed for less heat)
- 2 teaspoons minced garlic
- 2–4 tablespoons unsalted butter, melted

1. Preheat the oven to 375°F (190°C). Line a baking sheet with parchment paper.

2. In a large bowl, whisk together the flour, baking powder, salt, black pepper, and cayenne to combine. Add the cheese, sausage, onion, jalapeño, and garlic, and toss to coat. Add just enough of the melted butter until the mixture comes together but is still quite crumbly.

3. Form the dough into 1-inch balls. Bake on the prepared baking sheet for 20 to 25 minutes, until golden and cooked through. Serve warm.

NOTE: *These are great make aheads: Freeze unbaked balls on the baking sheet for 1 hour, or until firm, then pop into freezer bags and freeze for up to 3 months. Bake straight from the freezer for several minutes longer than the recipe specifies.*

LIVERWURST PÂTÉ

If you love liverwurst and also pâté, you're doubly in luck: This is a flavorful (and less expensive) alternative to foie gras, and apparently even James Beard himself had a favored formula, which incorporated cream cheese. Here, the liver sausage is combined with more chicken livers; you could use Braunschweiger (page 108), another liver sausage, in place of liverwurst.

Make the pâté at least a day before serving to allow the flavors to fully develop. Serve as you would any pâté — with mustards (see page 293), cornichons, and crackers or crusty bread.

Serves 8 to 10

- 3 tablespoons unsalted butter
- ¼ pound chicken livers
- 2 garlic cloves, minced
- 2 scallions, minced
- 1 tablespoon chopped fresh basil
- ¼ pound Liverwurst (page 96), diced
- 1 tablespoon cognac or other brandy
  Kosher salt and freshly ground black pepper
  Vegetable oil, for greasing

1. Heat the butter in a large skillet over medium heat. Add the chicken livers and sauté until they are uniformly colored, about 15 minutes. Add the garlic, scallions, and basil, and sauté until the garlic and scallions are softened but not browned, 1 to 2 minutes. Stir the liverwurst into the liver mixture. Remove the skillet from the heat.

2. Process the mixture in a food processor until it is puréed, then add the cognac and pulse until blended. Season with salt and pepper to taste.

3. Lightly oil a small crock or loaf pan. Pack the mixture into it firmly. Cover and refrigerate for at least 4 hours or preferably 1 day. The pâté will keep for up to 1 week, covered well with plastic wrap, in the refrigerator. Let come to room temperature before serving.

BRAUNSCHWEIGER BITES

This is a quick-assembly party dish that can be prepared a day or two ahead and refrigerated until needed.

Makes about 2 dozen

- ½ pound Braunschweiger (page 108)
- 2 teaspoons grated onion
- ½ teaspoon finely minced garlic
  Dash of hot pepper sauce, such as Tabasco
  Kosher salt and freshly ground black pepper
- 2 tablespoons chopped fresh flat-leaf parsley

1. In a medium bowl, mash the braunschweiger with a fork and add the onion, garlic, hot sauce, and salt and pepper to taste. Mix well.

2. Form the mixture into 24 little meatballs and roll each one in the parsley to coat.

3. Insert a toothpick into each ball and refrigerate for at least 2 hours before serving.

SAUSAGE-STUFFED MUSHROOMS

Talk about convenient: These tasty morsels can be ready in just half an hour and take well to being made in advance: Once cooked, allow to cool completely, then refrigerate, covered, for up to 1 day and reheat just before serving. This recipe can easily be made vegetarian by substituting any vegetarian sausage for the Italian sausage. The Pimentón Lentil Patties (see page 258) and the Vegetarian Breakfast Sausage (see page 245) are both excellent choices. *See photo, opposite.*

Serves 6 to 8

- 2 tablespoons extra-virgin olive oil, plus more for greasing
- 18 large mushrooms (at least 1½ inches in diameter)
- 2 tablespoons unsalted butter
- ¼ pound Sweet Italian Sausage (page 64) or Country-Style Breakfast Sausage (page 56), removed from casing
- 2 tablespoons finely minced onions
- ¼ cup dried bread crumbs
- 1 garlic clove, very finely minced
- 1 tablespoon chopped fresh flat-leaf parsley
- ½ teaspoon dried oregano
- 2 tablespoons dry sherry
- Kosher salt and freshly ground black pepper
- 4 ounces mozzarella cheese, grated

1. Preheat the broiler. Oil a baking sheet.

2. Clean the mushrooms and remove the stems. Chop the stems finely and set aside.

3. Melt the butter in a large skillet over medium heat. Add the mushroom caps and gently sauté just until they are slightly golden but not noticeably shrunken, 2 to 3 minutes. Remove them with a slotted spoon and drain on paper towels.

4. Add the sausage and onions to the skillet and sauté, crumbling the sausage with the back of a spoon, until it is cooked through and lightly browned and the onions are crisp-tender, about 10 minutes. Stir in the oil and chopped mushroom stems, and sauté for 2 minutes.

5. Remove the skillet from the heat. Add the bread crumbs, garlic, parsley, oregano, sherry, and salt and pepper to taste; mix well. Stir in the cheese.

6. Divide the mixture evenly among the mushroom caps and arrange the caps on the prepared baking sheet, filling side up.

7. Broil for 1 to 2 minutes, until the cheese bubbles. Serve hot.

Sausage-Stuffed
Mushrooms

MEZZE UPDATE

For summer entertaining, nothing beats the ease — and robust flavors — of a Mediterranean mezze platter, with its vibrant flavors, colors, and textures. Lamb, Rosemary, and Pine Nut Sausage (page 121) makes an exceptional stand-in for traditional kibbe (lamb and pine nut patties) and served with the usual go-withs. You may also want to make a batch of marinated red peppers (see page 286) and pick up some olives (such as Kalamatas or Castelvetranos), a good-quality feta, and pita breads.

Pita Crisps

Cut pita bread into wedges, brush with extra-virgin olive oil, sprinkle with salt and Aleppo pepper or freshly ground black pepper, and toast in a 350°F (190°C) oven for about 10 minutes, until golden and slightly crisp.

Tzatziki

Strain 1½ cups plain Greek yogurt in a fine sieve over a bowl for 2 hours. Meanwhile, place 2 coarsely grated English cucumbers in a square of cheesecloth (or a clean kitchen towel) and squeeze to remove as much liquid as possible. In a bowl, stir together the yogurt and cucumber with 2 tablespoons extra-virgin olive oil, 2 minced garlic cloves, 1 tablespoon fresh lemon juice, and 1 tablespoon each finely chopped fresh dill and mint. Season with kosher salt and freshly ground black pepper to taste. The tzatziki can be refrigerated, covered, until ready to serve, up to a few hours. Stir and garnish with more dill or mint to serve. Makes about 3 cups.

Hummus with Sumac

In a food processor, pulse to combine 1½ cups drained cooked chickpeas (or one 15-ounce can, rinsed and drained), 3 peeled garlic cloves, and ⅓ cup fresh flat-leaf parsley (leaves and tender stems) until finely chopped. Add ¼ cup tahini, 1 tablespoon ground sumac, the zest and juice of 1 lemon, and ¼ cup extra-virgin olive oil, and process to the desired consistency. Season with kosher salt and more sumac to taste. The hummus can be refrigerated, covered, until ready to serve, up to 3 days. Stir and garnish with more sumac and/or parsley to serve. Makes about 3 cups.

Baba Ghanoush

Preheat the oven to 400°F (200°C). Roast 2 eggplants (about 3 pounds) and 3 garlic cloves (unpeeled) on a baking sheet for about 30 minutes, or until very tender. Remove from the oven. Place the eggplant in a bowl, cover, and let stand for about 10 minutes. Scoop out the flesh into a food processor. Once the garlic is cool enough to handle, squeeze out the cloves into the processor bowl. Pulse until combined. Add ¼ cup fresh lemon juice, ⅓ cup tahini, 2 tablespoons plain Greek yogurt, and 2 tablespoons chopped fresh flat-leaf parsley, and purée to combine. Season with kosher salt and cayenne pepper, plus more lemon juice, to taste. The baba ghanoush can be refrigerated, covered, until ready to serve, up to 3 days. Stir and garnish with more parsley and extra-virgin olive oil to serve. Makes about 3 cups.

Tabbouleh

In a small bowl, combine ¼ cup fine bulgur wheat with 1 cup boiling water. Let soak for about 20 minutes, then drain in a sieve, pressing on the bulgur to extract as much liquid as possible. In a large bowl, combine the bulgur with 2 minced garlic cloves, the zest and juice of 2 lemons, 2 cups finely chopped fresh flat-leaf parsley, ½ cup chopped fresh mint, 8 ounces finely chopped tomatoes, and 1 bunch scallions, trimmed and very finely chopped. Toss to combine, then season with kosher salt to taste. Cover and refrigerate for at least 2 hours (or up to overnight). To serve, add 3 tablespoons extra-virgin olive oil and toss to combine. Adjust the seasoning with more salt and/or lemon juice to taste. Makes about 3 cups.

Panzanella with Salami, page 306

SAUSAGE FOR LUNCH OR DINNER

There is no love sincerer than the love of food.

— George Bernard Shaw

One of the marvelous things about sausage is that it is so eminently adaptable. There is hardly a vegetable or grain that doesn't go well with one type of sausage or another. Part of sausage's versatility stems from the fact that it can be prepared in so many different ways. There isn't a single cuisine in the civilized world that doesn't include as part of its repertoire at least some kind of sausage.

For every idea about how to use sausage, there is an infinite number of variations around that theme. One needn't be a culinary genius to come up with one's own variations.

In this chapter, we celebrate sausage as the main ingredient in a meal. The recipes range from the most simple to quite elaborate, and also span the seasons and the globe. But they all have two things in common: They are complete one-dish meals, and they are undeniably delicious. Just as you are encouraged to adapt the sausage recipes in the earlier chapters, so too should you make these recipes your own. Try them our way first, and then experiment to suit your own taste. Consider them our gift to you, as a fellow *wurstmacher*.

PANZANELLA WITH SALAMI

How to make any salad more satisfying? Add bread for more heft. This is also a great way to use up any remnants of bread from the day before. Panzanella is the Italian expression of this idea, and its appeal lies in the way the bread soaks up the oil-and-vinegar dressing and vegetable juices without becoming too sodden. Using a mix of fresh herbs, such as parsley, oregano, basil, summer savory, and thyme, is also essential to achieving the characteristic vibrancy. Salami ups the ante even more, making the salad the centerpiece of a summery meal. *See photo on page 304.*

Serves 4

- 2¼ cups extra-virgin olive oil
- ¾ cup red wine vinegar
- 1 small garlic clove, minced
  Kosher salt and freshly ground black pepper
- 2 large tomatoes, cored
- 2 medium cucumbers
- 2 medium zucchini or summer squash
- ½ pound Salamette (page 109)
- 1 cup mixed fresh herbs
- ½ cup Kalamata olives, pitted and chopped
- 6 thick slices rustic bread (about ¾ pound)

1. Make the dressing by whisking 2 cups of the oil, the vinegar, garlic, and salt to taste together in a medium bowl. Set aside.

2. Chop the tomatoes, cucumbers, and zucchini into bite-size pieces. Place in a large salad bowl.

3. Cut the salami into thin slices and then into thin strips. Place in the bowl and add the herbs and olives.

4. Grill or toast the bread until charred and crisp. Drizzle with the remaining ¼ cup oil, and cut or tear into bite-size pieces. Add to the salad bowl.

5. Add a generous amount of dressing, and toss to combine. The bread shouldn't be soggy and the vegetables should be well coated with dressing. Add more dressing if needed and season with salt and pepper.

VEGETARIAN SAUSAGE AND KALE HASH

Hearty enough for a satisfying supper, this meat-free hash would also be lovely as a brunch dish. Either way, you may want to top it with eggs: Once you've added the kale and tossed to wilt it, crack an egg into a cup, slide it onto the hash, and repeat with three more eggs. Cover and cook over moderate heat just until the whites are set but the yolks are still runny. Choose among Country-Style Breakfast Sausage (page 56), Portobello Sausage (page 252), or Italian Bulgur and Bean Sausage Rolls (page 250). *See photo, opposite.*

Serves 4

- 4 tablespoons extra-virgin olive oil
- 1 pound vegetarian sausage
- 1 onion, chopped
- 2 garlic cloves, minced
- ½ teaspoon paprika
- ¼ teaspoon ground cumin
  Kosher salt and freshly ground black pepper
- 1 large sweet potato (about ¾ pound), peeled and cut into ¼-inch dice
- 1 large russet potato (about ½ pound), peeled and cut into ¼-inch dice
- 1 cup vegetable broth
- 1 bunch kale, stemmed and cut into bite-size pieces
  Lime wedges, for serving
  Hot pepper sauce, such as Tabasco, for serving

recipe continued on page 308

Vegetarian Sausage with
Kale Hash

1. In a large skillet, heat 2 tablespoons of the oil over medium-high heat. Cook the sausage until browned and crisp. Remove from the skillet and cut into bite-size pieces.

2. Add the remaining 2 tablespoons oil to the pan and cook the onion and garlic, stirring occasionally, until tender, about 8 minutes. Add the paprika and cumin, season with salt and pepper, and cook until fragrant, about 1 minute.

3. Add the sweet potato and russet potato to the skillet. Stir to coat with the spices and onion. Add ½ cup of the vegetable broth, reduce the heat, and simmer. When the broth has boiled almost completely away, add the remaining ½ cup broth and repeat. If the potatoes are not fork-tender, add a little water and continue cooking.

4. Raise the heat to medium-high and return the sausage to the pan. Add the kale, tossing gently to incorporate and wilt it.

5. Serve immediately with lime wedges and hot sauce.

SAUSAGE, RICOTTA, AND FENNEL PIZZA

Skip the tomato sauce to make a white pizza that is every bit as authentic. Ricotta cheese is a creamy counterpart to spicy Italian sausage; use pork (page 64), turkey (page 224), or wild boar (page 184). Sliced fennel marries well with the sausage's seasonings. More fennel seeds are sprinkled over the top before baking. It's likely to become your new favorite pie.

Serves 6 to 8

1 tablespoon extra-virgin olive oil, plus more for brushing
½ pound hot Italian-style sausage, removed from casing
1 small fennel bulb, trimmed and thinly sliced, fronds reserved for garnish
All-purpose flour, for dusting
Basic Pizza Dough (see opposite page)
1½ cups ricotta cheese
1 garlic clove, minced
1 tablespoon fennel seeds
Crushed red pepper
Kosher salt and freshly ground black pepper

1. Heat the oil in a skillet over medium-high heat. Add the sausage and sauté, crumbling it with the back of a spoon, until it is cooked through and browned, about 10 minutes. Remove the sausage with a slotted spoon and place in a large bowl.

2. Add the fennel to the drippings in the pan and cook just until tender, about 3 minutes. Transfer to the bowl with the sausage.

3. Preheat the oven to 425°F (220°C).

4. On a floured surface, roll out the dough to form a rough 14-inch oval. Transfer the dough to a baking sheet or perforated pizza pan. Brush with oil. Sprinkle the sausage and fennel evenly over the dough, leaving a 1-inch border.

5. Mix together the ricotta and garlic, then spoon dollops over the sausage. Sprinkle with the fennel seeds, crushed red pepper, and salt and black pepper to taste.

6. Bake for about 15 minutes, or until the crust is golden and the cheese is bubbling. Serve hot.

GREEK PIZZA

Lamb, olives, and feta cheese give a decidedly Greek accent to this pizza. The Lamb, Rosemary, and Pine Nut Sausage (page 121) is a good choice here, as is Loukanika (page 152).

Serves 6 to 8

- 2 tablespoons extra-virgin olive oil
- ½ pound fresh lamb sausage, removed from casing
- 1 small onion, chopped
- 2 garlic cloves, minced
- ½ cup chopped ripe olives (preferably brine-cured)
- All-purpose flour, for dusting
- Basic Pizza Dough (see right)
- 8 ounces mozzarella cheese, shredded
- 1 cup crumbled feta cheese
- 1 tablespoon chopped fresh basil
- 1 tablespoon chopped fresh oregano
- Kosher salt and freshly ground black pepper

1. Heat the oil in a large skillet over medium-high heat. Add the sausage and sauté, crumbling it with the back of a spoon as it cooks, until it is cooked through and browned, about 10 minutes. Remove with a slotted spoon and set aside in a large bowl.

2. Drain off all but 1 tablespoon of the drippings, add the onion and garlic to the pan, and sauté until they are crisp-tender, 2 to 3 minutes. Remove the vegetables with a slotted spoon and add to the sausage. Discard the drippings.

3. Preheat the oven to 425°F (220°C).

4. Add the olives to the sausage mixture. Mix well.

5. On a floured surface, roll out the dough to form a rough 14-inch oval. Transfer the dough to a baking sheet or perforated pizza pan. Sprinkle a thin layer of mozzarella on the dough. Distribute the sausage mixture evenly over the cheese. Dot with the crumbled feta. Sprinkle with the basil, oregano, and salt and pepper to taste. End with a layer of mozzarella.

6. Bake for about 25 minutes, or until the top is golden and bubbling. Serve hot.

BASIC PIZZA DOUGH

To save time, you may want to double this recipe each time you prepare it, then freeze half of the dough for another night. It will keep for up to 3 months, wrapped well in plastic wrap; thaw overnight in the refrigerator before using.

Makes 1 pizza crust

- 1 package (¼ ounce) active dry yeast
- ¾ cup plus ½ cup warm water (105 to 115°F; 41° to 46°C)
- 1 teaspoon kosher salt
- 3 cups all-purpose flour, plus more for dusting
- 1 tablespoon extra-virgin olive oil, plus more for greasing

1. Mix the packet of yeast into ¾ cup of the warm water. Let stand for 15 minutes, or until bubbles form.

2. In a large bowl, stir the salt into the flour. Gradually add the yeast mixture and the oil. Turn the dough out onto a floured surface. Gradually add the remaining ½ cup warm water while kneading the dough, and continue to knead for about 10 minutes. The dough should be smooth and elastic.

3. Transfer the dough to a lightly greased bowl, turning to coat. Cover the bowl and let rise in a warm place until the dough doubles in bulk, about 1 hour.

Black Bean Sausage and
Sweet Potato Tacos

BLACK BEAN SAUSAGE AND SWEET POTATO TACOS

These vegetarian tacos are among the best reasons to make the Black Bean and Smoked Corn Sausage on page 246. Sautéed sweet potato adds even more substance, while the "lime-pickled" cabbage is made quickly by tossing shredded cabbage (and sliced jalapeños, for a kick of heat) with lime juice and cilantro, and then rubbing the mixture to tenderize the vegetables. Try it on all of your favorite Mexican-style dishes. *See photo, opposite.*

Serves 4

2 cups shredded green cabbage
2–3 jalapeño chiles, sliced into thin rings (ribs and seeds removed for less heat)
¼ cup fresh lime juice
2 tablespoons minced fresh cilantro, plus whole leaves for serving
1 tablespoon kosher salt
3 tablespoons vegetable oil
¼ pound Black Bean and Smoked Corn Sausage (page 246)
1 medium sweet potato, peeled and cut into ¼-inch dice
2 garlic cloves
1 tablespoon chili powder
1 teaspoon ground cumin
¼ teaspoon cayenne pepper
8 corn tortillas (6-inch size)
   Sour cream, for serving
   Lime wedges, for serving
   Hot pepper sauce, such as Tabasco, for serving

1. In a medium bowl, combine the cabbage, chiles, lime juice, minced cilantro, and 1 teaspoon of the salt. Using your hands, massage the ingredients together to wilt the cabbage and soften the chiles. Set aside.

2. Heat 1 tablespoon of the oil in large skillet, add the sausage, and cook until browned and cooked through. Remove from the pan and cut into bite-size pieces.

3. Add the remaining 2 tablespoons oil to the pan and heat over medium heat. Add the sweet potatoes and cook, stirring occasionally, until browned and tender, about 10 minutes.

4. Add the garlic, chili powder, cumin, cayenne, and remaining 2 teaspoons salt to the pan and cook until fragrant. Add 1 cup water and the sausage, bring to a simmer, and cook until the water has nearly evaporated. Remove from the heat.

5. Layer each tortilla with the sweet potato–sausage mixture and then with the pickled cabbage. Top with sour cream and cilantro leaves and serve with lime wedges and hot sauce.

SAUSAGE-STUFFED ARTICHOKES

An artichoke is eminently stuffable, often with prosciutto or pancetta in the filling. Here, we swapped out those cured meats for fresh sausage. Good ones to try: Hot Italian Sausage (page 64), Luganega (page 66), or Italian-Style Wild Boar Sausage (page 184).

Serves 4

- 2 lemons, 1 halved and 1 quartered
- 4 large artichokes
- 6 garlic cloves; 4 crushed and 2 finely minced
- 4 bay leaves
- ¼ cup plus 1 teaspoon extra-virgin olive oil
- 1 pound fresh sausage, removed from casing
- ¼ cup minced onion
- ½ cup dried bread crumbs
- ¼ cup grated Parmesan cheese
- 1 teaspoon dried thyme
  Pinch of cayenne pepper
- 1 tablespoon chopped capers
- 1 large egg, well beaten
- ½ cup dry white wine
  Kosher salt and freshly ground black pepper
- ½ cup fresh lemon juice

1. Bring a medium pot of water to a boil. Squeeze the lemon halves into a large bowl of water.

2. To prepare the artichokes, cut off the stems, leaving a flat base on each. Reserve the stems. Remove any bruised outer leaves. Cut about 1 inch off the top of each artichoke with a sharp knife. With a pair of kitchen scissors, snip off the tip of each outer leaf. Place each trimmed artichoke in the bowl of lemon water to keep it from discoloring while you prepare the rest.

3. Add the artichokes, reserved artichoke stems, lemon quarters, crushed garlic, and bay leaves to the pot and boil, covered, until the leaves are tender at their thickest part, about 30 minutes. Drain and let cool, then finely chop the stems.

4. While the artichokes are cooking, heat 1 teaspoon of the oil in a large skillet over medium-high heat. Add the sausage and sauté, crumbling it with the back of a spoon, until it is cooked through and browned, about 10 minutes. Remove with a slotted spoon and set aside.

5. In the sausage drippings remaining in the skillet, sauté the onion until it is translucent, about 5 minutes.

6. In a medium bowl, combine the onion, sausage, bread crumbs, Parmesan, thyme, cayenne, capers, egg, and wine. Mix well. Season with salt and black pepper. Stir in the artichoke stems.

7. Preheat the oven to 350°F (175°C).

8. Pull back the leaves from the artichokes and remove the inner choke. Stuff the sausage mixture into the artichokes. If you have enough stuffing, also place some between the large leaves at the base. Arrange the artichokes in a large baking pan just large enough to hold them.

9. In a small bowl, whisk together the lemon juice and remaining ¼ cup oil; stir in the minced garlic. Pour the lemon juice mixture over the artichokes.

10. Cover and bake for about 20 minutes, or until the base leaves are very tender. Serve hot with the pan juices for dipping.

SAUSAGE-FILLED PIROSHKI

If you happen to have a Russian *babushka*, you're no doubt already familiar with piroshki (literally "little pie") made with a soft, eggy yeast dough and stuffed with savory or sweet fillings. If not, you may want to give the delectable dumplings a shot.

Ground beef, with or without cabbage, is among the most traditional fillings; here, we replace that with fresh sausage meat. You could use most any fresh sausage in the filling, including Luganega (page 66), Pork and Apple Sausage (page 71), Weisswurst (page 145), and Fresh Kielbasa (page 144). In Russia, piroshki are often fried or baked, but we opted to boil them instead for a dish that's more akin to dumplings — just right for tossing with melted butter to serve.

Serves 4 to 6

DOUGH
    1 package (¼ ounce) active dry yeast
    ½ cup warm water (105 to 115ºF; 41 to 46ºC)
    2½ cups sifted all-purpose flour, plus more
        for dusting
    1 large egg, well beaten
    2 tablespoons vegetable oil, plus more for
        greasing

FILLING
    1 tablespoon olive oil or vegetable oil
    1 pound fresh sausage, removed from casing
    1 small onion, chopped
    1 large egg, well beaten

TO SERVE
    Melted unsalted butter
    Chopped fresh flat-leaf parsley
    Kosher salt and freshly ground black pepper

Make the dough

1. Mix the yeast with the warm water in a bowl and let it rest for 15 minutes, or until it becomes foamy.

2. In a large bowl, mix the flour, egg, and oil together. Stir the yeast mixture into the flour mixture to make a soft dough.

3. Place the dough in a greased bowl, cover, and let rise until doubled in bulk, about 1 hour.

Make the filling

1. While the dough is rising, heat the oil in a large skillet over medium-high heat. Add the sausage and sauté, crumbling it with the back of a spoon, until it is cooked through and browned, about 10 minutes. Remove the sausage with a slotted spoon and set aside.

2. In the drippings remaining in the skillet, sauté the onion until tender, about 5 minutes. Remove with a slotted spoon and add to the sausage.

3. Add the egg to the sausage mixture; blend thoroughly.

Prepare the piroshki

1. Bring a large pot of salted water to a boil.

2. Punch down the dough and roll it out on a floured surface until it is about ⅛ inch thick. With a biscuit cutter or an inverted glass, cut the dough into circles about 3 inches in diameter.

3. Place about 1 tablespoon of the sausage mixture on each circle of dough; fold over to form half-moons. Pinch the edges tightly to seal.

4. Slip the piroshki, a few at a time, into the boiling water and cook until the piroshki float, 3 to 4 minutes. Transfer with a slotted spoon to a large bowl and cover with foil to keep warm until all piroshki are cooked.

5. Toss the piroshki with butter, parsley, and salt and pepper to taste. Serve warm.

NOTE: *To bake the piroshki, arrange them snugly in a baking dish, brush with melted butter, and bake in a 400°F (300°C) oven for about 25 minutes, until golden. Serve warm.*

KOREAN LETTUCE WRAPS (BO SSAM)

If you're looking for a fun, easy, delicious, and easily doubled dish for a casual dinner party, here's your answer: Set out all the components and let your guests assemble their own wraps, exactly to their liking. Fresh pork sausage meat is added to the filling as a shortcut for the traditional pork belly. If you can find them, fresh oysters are a traditional addition to the spread, or you can use the Oyster Sausage on page 235 in the same spirit. *See photo, opposite.*

Serves 4

- 1 cup shredded carrots
- ¼ cup shredded daikon radish
- 1 teaspoon kosher salt
- 1 cup rice vinegar
- 2 tablespoons gochujang (spicy Korean chili paste)
- 1 garlic clove, minced
- 1 pound fresh sausages
- 1 head Boston or Bibb lettuce
- 4 cups cooked white rice
- 2 cups Kimchi (page 289)
- 1 cup cilantro sprigs
- ½ cup sliced scallions

1. Preheat the oven to 400°F (200°C).

2. In a medium bowl, mix the carrots, daikon, salt, and ½ cup of the vinegar together. Set aside.

3. In a small bowl, whisk the rest of the vinegar with the gochujang and garlic.

4. Rub the sausages with ¼ cup of the gochujang mixture. Set the rest aside.

5. Place the sausages on a baking sheet and roast for about 10 minutes, until browned and cooked through. Cut the sausages diagonally into ovals.

6. Separate the lettuce into individual leaves. Prepare a serving platter with a mound of rice in the center, and the kimchi, pickled carrots and daikon, and sausages arranged around it. Sprinkle with the cilantro and scallions and serve with lettuce leaves for wrappers.

NOTE: *For a pan-Asian variation, you could also use any of the following sausages with complementary flavor profiles: Curry Sausage with Cilantro and Lime (page 82), Vietnamese Pork and Lemongrass Sausage (page 70), Sai Krok Isan (page 83), or Thai Chicken Sausage (page 219).*

CURRIED SAUSAGE STRUDEL

Here's a melting-pot dish that marries flaky phyllo dough with a savory, wonderfully aromatic filling of sausage, leek, apple, and mushrooms that's simmered in a coconut curry sauce. For best results, use a mild-mannered sausage such as Country Chicken Sausage (page 206) or one with matching flavors, including Curry Sausage with Cilantro and Lime (page 82). This recipe calls for ready-made phyllo dough from your grocer's freezer section, but by all means feel free to prepare strudel dough from scratch if you are so inclined.

Serves 4 to 6

- 1 tablespoon vegetable oil, plus more for greasing
- ¾ pound fresh sausage, removed from casing
- ¼ pound cremini mushrooms, finely chopped
- 1 small leek (about 6 ounces), trimmed, sliced crosswise ⅛ inch thick, washed well and drained
- 1 small green apple, cored and diced
- 2 garlic cloves, finely minced
- 1 tablespoon minced fresh ginger
- 1 tablespoon curry powder
- ¾ teaspoon ground cumin

recipe continued on page 316

Korean Lettuce Wraps

½ cup coconut milk

¼ cup finely minced fresh flat-leaf parsley

2 tablespoons minced fresh cilantro

4 sheets (12 by 16 inches each) phyllo dough,
    thawed according to package directions

2 tablespoons unsalted butter, melted

¼ cup dried bread crumbs

1. In a large skillet, heat the oil over medium-high heat. Add the sausage and sauté, crumbling it with the back of a spoon, until it is cooked through and browned, about 10 minutes. Remove the sausage with a slotted spoon and set aside in a large bowl.

2. In the same skillet, sauté the mushrooms, leek, apple, and garlic in the drippings until soft, about 10 minutes. Add the ginger, curry powder, cumin, and coconut milk, and simmer until the coconut milk has been absorbed. Transfer the mixture to the bowl with the sausage. Add the parsley and cilantro, and mix well.

3. Preheat the oven to 400°F (200°C). Lightly grease a baking sheet.

4. Lay one sheet of phyllo on a work surface. Cover the remaining phyllo sheets so they won't dry out. Brush the phyllo sheet with some of the melted butter, then sprinkle with some of the bread crumbs. Add another sheet on top of the first, brush with butter, and sprinkle with crumbs. Continue in the same manner with the third and fourth sheets.

5. Spread the sausage mixture evenly over the top sheet of phyllo, leaving a 1½-inch border all around. Tuck in the sides. Carefully roll up the phyllo, jelly-roll fashion. Place the roll seam side down on the prepared baking sheet.

6. Bake for about 15 minutes, or until the strudel is evenly browned. Let cool slightly before slicing and serving.

RISOTTO WITH LUGANEGA AND RED WINE

Sausage, mascarpone, and Parmesan combine to make this an ultra-luxe version of eternally beloved risotto. Red wine tints the dish a lovely shade of pink, making it especially festive; radicchio adds welcome bitterness to cut through the richness (plus even more rosy color). *See photo, opposite.*

Serves 8 to 10

2 tablespoons unsalted butter

2 tablespoons extra-virgin olive oil

2 shallots, minced
    Kosher salt and freshly ground black pepper

2 cups Arborio or carronoli rice

½ pound Luganega (page 66) or Sweet Italian Sausage (page 64), removed from casing

3 cups dry red wine

4 cups chicken broth, warmed

2 cups finely shredded radicchio

½ cup mascarpone cheese

½ cup grated Parmesan cheese, preferably Parmigiano-Reggiano

1. Heat the butter and oil in a large skillet over medium heat. Add the shallots and season with salt; cook, stirring, until softened.

2. Add the rice to the skillet and sauté, coating the grains with oil and "toasting" them: They will become translucent and the starchy core will become more prominent.

3. Add the sausage to the pan, crumbling it into the rice. When it has started to brown, deglaze the pan with 1 cup of the wine, stirring until the wine is almost evaporated. Season with salt.

4. Add a ladleful of broth and keep stirring, slowly. Reduce the heat to medium-low so the rice

recipe continued on page 319

Risotto with Luganega
and Red Wine

Sausage and Beer
Hand Pies

mixture is at a gentle simmer. Alternate adding chicken broth and wine, continuing to stir the rice. The rice will start to release its starch and the risotto will become creamy.

5. Once the rice is just tender, remove from the heat and stir in the radicchio, mascarpone, and Parmesan. Season with pepper and more salt, as needed. Serve immediately.

SAUSAGE AND BEER HAND PIES

This recipe was inspired by one that ran in *The American Craft Beer Cookbook* by John Holl. It was contributed to that book by Fullsteam Brewery's spin-off bakery, Bullytown, in Durham, North Carolina. They make it with porter (or smoked porter), but you can use brown ale or other dark beers, such as stout. The original recipe called for ground pork in addition to pork sausage, but here we just double the sausage. Country-Style Breakfast Sausage (page 56) and Pork and Apple Sausage (page 71) are your best bets. *See photo, opposite.*

Makes 10 hand pies

DOUGH

- 2 cups all-purpose flour, plus more for dusting
- ½ cup fine cornmeal
- 1 teaspoon kosher salt
- 1 cup cold unsalted butter, cut into small pieces
- ¼ cup ice water, plus more as needed

FILLING

- 1 tablespoon extra-virgin olive oil
- 1 small onion, finely chopped
- 2 garlic cloves, minced
- 1 pound fresh pork sausage, removed from casing
- 3 Granny Smith apples, peeled, cored, and cut into ½-inch dice
- 2 medium sweet potatoes, peeled and cut into ½-inch dice
- ½ cup dark beer, such as porter or stout (preferably smoked)
- 1 tablespoon finely chopped mixed fresh herbs, such as thyme, sage, and summer savory
  Kosher salt and freshly ground black pepper
- 1 large egg, lightly beaten

Make the dough

1. Pulse the flour, cornmeal, and salt in a food processor to combine, then add the butter and pulse until the mixture resembles coarse meal with pea-size pieces of butter. Add ¼ cup ice water and pulse until the dough just comes together, adding more ice water, 1 tablespoon at a time, as needed. Do not overprocess.

2. Turn the dough out onto a piece of plastic wrap, then divide in half and shape each into a disk. Wrap separately in plastic wrap and refrigerate until firm, at least 1 hour or up to 2 days. (The dough can be frozen in a resealable plastic bag for up to 3 months; thaw overnight in the refrigerator before proceeding.)

Make the filling

1. Heat the oil in a large skillet over medium-high heat. Add the onion and cook, stirring occasionally, until soft, 3 to 4 minutes. Add the garlic and cook for 1 minute longer.

2. Add the pork sausage to the skillet and cook, breaking the meat up with the back of a spoon, until browned, 3 to 5 minutes.

3. Add the apples, sweet potatoes, beer, and herbs to the skillet. Cover and cook until the sweet potatoes and apples are tender, 15 to 20 minutes.

recipe continued on next page

Season with salt and pepper to taste. Transfer to a bowl and let cool.

Prepare the pies

1. On a lightly floured surface, roll out one half of the dough to ⅛ inch thick. Use a biscuit cutter to cut out ten 4-inch rounds, gathering together scraps and rerolling once, as needed. Arrange the rounds on a parchment-lined baking sheet. Cover and place in the refrigerator to chill.

2. Repeat the process with the remaining dough, this time cutting out ten 4¼-inch rounds (you can use a plate or inverted bowl and a paring knife if you don't have a cutter of that dimension). Chill all dough for at least 1 hour or up to 1 day, covered.

3. Preheat the oven to 450°F (230°C).

4. Divide the filling among the smaller rounds of dough (about ¼ cup each, or ⅛ cup for half-moon shapes), leaving a ½-inch border. Brush the edges with the beaten egg and press a larger round of dough over the top of each pie (or fold in half), pressing firmly around the edges to seal.

5. Crimp the edges of the dough with the tines of a fork and cut three small air vents in the top of each pie. Brush the tops with the beaten egg. Bake for 20 to 25 minutes, until golden brown and the filling is bubbling. Let cool slightly before serving.

SAUSAGE AND POTATO HOBO PACKS

Wrapped in aluminum foil and roasted over hot coals, these parcels are just right for serving at a campfire dinner — or a backyard barbecue. They're made with a winning combination of potatoes (two types: little red ones and also sweet potatoes) and sausage. Use any fresh sausage you happen to have already made: breakfast sausage, Italian sausage, chorizo, merguez . . . they all fit the bill terrifically.

Serves 4

- 1 pound small red-skinned potatoes, sliced ¼ inch thick
- ½ pound sweet potatoes, sliced ¼ inch thick
- 2 zucchini, sliced ¼ inch thick
- 1 onion, thinly sliced
- 2 garlic cloves, minced
- 1 teaspoon celery salt
- ½ teaspoon freshly ground black pepper
- ½ teaspoon paprika
- 1 pound fresh sausage, removed from casing
- 2 tablespoons unsalted butter, cut into small pieces
- 2 tablespoons chopped fresh flat-leaf parsley

1. Prepare a charcoal grill (or campfire) and let the fire burn down to coals covered with a layer of gray ash.

2. Combine both kinds of potatoes and the zucchini in a large bowl. Add the onion, garlic, celery salt, pepper, and paprika. Toss to coat everything evenly.

3. Cut eight 16-inch lengths of aluminum foil. Distribute the vegetables, then the sausage, evenly among half of the sheets. Dot each portion with butter and sprinkle with the parsley.

4. Crimp the foil around the ingredients to create tightly sealed packets. Wrap each packet in a second layer of foil.

5. Lay the packets directly on the coals, mounding them up and around the sides of the packets. Cook for 30 minutes. Remove one packet and check (opening carefully) to see if the vegetables are tender. If not, rewrap and cook for another 5 to 10 minutes.

NOTE: *If you don't have access to a grill, the packets can be cooked in a 400°F (200°C) oven for about 30 minutes.*

CHEESE RAVIOLI WITH PORTOBELLO SAUSAGE SAUCE

Ravioli is one of the easiest forms of pasta to make because you roll out the dough in large sheets rather than cutting or shaping it into trickier styles. If you have a rolling pin, you are in business. Be sure to use real semolina flour for the pasta. Here, a vegetarian sausage made from portobello mushrooms flavors the sauce. The ricotta needs to drain for at least 8 hours, so plan accordingly.

Serves 8 to 10

DOUGH

2½–3 cups semolina flour
1 teaspoon kosher salt
2 large eggs, well beaten
About ¼ cup water

FILLING

1 pound ricotta cheese, drained for at least 8 hours in a cheesecloth-lined colander
¼ cup grated pecorino Romano cheese
1 tablespoon minced fresh basil
1 tablespoon minced fresh flat-leaf parsley
1 tablespoon minced fresh chives
½ teaspoon kosher salt
½ teaspoon freshly ground black pepper
¼ teaspoon freshly grated nutmeg
1 garlic clove, minced
2 large eggs, well beaten

SAUCE

1 tablespoon olive oil or vegetable oil
1 pound Portobello Sausage (page 252)
4 cups Marinara Sauce (page 251)

ASSEMBLY

½ cup grated Parmesan cheese
½ cup grated pecorino Romano cheese
2 tablespoons chopped fresh flat-leaf parsley

Make the dough

1. Mix the semolina and salt, then mound 2½ cups on a pastry board or work surface and make a well in the center. Pour the eggs into the well and gradually mix them into the flour using a fork. Add the water, a few tablespoons at a time, until you have a dough that can be shaped into a ball. Knead it for 5 to 7 minutes, adding up to ½ cup more semolina a little at a time, if necessary, to prevent sticking. The dough should be firm but pliable.

2. Place the dough in a bowl, cover it with a kitchen towel or plastic wrap, and keep at room temperature until you are ready to roll it out.

Make the filling

1. In a large bowl, combine the ricotta, Romano, basil, parsley, chives, salt, pepper, nutmeg, garlic, and eggs. Mix well, using your hands. Refrigerate until you are ready to make the ravioli.

Make the sauce

1. Heat the oil in a large skillet over medium heat. Add the sausage and cook until it is cooked through and browned, about 10 minutes. Slice the sausage into bite-size pieces.

2. Add the marinara sauce to the pan, and gently simmer for about 10 minutes.

Prepare the ravioli

1. Roll out a piece of dough about the size of a lemon on a floured board to a rectangle that is between ⅛ and 1/16 inch thick. Lay the dough across the serrated plate of a ravioli maker and roll to form

recipe continued on next page

shapes. (If you don't have a ravioli maker, mark a 1½-inch square grid with the dull side of a chef's knife or with a bench scraper.) Spoon 1 teaspoon filling on top of each square.

2. Roll out a second sheet of dough to the same shape and thickness and lay it over the filling. Press down around the sides of each mound of filling. With a serrated pastry cutter, cut the squares apart. Set them on a lightly floured surface.

3. Repeat with the remaining dough to make about 6 dozen ravioli.

4. Bring a large pot of salted water to a boil. Preheat the oven to 400°F (200°C).

5. Cook the ravioli in batches in the rapidly boiling water until the ravioli float to the surface, 4 to 5 minutes. Gently remove with a slotted spoon and let drain.

6. In a large baking dish or individual ramekins, layer the sauce, ravioli, and Parmesan and Romano cheeses. Bake for about 15 minutes, or until the sauce is bubbly. Sprinkle with the chopped parsley and serve.

MIXED SEAFOOD SAUSAGE WITH FETTUCCINE

Linguine or fettuccine is the traditional pasta shape for pairing with seafood sauces. Use either in this recipe, but the fettuccine, because it is slightly wider, cradles the sauce in the most delicious manner.

Serves 6 to 8

    Kosher salt and freshly ground black pepper
2 tablespoons extra-virgin olive oil, plus more for tossing

2 garlic cloves, crushed
¼ cup finely chopped shallots
¼ cup finely minced fresh flat-leaf parsley
2 pounds Mixed Seafood Sausage with Shrimp, Crabmeat, and Scallops (page 237)
1 cup dry white wine
3 cups Marinara Sauce (page 251)
1 pound fettuccine

1. Bring a large pot of water to a boil and add a generous amount of salt.

2. Meanwhile, in a large skillet, heat the oil over medium-high heat. Add the garlic and sauté until lightly browned, about 2 minutes. Discard the garlic.

3. Sauté the shallots in the same pan until they are softened, about 5 minutes. Add the parsley and sausage and sauté, turning the sausages until they are lightly browned all over, about 10 minutes.

4. Stir in the wine and marinara sauce, then season with salt and pepper to taste. Simmer for 10 minutes.

5. Meanwhile, cook the fettuccine until al dente in the boiling water. Drain, toss with oil, and arrange on a platter.

6. Pour the sauce over the fettuccine and arrange the sausages on top. Serve immediately.

CHICKEN AND SAUSAGE CACCIATORE

Cacciatore translates as something prepared "in the style of the hunter," and there are as many cacciatore recipes as there are hunters. This hearty dish is rich in flavor, easily assembled, and good for family meals and potluck dinners.

Serves 6 to 8

¼ cup olive oil or vegetable oil

1 whole roaster chicken (about 3 pounds), cut into 6–8 serving pieces

1 pound Hot or Sweet Italian Sausage (page 64) or Italian-Style Wild Boar Sausage (page 184)

1 onion, thinly sliced

1 red, orange, or yellow bell pepper, cut into ¼-inch strips

2 garlic cloves, minced

4 cups peeled, cored, seeded, and chopped plum tomatoes (see page 251)

2 teaspoons dried oregano

1 teaspoon dried basil

1 bay leaf

1 whole dried hot red pepper, or 1 teaspoon crushed red pepper

Kosher salt and freshly ground black pepper

1 pound spaghetti or spaghettini

1 tablespoon unsalted butter

Grated Parmesan cheese

Chopped fresh flat-leaf parsley, for garnish

1. Heat the oil in a large skillet over medium-high heat. Brown the chicken on all sides, about 20 minutes. Transfer the chicken to a platter; cover to keep warm.

2. Panfry the sausage in the same skillet until cooked through and browned, about 20 minutes. Remove and keep warm.

3. Discard all but 2 tablespoons of the drippings. Add the onion, bell pepper, and garlic; sauté until crisp-tender, about 5 minutes.

4. Add the tomatoes, oregano, basil, bay leaf, red pepper, and salt and black pepper to taste. Bring the sauce to a simmer and return the chicken and sausage to the sauce. Simmer gently.

5. Meanwhile, bring a large pot of salted water to a boil. Cook the pasta until al dente; drain, then toss with the butter.

6. Arrange the pasta on a large platter. Using tongs, take the chicken and sausage out of the sauce and arrange on top of the pasta. Pour the sauce over all. Sprinkle with Parmesan, garnish with parsley, and serve.

SAUSAGE AND KALE RAGÙ

Removing the sausages from their casings allows the meat to better combine with the sauce. If you prefer a bit more texture, keep them in their casings while cooking in the sauce, and then cut into bite-size pieces before serving. This sauce freezes well, so double the quantities and stock up. Use it in lasagna (see Note, page 324), or toss with penne or other short, tubular pasta shapes.

Makes about 1 quart, enough to generously sauce 1 pound of pasta

2 tablespoons extra-virgin olive oil

1 pound Hot or Sweet Italian Sausage (page 64), removed from casing

1 small onion, finely chopped

4 garlic cloves, minced

1 bunch kale, stems removed, leaves chopped

Kosher salt

1 can (28 ounces) crushed tomatoes in purée

2 teaspoons dried oregano

Pinch of crushed red pepper

1. Heat the oil in a heavy saucepan over medium-high heat. Add the sausage, onion, and garlic. Sauté, stirring often and breaking up the meat with the back of a spoon, until the onion begins to turn golden.

2. Add the kale, season with salt to taste, and stir to coat with oil. Cook, tossing occasionally, until the kale wilts.

3. Add the tomatoes and purée, oregano, and crushed red pepper to the pan. Bring to a boil

recipe continued on next page

and then reduce the heat to a simmer. Cook until the sauce thickens slightly, about 10 minutes.

NOTE: *For a quick path to lasagna, make a double batch and layer the ragu and a generous sprinkling of mozzarella and Parmesan between layers of no-boil lasagna noodles in a baking dish, then top with more cheeses. Bake at 375°F (190°C) for 30 minutes, covered with aluminum foil, then remove the foil and continue baking until the sauce is bubbling and the topping is golden, about 10 minutes more.*

COQ AU VIN WITH SAUSAGE

This dish is a robust variation on the venerable French dish, with sausage adding even more heft. The thyme and sherry vinegar are nontraditional, but they add bright, fresh notes. Serve it with lots of crusty bread to sop up the juices.

Serves 6 to 8

- ¼ cup olive oil
- 1 whole roaster chicken (about 3 pounds), cut into 6 to 8 serving pieces
- 1 pound French Garlic Sausage (page 114)
- 18 small red potatoes
- 2 carrots, cut into 1-inch pieces
- 1 large onion, chopped
- 2 celery stalks, finely chopped
- 4 garlic cloves, crushed
- 2 cups dry red wine
- 1 cup chicken stock
  Bouquet garni (see right)
- 1 dried bay leaf
  Kosher salt and freshly ground black pepper
- 1 teaspoon fresh thyme, minced
- 2 teaspoons sherry vinegar

1. Heat the oil in a Dutch oven or heavy pot over medium-high heat. Brown the chicken pieces on all sides, about 20 minutes. With a slotted spoon, transfer the chicken pieces to a large bowl; cover to keep warm.

2. Add the sausage to the Dutch oven and cook until browned, about 20 minutes. Transfer the sausage to the bowl with the chicken.

3. Sauté the potatoes, carrots, onion, celery, and garlic in the Dutch oven until the vegetables are browned, about 10 minutes.

4. Return the sausage and chicken to the pot. Stir in the wine, chicken stock, bouquet garni, bay leaf, and salt and pepper to taste.

5. Bring to a simmer and cook, covered, for about 1 hour, or until the chicken is tender; an instant-read thermometer inserted into the thickest part of a chicken thigh should read 165°F (74°C). Before serving, remove the bay leaf and the bouquet garni and stir in the thyme and vinegar.

BOUQUET GARNI

This bundle of fresh herbs is the traditional way to impart subtle herbal flavor to all manner of preparations, from sauces and stocks to stews and braises, such as coq au vin. The classic combination, depending on whom you ask, will be a sprig each of fresh flat-leaf parsley or basil and thyme and maybe marjoram, plus a few bay leaves, but you'll find variations with rosemary or chervil, and with other aromatics such as orange peel, peppercorns, cloves, or even fennel fronds. Basically, use what's in your crisper — or herb garden — to skew to one cuisine or another.

To make one: Cut a piece of cheesecloth into an 8-inch square and pile the aforementioned herbs in the center, then gather together the corners and tie into a pouch with butcher's twine. If you aren't using any spices, you can just tie the herbs with twine and skip the cheesecloth. *Et voilà!*

ROAST DUCKLING WITH SAUSAGE

With its citrusy flavor, luganega sausage would be a good choice to use here, in combination with the duck. For best results, use a full-bodied red wine, such as Sangiovese, Malbec, or Shiraz, in the sausage mixture — and for serving with the meal.

Serves 4

> 1 whole duckling (4–5 pounds), with giblets
> ¾ cup water
> ¾ pound mushrooms, chopped
> 1 small onion, chopped
> 2 garlic cloves, minced
> 2½ cups dry red wine
> Kosher salt and freshly ground black pepper
> 1 pound Luganega (page 66)

1. Preheat the oven to 425°F (220°C).

2. Set aside the giblets. Cut the duckling into quarters. Cut off as much fat from the tail section as possible. Prick the skin in several places to allow fat to escape as the duck cooks.

3. Arrange the pieces of duckling on a rack in a shallow roasting pan and roast for 30 minutes.

4. In a small saucepan, bring the giblets and the water to a boil. Reduce the heat to maintain a low boil and cook until the liquid is reduced to ½ cup, about 30 minutes. Discard the giblets and reserve the liquid.

5. Remove the duckling from the oven. Using tongs, remove the pieces from the pan. Drain off the accumulated drippings and remove the rack from the pan. Return the duck to the roasting pan.

6. Combine the mushrooms, onion, garlic, wine, and giblet liquid. Season with salt and pepper. Pour the mixture over the duck pieces. Add the sausage to the roasting pan.

7. Lower the oven temperature to 375°F (190°C) and roast for 45 minutes, or until the sausages and duck pieces are well browned and the liquid is bubbling.

8. Serve the duck and sausage on a platter with the wine sauce poured over them.

STUFFED ROAST CAPON

Sunday supper, solved: Nothing is quite as satisfying as a well-roasted chicken, and that's particularly true when you enhance it with a savory stuffing. Here are three ideas for stuffings where sausage is the star. You can use a whole roaster chicken if you can't find a capon.

Serves 8 to 10

> 1 roasting capon or hen (4–5 pounds)
> Kosher salt
> 1 recipe stuffing of choice (recipes follow)
> 2 tablespoons olive oil or melted unsalted butter

1. Preheat the oven to 450°F (230°C). Sprinkle the capon inside and out with salt.

2. Stuff the bird as desired, then truss the cavity to close (or pin with toothpicks).

3. Set the stuffed bird on a rack in a roasting pan, breast side down. Rub the top of the bird with the oil.

4. Put the roasting pan in the oven, then immediately lower the temperature to 350°F (175°C). Roast for about 20 minutes per pound, basting frequently with the juices after the first 30 minutes, until the thigh juices run clear when pierced with a sharp knife; an instant-read thermometer inserted into thigh should read 165°F (74°C).

6. Transfer the chicken to a platter to rest for 15 minutes before carving. Use the drippings to make gravy, if desired.

recipe continued on next page

HERB AND SAUSAGE STUFFING

Makes 2 to 3 cups

>   1 teaspoon olive oil or vegetable oil
>   ½ pound fresh sausage, removed from casing
>   2 cups fresh bread crumbs (see page 245)
>   ¼ cup finely chopped celery, including inner leaves
>   1 tablespoon grated Parmesan cheese
>   2 teaspoons chopped fresh basil
>   2 teaspoons chopped fresh oregano
>   2 teaspoons chopped fresh flat-leaf parsley
>   ½ cup white wine or chicken stock, or more as needed
>   Freshly ground black pepper

1. Heat the oil in a medium skillet over medium-high heat. Add the sausage and sauté, crumbling it with the back of a spoon, until it is cooked through and browned, about 10 minutes. Drain on paper towels, if necessary.

2. In a large bowl, toss together the sausage, bread crumbs, celery, Parmesan, basil, oregano, parsley, and wine. Add more wine, if needed, to make the mixture moist. Season with pepper.

SAUSAGE, CORN BREAD, AND CRANBERRY STUFFING

Makes 2 to 3 cups

>   1 teaspoon olive oil or vegetable oil
>   ½ pound fresh sausage, removed from casing
>   1 apple, peeled and chopped
>   1 small onion, chopped
>   2 cups day-old corn bread (cut into cubes)
>   ½ cup chopped fresh cranberries
>   ½ cup white wine or chicken stock, or more as needed
>   Freshly ground black pepper

1. Heat the oil in a medium skillet over medium-high heat. Add the sausage and sauté, crumbling it with the back of a spoon, until it is cooked through and browned, about 10 minutes. Add the apple and onion and cook until they are softened, about 5 minutes. Drain on paper towels, if necessary.

2. In a large bowl, toss together the sausage mixture, corn bread, cranberries, and wine. Add more wine, if needed, to make the mixture moist. Season with pepper.

WILD RICE STUFFING WITH SAUSAGE AND CHESTNUTS

Makes 2 to 3 cups

>   1 teaspoon olive oil or vegetable oil
>   ½ pound fresh sausage, removed from casing
>   1 tablespoon unsalted butter
>   1 cup roasted chestnuts (see page 221)
>   ¼ cup chopped celery
>   1 tablespoon chopped shallots
>   2 cups cooked wild rice
>   ½ cup chicken stock, as needed
>   Freshly ground black pepper

1. Heat the oil in a medium skillet over medium-high heat. Add the sausage and sauté, crumbling it with the back of a spoon, until it is cooked through and browned, about 10 minutes. Drain on paper towels, if necessary, and wipe out the skillet.

2. Melt the butter in the skillet. Add the chestnuts, celery, and shallots, and sauté for about 5 minutes to soften them.

3. In a large bowl, combine the sausage, chestnut mixture, and wild rice. Add the chicken stock, a little at a time, as needed, to moisten the stuffing. Season with pepper.

BLACK BEANS AND RICE WITH CHORIZO

Moros y Cristianos is widely considered the national dish of Cuba, where it is traditionally made by cooking dried black beans with smoked ham hocks — a step that is highly encouraged for the most desirable result (although canned beans can certainly be substituted). In this update, chorizo is cooked along with the rice for an added dose of authentic flavors, and then some. If you prefer to eat the rice topped with the beans, following in the heels of some Cubans, then do so: fluff the rice in step 5, but do not add to the pan.

Serves 4 as a main dish or 8 as a side dish

- 1 pound Spanish-Style Chorizo (page 110), cut into thin slices
- 1 cup long-grain white rice
- 1 tablespoon olive oil or vegetable oil
- 1 small onion, chopped
- 1 green bell pepper, ribs and seeds removed, thinly sliced
- 1 red bell pepper, ribs and seeds removed, thinly sliced
- 1 yellow bell pepper, ribs and seeds removed, thinly sliced
- 3 garlic cloves, minced
- ¼ teaspoon cayenne pepper
- ¼ teaspoon ground cumin
  Kosher salt and freshly ground black pepper
- 3½ cups cooked black beans, drained, or 2 cans (15 ounces each) black beans, rinsed and drained
- 2 tablespoons fresh lime juice
- 1 teaspoon dried oregano, preferably Mexican oregano
  Chopped fresh cilantro, for garnish
  Lime wedges, for garnish

1. In a medium saucepan over medium-high heat, cook the chorizo. After the chorizo has started to crisp, stir in the rice to coat with the rendered fat. Cook, stirring, for 8 minutes to toast the rice.

2. Add 1½ cups water to the saucepan and bring to a boil. Reduce the heat, cover, and simmer until the rice is tender and the liquid is absorbed, about 15 minutes. Remove from the heat and let stand, covered, for 10 minutes.

3. Heat the oil in a large skillet over medium-high heat. Add the onion, bell peppers, garlic, cayenne, and cumin, and season with salt and black pepper to taste. Cook, stirring occasionally, until the onion is golden and the peppers have softened, about 8 minutes.

4. Stir the beans, lime juice, and oregano into the vegetables and cook to heat through.

5. Fluff the rice and add to the pan. Stir gently to combine. Garnish with cilantro and lime wedges, and serve.

Chicken Sausage
Tandoori

CHICKEN SAUSAGE TANDOORI

To be truly authentic, this dish would have to be cooked in a tandoori oven — a rounded-top, brick and clay oven that's used throughout India for cooking meats super quickly. But this recipe is still delicious, regardless of where it is baked. Round out the meal with naan, now widely available at most supermarkets. *See photo, opposite.*

Serves 6

- 1 tablespoon plus 1 teaspoon ground cumin
- 2 teaspoons cayenne pepper
- 2 teaspoons ground coriander
- 2 teaspoons paprika
- 1 tablespoon finely minced fresh ginger
- 3 garlic cloves, finely minced
- 1 cup plain yogurt
  Kosher salt and freshly ground black pepper
- 2 pounds chicken sausage
  Sliced cucumbers, for serving
  Hot red peppers, for serving
  Cilantro springs, for serving
  Naan or other flatbread, for serving

1. In a medium bowl, mix together the cumin, cayenne, coriander, and paprika. Stir in the ginger, garlic, and yogurt. Season with salt and pepper to taste.

2. Put the sausage in a casserole dish, then pour the yogurt mixture over the sausage, coating it well. Cover and marinate in the refrigerator for 1 day.

3. Preheat the broiler or prepare the grill for high heat.

4. Lift the sausage out of the marinade. Discard the marinade. Broil or grill the sausage for about 20 minutes, turning frequently, until evenly browned and cooked through. Serve with cucumber, peppers, cilantro, and flatbread.

CHICKEN AND SAUSAGE JAMBALAYA

We have the "holy trinity" to thank for making so many of the iconic Cajun dishes so distinctively delicious. The simple combination of chopped onion, green bell pepper, and celery is behind the prowess of many a New Orleans cook, and it's the foundation of jambalaya — certainly a contender for most-beloved dish, along with gumbo and étouffée. Here, shrimp and chicken are aided and abetted by the addition of andouille — another mainstay of Louisiana cooking.

Serves 4

- ½ pound medium shrimp, peeled, deveined, and chopped
- ¼ pound boneless, skinless chicken thighs, cut into ½-inch pieces
- 2 teaspoons paprika
- ½ teaspoon dried oregano
- ½ teaspoon dried thyme
- ¼ teaspoon cayenne pepper
  Kosher salt and freshly ground black pepper
- 2 tablespoons olive oil
- ¼ cup chopped onion
- ¼ cup chopped green bell pepper
- ¼ cup chopped celery
- 4 garlic cloves, minced
- ¾ cup long-grain rice
- ½ cup chopped tomatoes
- 1 dried bay leaf
- 2 teaspoons Worcestershire sauce
- 1 teaspoon hot pepper sauce, such as Tabasco, plus more for serving
- 2 cups chicken stock
- 5 ounces andouille sausage

recipe continued on next page

1. In a large bowl, combine the shrimp, chicken, paprika, oregano, thyme, cayenne, and salt and black pepper to taste. Toss together to coat the shrimp and chicken with the spices. Set aside.

2. Heat the oil in a large skillet over medium-high heat. Add the onion, bell pepper, celery, and garlic, and cook, stirring occasionally, until the vegetables are tender and starting to brown. Add the rice, tomatoes, bay leaf, Worcestershire, and hot sauce, and cook, stirring occasionally, for 5 minutes.

3. Add the chicken stock to the pan and increase the heat to bring it to a simmer. Reduce the heat to medium-low and cook until the rice is nearly tender, about 15 minutes.

4. Add the shrimp and chicken mixture to the pan, along with the andouille, stir once, and cook until the meat is cooked through, 10 minutes. Serve with additional hot sauce on the side.

BRACIOLA

Braciola is an Italian specialty traditionally made from flank steak, a versatile and flavorful cut that benefits from tenderizing and braising. It can also be prepared with thinly sliced round steak, though the flavor won't be quite the same. Regardless, lots of black pepper is one of the secrets of this dish.

Serves 4 to 6

- 1 flank or bottom round steak (about 1 pound)
- 1 teaspoon freshly ground black pepper
- ½ teaspoon dried oregano
- ½ teaspoon crushed red pepper
- ¼ cup chopped fresh flat-leaf parsley
- 1 tablespoon chopped fresh basil
- 1½ teaspoons chopped fresh mint
- 2 tablespoons chopped onion
- 2 tablespoons grated Parmesan or Romano cheese
- 1 garlic clove, minced
- 2 links Hot Italian Sausage (page 64)
- 2 tablespoons olive oil
- 2–3 cups Marinara Sauce (page 251)

1. With the side of a meat cleaver or a meat mallet, flatten the steak to ¼ inch thick. The meat should be rectangularly shaped.

2. In a small bowl, mix together the black pepper, oregano, crushed red pepper, parsley, basil, mint, onion, Parmesan, and garlic. Sprinkle the mixture evenly over the meat.

3. Lay the sausages end to end across the narrower side of the steak and roll it up jelly-roll fashion. As you get toward the end, tuck the sides in. With butcher's twine, tie the roll tightly at 2-inch intervals.

4. Heat the oil in a large skillet over medium-high heat, then brown the roll on all sides, about 10 minutes.

5. Reduce the heat to medium-low. Pour the marinara sauce over the roll and simmer, partially covered and gently turning occasionally, until the meat is tender, about 1½ hours.

6. To serve, cut the roll into ½-inch-thick slices and spoon the sauce over the top.

DOUBLE-THICK STUFFED PORK CHOPS

Here's a good way to stuff a pork chop so that the stuffing stays put: With a very sharp paring or slicing knife, pierce the chop in the center of the edge opposite the bone, then thrust the knife all the way to the bone. Without making that opening any larger, work a pocket into the chop by moving the knife first in one direction and then, turning it over, the other. Be careful not to poke a hole in the wall of the chop and to leave about a ¼-inch border along the chop's outer edge.

A pastry bag (or plastic bag with a corner snipped off) makes fast and neat work of stuffing the chops, or you can just spoon in the filling. We especially like to use the Pork and Apple Sausage (page 71) in this recipe, and then serve with Oven-Roasted Applesauce (recipe follows) and wilted Swiss chard (with or without wine-plumped raisins added) alongside.

Serves 6

- 6 double-thick (about 1½ inches) center-cut loin pork chops
- ¼ pound fresh sausage, removed from casing
- ¼ cup diced bits of any dried red sausage, such as Pepperoni (page 170) or Salamette (page 109)
- ½ cup dried bread crumbs
- 1 tablespoon minced onion
- 1 teaspoon ground ginger
- 1 teaspoon chopped fresh flat-leaf parsley
- 2 tablespoons dry white wine
- 2 tablespoons olive oil
- Kosher salt and freshly ground black pepper

1. Prepare the pockets in the chops, as described in the headnote.

2. Preheat the oven to 375°F (190°C).

3. In a medium bowl, mix together the fresh and dried sausage, bread crumbs, onion, ginger, parsley, and wine. Divide the mixture into six portions. With a pastry bag or spoon, fill each chop with one portion of the stuffing.

4. Brush the chops with the oil, then season with salt and pepper.

5. Arrange the chops on a rack in a roasting pan and roast for about 45 minutes, or until browned and tender. Baste frequently with the pan juices to prevent the chops from drying out. Serve hot.

OVEN-ROASTED APPLESAUCE

Although applesauce is typically prepared on the stovetop, roasting all the ingredients results in a "sauce" that's deeply flavorful. Using a mix of apples also helps.

To make it: In a roasting pan, combine about 4 pounds apples (preferably a combination of Granny Smith, McIntosh, Braeburn, and/or Cortland), peeled, cored, and cut into quarters, with the zest and juice of 1 lemon, ¼ cup apple cider (unsweetened), and a generous pinch each of ground ginger and allspice. Toss well and spread evenly. Dot with 2 tablespoons unsalted butter, cut into pieces (or, better yet, grated directly over the pan), and sprinkle with up to 2 tablespoons brown sugar (depending on the tartness of the apples).

Roast at 375°F (190°C) for about 30 minutes, until the apples are very tender and golden. Mash with a potato masher for a chunky consistency, or process in a food processor or pass through a food mill for a smoother texture. Season with kosher salt to taste, and serve, drizzled with a bit of apple cider vinegar, if desired.

SAUSAGE AND BEER
A MATCH PERFECTED

Now that you've gone to all the effort to prepare your own sausages from scratch, you'll want to give what to pair it with your due consideration. It's not so much that there's a *wrong* match per se for any particular sausage; it's more that there are definitely some *right* ones that can make your sausages taste that much better. Plus, it's fun to explore all the wonderful varieties of beer that are available everywhere these days. We tapped into the expertise of Jeff Alworth, author of *The Beer Bible* and *Brewing the World's Classic Styles: Advice from the Pros*, as well as his regular blog, *Beervana*. He also happens to reside in Portland, Oregon, where beer is always on tap for some spirited discussion.

"Because of its broad flavor spectrum, beer is a more versatile partner at the dinner table than wine — but that range also means pairing combinations threaten to spiral out of control. Fortunately, there are a few handy guidelines to help you find the perfect match," says Jeff. When approaching a particular pairing, he advises to recognize that beer can either complement or contrast, and both have their advantages. "Light American wheat beers have a sweetness that draws out a similar sweet note in pork, but another classic is stout and seafood, where the acrid bitterness in the beer also helps frame and enliven oysters (the quintessential pairing). Acid and carbonation have a cutting quality that helps balance salt and fat. Also look to match intensities; if you're working with a delicately spiced sausage, don't overwhelm it with a gale-force IPA; similarly, don't send a frail pilsner into battle with a searing andouille sausage."

Some Style Pairing Guidelines

CLASSIC GERMAN AND POLISH. Oktoberfest/märzen beers have lovely warming tones, with soft herbal spicing, and harmonize perfectly with these kinds of sausages.

EXAMPLES: *Bratwurst, Weisswurst, Kasekrainer, Liverwurst, Fresh or Smoked Kielbasa*

ITALIAN. For fiery links, a rich bière de garde will stand up to the flames and draw out the meaty flavors underneath; for salty and earthy sausages, try a lightly herbal English cask ale.

EXAMPLES: *Hot or Sweet Italian Sausage, Luganega, Cotechino, Salamette, Genoa Salami*

FRENCH. The sweet-sour-tart ales made in Belgian Flanders are a superb match for earthy, bloody sausages, but for more delicate, lightly herbal sausages, try a spicy saison.

EXAMPLES: *French Garlic Sausage, Boudin Noir, Boudin Blanc, Herbed Game Sausage, Herbes de Provence Chicken Sausage*

MIDDLE EASTERN. The wheat beers of Bavaria, with their fruit-and-clove palate, are a great match for these rich, warming sausages.

EXAMPLES: *Merguez, Loukanika*

LATIN (SPANISH/MEXICAN/PORTUGUESE). The intense, salty, and fatty sausages of these countries require a flavor powerhouse. A gueuze lambic, with a spine of acidity and rich, earthy flavors, is an excellent choice.

EXAMPLES: *Fresh Chorizo, Spanish–Style Chorizo, Linguiça*

CAJUN. The German schwarzbier, a black lager, can do double duty here, offering a bit of roastiness to complement the sausage's earthiness, along with a crisp, fire-dampening finish.

EXAMPLE: *Andouille*

ASIAN. No beer complements Asian spices better than the lightly spiced Belgian witbier; they were practically designed for each other.

EXAMPLES: *Curry Sausage with Cilantro and Lime, Vietnamese Pork and Lemongrass Sausage, Ginger-Scallion Seafood Sausage*

SEAFOOD. Dry Irish stouts help pull out fish's sweeter notes, and the briny/salty flavors are surprisingly compatible with the dry roastiness of the beer.

EXAMPLES: *Oyster Sausage; Mixed Seafood Sausage with Shrimp, Crabmeat, and Scallops; Clam Dogs; Chesapeake Bay Sausage*

SWEET. Contrast is the way to go with sweet sausages, and a crisp, effervescent pilsner will make their rounder, heavier flavors pop while providing a palate-cleansing snap at the finish.

EXAMPLES: *Turkey and Cranberry Sausage; Lamb, Ginger, and Dried Apricot Sausage; Pork and Apple Sausage; Maple Breakfast Sausage; Cider and Sage Breakfast Sausage*

Seafood Sausage
Cioppino

SEAFOOD SAUSAGE CIOPPINO

Ever frugal, the Italian fishermen who settled in San Francisco created a way to use up whatever fish was left over after selling their catch of the day. They called it cioppino ("chopped," as the ingredients are prepared). The stew is typically made with a variety of fish and shellfish, similar to bouillabaisse. Here, the sausage provides the variety of fish, with clams adding an extra briny kick. It's excellent served with slabs of garlic-rubbed grilled bread on the side. *See photo, opposite.*

Serves 4

- 3 tablespoons olive oil
- 1 onion, chopped
- 1 large russet potato, diced
- ½ bulb fennel, chopped, fronds reserved for garnish
- 4 garlic cloves, minced
  Kosher salt
  Pinch of saffron
- 1 cup dry white wine
- ½ cup tomato purée
- 4 cups seafood broth
- 1 pound Mixed Seafood Sausage with Shrimp, Crabmeat, and Scallops (page 237)
- 16 littleneck clams, cleaned

1. Heat 2 tablespoons of the oil in a 3-quart saucepan over medium-high heat. Add the onion, potato, fennel, and garlic, and season with salt. Sauté until the onion and fennel are tender and starting to brown.

2. Add the saffron to the pan and stir. Add the wine and cook until it has nearly evaporated.

3. Add the tomato purée and seafood broth to the pan and bring to a simmer. Cook until the potatoes are tender and starting to fall apart.

4. While the broth is simmering, heat the remaining 1 tablespoon oil in a large skillet over medium heat. Cook the sausages until cooked through and browned all over, about 8 minutes. Remove from the pan. Cut into bite-size pieces.

5. Add the cooked sausage and clams to the simmering broth. Cook, covered, until the clams open. Discard any clams that remain closed.

6. To serve, portion the stew into bowls or soup plates, garnish with the fennel fronds, and season with pepper.

BRATS IN BEER

It's unclear why bratwurst cooked in beer seems better than brats cooked any other way. It could be because the beer adds something essential to the meat, or simply because the vapors rising from the pot whet the appetite. Suffice it to say that it is better, period.

Serves 4

- 1 bottle (12 ounces) beer or ale
- 1 cup water
- 1 pound Bratwurst (page 142)
- 2 cups prepared sauerkraut, rinsed and drained
- ½ teaspoon caraway seeds
- 4 hard rolls or hot dog buns

1. In a large saucepan, pour the beer and water over the bratwurst. Bring to a boil, then reduce the heat so that the liquid barely simmers. Cook for 10 minutes.

2. Set a colander or vegetable steamer over the pan with the bratwurst, then spoon the sauerkraut into it. Sprinkle the caraway seeds on the sauerkraut. Cover and allow the steam to heat the sauerkraut, about 10 minutes.

3. To serve, make a bed of sauerkraut in each roll and place a brat on top.

CURRYWURST

What business does curry powder have playing a role in a Berliner sausage phenomenon? Like so many popular-food histories, this one is a bit storied: In 1949, Herta Heuwer, then running a food kiosk in West Berlin, managed to obtain some ketchup, Worcestershire sauce, and curry powder from British soldiers, and then had the presence of mind to mix them together and serve the concoction over the local sausages — instant success! It was just after World War II, and food and money were scarce.

Legend has it that Herta sold more than 10,000 orders each week during the peak of her success, which was thwarted when competitors stepped in and created the craze for currywurst that still exists in Berlin today, where you'll likely be served it with a heap of French fries on top.

Serves 4

2 tablespoons vegetable oil
2 tablespoons curry powder
1 tablespoon hot paprika
1 onion, finely chopped
Kosher salt
1 can (16 ounces) whole peeled tomatoes
½ cup sugar
¼ cup red wine vinegar
½ teaspoon anise seed
¼ teaspoon freshly grated nutmeg
1 pound Weisswurst (page 145) or Bratwurst (page 142), cooked as desired

1. Heat the oil in a medium saucepan over medium heat. When hot, add the curry powder and paprika. Cook, stirring frequently, until fragrant, 1 minute.

2. Add the onion to the pan and stir to cover with the spices. Lower the heat and cook until the onion is softened but not browned.

3. Pour the tomatoes and their juice into a small bowl and crush with your hands until broken up. Add to the saucepan along with the sugar, vinegar, anise seed, and nutmeg, and stir to combine. Simmer (increase the heat if necessary) until the sauce is thickened, about 20 minutes.

4. Let the sauce cool for about 10 minutes, then purée in a blender until smooth. (The sauce can be made up to 5 days ahead and kept in the refrigerator, covered.)

5. Rewarm the sauce and serve over the cooked sausages.

STEAMED MUSSELS WITH CHORIZO

It's a wonder why more people don't prepare mussels at home. They are inexpensive, fast-cooking, and singularly delicious, making them just right for busy weeknight meals but also special enough for a casual dinner party. Chorizo is traditionally paired with mussels (and clams) in Spanish cooking, its spicy, smoky flavor working its magic with the briny shellfish. This brothy dish demands thick slices of crusty bread, preferably charred in spots. *See photo, opposite.*

Serves 4 as a main dish or 8 to 10 as a starter

2 tablespoons olive oil
2 ounces Spanish-Style Chorizo (page 110), removed from casing and sliced thin
1 small onion, halved and thinly sliced
4 garlic cloves, thinly sliced
1 tablespoon tomato paste
4 pounds mussels, cleaned and beards removed (see box, page 338)
1 cup dry white wine
Pinch of red pepper flakes (optional)
Kosher salt and freshly ground black pepper

recipe continued on page 338

Steamed Mussels
with Chorizo

1. Heat the oil in a Dutch oven or large heavy pot over medium-high heat. Add the chorizo, onion, and garlic. Sauté until the chorizo is crisp and the onion and garlic have softened, about 8 minutes. Add the tomato paste and stir to combine and coat the ingredients.

2. Add the mussels, wine, and red pepper flakes (if desired) to the pan, then season with salt and black pepper.

3. Cover the pot and cook until mussels have opened, about 10 minutes. Discard any mussels that remain closed. Serve immediately.

CLEANING MUSSELS

Those straggly beards? Not harmful, just unpleasant. Mussels need only a thorough rinsing under cold water and some tugging to remove the strings before cooking. Because most mussels are farmed today, you don't really even need to soak them to remove the sand. Do discard any that are cracked and/or remain open before cooking.

COD AND CLAMS WITH LINGUIÇA

In Portugal, so the saying goes, there are more codfish (*bacalau*) recipes than there are days in the year. The flaky white fish is a staple there (and also in Portuguese enclaves in the United States — for instance, New Bedford, Massachusetts, and surrounding communities), where the sausage is added to soups, stews, and plenty of other national treasures. In this oven-to-table stew, the fish is baked along with littleneck clams and linguiça for a quick, hands-free version of *cataplana*.

Serves 4

- 1½ pounds cod fillet
- 3 tablespoons olive oil
  Kosher salt and freshly ground black pepper
- 1 onion, halved, sliced into thin rings
- 1 lemon, sliced into thin rings
- ¼ cup chopped fresh flat-leaf parsley
- 3 garlic cloves, very thinly sliced
- 2 tablespoons pimentón (smoked paprika)
- ¾ pound linguiça, diced
- 12 littleneck clams, cleaned
- 1 cup white wine
- 3 tablespoons unsalted butter, cut into small cubes

1. Preheat the oven to 450°F (230°C).

2. Rub the cod with 1 tablespoon of the oil and ¼ teaspoon salt. Place the cod in a 9- by 13-inch baking dish.

3. In a medium bowl, combine the remaining 2 tablespoons oil with the onion, lemon, parsley, garlic, and pimentón. Season with salt and pepper. Toss to coat everything with the oil. Spoon over and around the cod.

4. Arrange the linguiça and clams in the baking dish. Pour the wine into the dish. Bake for about 15 minutes, or until the fish is cooked through and the clams have opened. Discard any clams that remain closed. Dot with the butter and serve.

VENISON SAUSAGE CHILI

Venison has a long history in so-called "cowboy cuisine": Ranchers (or hunters) would use the fresh meat from the day's hunt to cook a hearty stew over the open campfire, and nothing would warm them up like a bowl of spicy chili. This recipe gets added flavor from chorizo. In place of venison sausage, you could use sausage made from other game meat, such as elk, moose, or antelope.

Serves 8

    6 ounces Fresh Chorizo (page 67), removed from casing
    3 pounds Herbed Game Sausage (page 198) made with venison, removed from casing
    2 onions, finely chopped
    4 garlic cloves, minced
    2 jalapeño or serrano chiles, minced (ribs and seeds removed for less heat)
    ¼ cup chili powder
    ¼ cup unsweetened cocoa powder
    1 tablespoon ground cumin
    2 cans (28 ounces each) whole tomatoes in purée
    3 tablespoons unsulfured molasses
      Kosher salt
    4 cups cooked pinto or black beans, drained, or 2 cans (15 ounces each), rinsed and drained
      Chopped fresh cilantro, for serving
      Mexican crema (or sour cream), for serving
      Chopped scallions, for serving
      Diced avocado, for serving

1. Heat a heavy pot over medium heat. Add the chorizo and cook, stirring and breaking up the meat with a spoon, until the fat is rendered and the meat is browned, about 5 minutes. Add the venison sausage and cook, breaking up the meat with the back of a spoon, until browned, about 8 minutes.

2. Add the onions, garlic, and chiles. Cook until tender, stirring frequently, about 5 minutes. Stir in the chili powder, cocoa powder, and cumin, and cook, stirring, until toasted, about 1 minute.

3. Add the tomatoes to the pot, breaking them up with your hands as you go, then pour in the purée and stir to combine. Add 1 cup water and the molasses, then bring to a boil. Reduce the heat and simmer, partially covered, until thickened, about 30 minutes. Season with salt.

4. Add the beans to the pot and cook until the meat and beans are very tender and the chili has reached the desired consistency (we like it on the thicker side), about 30 minutes more, adding more water and stirring as necessary. Serve immediately with the suggested toppings.

SPLIT PEA SOUP WITH KIELBASA

Instead of ham hocks, the classic seasoning component, kielbasa provides the characteristic smoky flavor to this version of split pea soup.

Serves 8

    1 pound dried green split peas
    3 onions, chopped
    2 carrots, chopped
    2 medium russet potatoes, peeled and cubed
    1 teaspoon dried thyme
    1 pound Smoked Kielbasa (page 173)
      Kosher salt and freshly ground black pepper

recipe continued on next page

1. Rinse the peas and discard any discolored ones.

2. In a large soup pot, cover the peas with 8 cups water. Bring to a boil, reduce the heat, and simmer for 20 minutes.

3. Stir the onions, carrots, potatoes, and thyme into the pot, lower the heat to a gentle simmer, and cover. Cook for about 45 minutes, stirring frequently so the peas don't stick to the bottom.

4. Cut the kielbasa into ½-inch-thick slices and add to the soup. Season with salt and pepper to taste. Simmer for about 30 minutes longer.

5. Taste the soup and add more salt and pepper as needed. Serve hot.

CASSOULET

The name of this classic dish — hailing from the South of France — comes from the clay pot, or *cassole*, that it was traditionally prepared in. Dating back to medieval times, the vessel has a narrow base and wide top, all the better to brown those bread crumbs, but a Dutch oven works well, too. The preparation of this recipe is spread over two days, so plan accordingly. You can also stretch it to three days to lighten your load. It will be every ounce worth the effort — and extended wait.

Serves 8

- 2 pounds Tarbais, cannellini, or Great Northern beans
- 2 tablespoons kosher salt
- 6 ounces fresh pork skin (optional)
- 2 ounces salt pork
- ⅓ cup duck or chicken fat or olive oil
- 2 fresh ham hocks (about 1½ pounds)
- 1 pound boneless pork shoulder, cut into 1½-inch cubes
- ½ pound pancetta (on one piece), halved horizontally
- 3 carrots, thinly sliced
- 2 onions, diced
- 14 garlic cloves (peeled)
- 2 quarts chicken stock
  Freshly ground black pepper
- 4 flat-leaf parsley sprigs
- 3 small celery stalks
- 2 thyme sprigs
- 2 bay leaves
- 1 tablespoon olive oil
- 1 pound fresh pork sausages
- 6 duck confit legs
- ½ cup fresh bread crumbs

1. Place the beans in a large bowl with 1 tablespoon of the salt and cover with water by 4 inches. Swish the beans around to dissolve the salt. Let the beans soak overnight.

2. The next day, place the pork skin (if using) and salt pork in a medium saucepan. Cover with water and bring to a boil, then reduce the heat to a simmer. Cook until the skin is soft and flexible — a sharp knife should meet no resistance. Drain well, then let cool and refrigerate the salt pork. Cut the pork skin into four or five long strips. Roll each strip up and tie with butcher's twine.

3. Heat the duck fat in a large enameled cast-iron casserole or Dutch oven over medium-high heat. Pat any moisture from the ham hocks and pork shoulder with paper towels. Unroll the pancetta if possible. Fry the pork shoulder cubes in batches so as not to crowd the pan. When each batch is golden, transfer to a plate and continue with the rest of the pork cubes, then the hocks and the pancetta. Set the meats aside.

4. Add the carrots and onions to the same pan and cook, stirring frequently, until the onions are golden and tender, 5 to 10 minutes. Add 10 of the garlic cloves and cook for a few minutes longer, until fragrant. Add the chicken stock, pork skin, pork, hocks, pancetta (and any juices from the

plate), and 2 teaspoons of the salt. Season with pepper. Bring to a boil, then reduce the heat and simmer, covered, for 1½ hours.

5. Make a bouquet garni by wrapping the parsley sprigs, celery, thyme, and bay leaves in a square of cheesecloth and securing with butcher's twine.

6. Drain the beans and add, along with the bouquet garni, to the casserole dish with the pork. Add a little water if the mixture looks too dry. Simmer, covered, until the beans are tender, adding more water as necessary. Remove from the heat and let cool.

7. When the stew is cool enough, pick out the hocks, pancetta, skin bundles, and bouquet garni. Discard the bouquet garni. Pick all the meat off the hocks, discarding the bones, skin, and anything that isn't meat. Cut or tear the hock meat and pancetta into bite-size pieces.

8. Preheat the oven to 400°F (200°C). Bring the stew to a simmer.

9. Cut the salt pork into small pieces. Add to a food processor along with the remaining 4 garlic cloves. Process to a smooth paste, adding liquid from the stew if needed. Stir the paste into the stew and simmer over low heat for 15 minutes, stirring occasionally. Stir in the hock meat and pancetta. Taste and add the remaining 1 teaspoon salt if necessary.

10. Unroll the pork skin bundles, if using. Line the bottom of a 5- to 6-quart cassole or enameled cast-iron casserole with the pork skin.

11. Spoon the stew into the casserole. Add 1 to 2 cups water to the casserole, depending on how much liquid was left in the stew. The dish shouldn't be swimming in liquid, but it should reach the top of the pan's contents. Bake, uncovered, for 1½ hours.

12. Reduce the oven temperature to 275°F (135°C).

13. Heat the olive oil in a medium skillet. Add the sausages and cook over high heat until browned all over and the casings are starting to crisp; they will not be cooked through. Remove from the heat and, when cool enough to handle, cut into bite-size pieces. Push the pieces into the cassoulet. Arrange the confit duck legs on the top of the cassoulet, and push down on them gently so that the bottom part of the leg is submerged but the skin is exposed.

14. Sprinkle the cassoulet with ¼ cup of the bread crumbs. Bake for 1 hour, or until the crumbs are dark golden brown. Push the brown crumbs down a little to break the crust, sprinkle the remaining ¼ cup bread crumbs over the top, and bake for another 45 minutes, or until the crumbs are browned. Let rest for at least 15 minutes before serving.

Paella

PAELLA

A paella pan is not essential to preparing this Spanish favorite, but it does help ensure that the rice forms that desirable layer of crust, called *socarrat*, on the bottom (avoiding stirring at the end also helps).

In Spain, there are countless variations on paella, some with only seafood, others with rabbit, but all with saffron, which gives the rice its characteristic hue. This version leans toward the most famous, paella Valenciana, which features shrimp, mussels (or clams), chicken, and chorizo. *See photo, opposite.*

Serves 4 to 6

- 3 tablespoons olive oil, plus more for drizzling
- ¾ pound boneless, skin-on chicken thighs
- ¼ pound Spanish-Style Chorizo (page 110), Linguiça (page 107), or other highly spiced sausage, cut into ¼-inch thick slices
- 2 garlic cloves, minced
- 1 onion, finely chopped
- 1 carrot, chopped
- 2 teaspoons pimentón (smoked paprika)
- 1 red bell pepper, ribs and seeds removed, cut into thin strips
- 1 tablespoon tomato paste
  Pinch of saffron
- 1½ cups Bomba or Calasparra rice
- 5 cups chicken stock
  Kosher salt
- 1 cup fresh or frozen peas (thawed)
- ½ pound medium shrimp, peeled and deveined
- ½ pound mussels and/or clams, cleaned (see page 338)
- ¼ cup chopped fresh flat-leaf parsley
  Lemon wedges, for serving

1. Heat a paella pan or large ovenproof skillet over medium-high heat. Add the oil. When hot and shimmering, carefully lay the chicken thighs, skin side down, in the oil. Cook until the skin is rendering fat and starting to crisp. Turn the chicken thighs over and add the chorizo, garlic, onion, and carrot. Cook for 5 minutes longer, or until the onion starts to soften. Add the pimentón, bell pepper, tomato paste, and saffron. Stir to incorporate.

2. Add the rice to the pan and stir to coat with the tomato oil. Add the chicken stock and salt to taste and bring to a boil, then reduce the heat to a simmer and cover. Let simmer for 15 minutes, stirring occasionally.

3. Stir in the peas and shrimp and nestle the mussels and/or clams in the rice. Cover and cook for another 5 minutes, or until the shrimp are opaque throughout and the mussels have opened (discard any that remain closed).

4. Serve immediately, topped with the parsley and a drizzle of olive oil and with lemon wedges for spritzing.

CALDO VERDE

Kale (or collards, a suitable substitute used here) is the green in Portuguese *caldo verde* (literally "green broth"), while potatoes are what lend the stew body. Linguiça (page 107) is classic, but other smoked sausages, such as Andouille (page 106) or kielbasa (page 144), are also good options, as is Spanish-Style Chorizo (page 110). A splash of sherry vinegar is an optional last-minute addition, lending a welcome bright note to this ultra-hearty soup.

Serves 4 to 6

¼ cup olive oil
1 pound smoked sausage, cut into ¼-inch-thick slices
1 onion, chopped
4 garlic cloves, minced
2 quarts chicken or vegetable stock
2 medium russet potatoes, peeled and cut into ½-inch cubes
1 bunch collard greens or kale, stemmed and chopped
Kosher salt and freshly ground black pepper
Sherry vinegar, for serving (optional)

1. In a 6- to 8-quart saucepan, heat the oil over medium-high heat. Add the sausage and cook until crisp and golden. Remove from the pan.

2. Reduce the heat to medium and add the onion and garlic to the pan. Cook until tender, about 10 minutes.

3. Add the sausage, chicken stock, potatoes, collards, and salt and pepper to taste. Bring to a simmer and cook until the potatoes and collards are very tender, 30 to 45 minutes.

4. Serve hot, with a dash of sherry vinegar (if desired).

STUFFED ACORN SQUASH

Here's a meal-in-one dish that's also conveniently cooked in its own edible vessel, although simple braised greens such as chard, kale, or collards would make a worthy accompaniment. Use any of the Italian-style sausages found in the book, or the Pork and Pistachio Sausages on page 77. *See photo, opposite.*

Serves 4

2 large acorn squash
2 teaspoons dark brown sugar
1 pound fresh pork sausage, removed from casing
½ cup fresh bread crumbs
1 large egg, well beaten

1. Preheat the oven to 375°F (190°C).

2. Cut each squash in half horizontally and scrape out the seeds and pith. If needed to make the squash halves sit flat, cut a small slice from the bottom. Sprinkle the cavity of each squash with ½ teaspoon of the brown sugar.

3. Pour 1 inch of water in the bottom of a baking pan and arrange the squash, skin side down, in the pan. Bake, uncovered, for 20 minutes.

4. Meanwhile, in a large bowl, combine the sausage, bread crumbs, and egg. Mix well.

5. When the squash has baked for 20 minutes, remove it from the oven and stuff one-quarter of the sausage mixture into each of the squash halves.

6. Return the squash to the oven and bake for 30 minutes longer, or until the sausage is cooked through and crisp on top. Serve hot.

Stuffed Acorn
Squash

KAPUSTA

Kapusta is Polish for "cabbage," and that's what you'll find in the national dish of the same name. It's often served as a side for kielbasa, but here we combine the two in a hearty meal that's got "comfort food" written all over it. You could offer buttered boiled potatoes alongside (another classic go-with), but we find this version satisfying enough on its own.

Serves 4

- 2 tablespoons vegetable oil
- 1 large onion, chopped
- 1 pound Smoked Kielbasa (page 173), cut into 1-inch pieces
- 1 small head cabbage, shredded
- 1 cup tomato juice, preferably low-sodium
  Kosher salt and freshly ground black pepper

1. Heat the oil in a Dutch oven or large heavy pot over medium-high heat. Add the onion and sauté until it is translucent, about 10 minutes.

2. Add the kielbasa and sauté for 5 minutes. Stir in the cabbage, tomato juice, and salt and pepper to taste. Cover and cook, barely simmering, for 1 hour. Serve hot.

CHOUCROUTE GARNI

There are as many varieties of *choucroute garni* (French for "dressed sauerkraut") as there are Alsatian grandmothers. This version calls for several kinds of sausage, but feel free to use only one or two types, or try substituting other varieties.

Serves 8 to 10

- 1½ pounds smoked ham hocks
- ½ pound thick-sliced bacon strips, cut crosswise into 1-inch pieces
- 2 pounds fully cooked Fresh Kielbasa (page 144), cut diagonally into 1-inch pieces
- 1 pound fully cooked knockwurst, cut diagonally into 1-inch pieces
- 1 pound fully cooked Bratwurst (page 142), cut diagonally into 1-inch pieces
- 2 onions, chopped
- 3 garlic cloves, minced
- 1 teaspoon juniper berries
- 1 teaspoon whole black peppercorns
- 5 whole cloves
- 4 allspice berries
- 2 bay leaves
- 4 pounds (2 quarts) prepared sauerkraut, drained and squeezed dry
- 2 cups dry white wine
- 2 pounds small red potatoes, halved
- ½ cup chopped fresh flat-leaf parsley
  Dijon and Grainy mustards, for serving (see page 293)
  Prepared horseradish, for serving
  Cornichons, for serving

1. Place the ham hocks in a 4-quart saucepan. Add enough water to cover by 2 inches. Bring to a boil, and then reduce the heat to a simmer. Cover and simmer until the meat is very tender and falling from the bones, 1½ to 2 hours. Drain the hocks and place in a bowl, reserving the liquid.

2. Preheat the oven to 350°F (175°C).

3. Heat a Dutch oven or large heavy pot over medium heat. Add the bacon and cook until crisp. Transfer to the bowl with the hocks, leaving any fat in the pan. Add all of the sausages to the pan and brown them in the bacon fat. Remove the sausage when browned and add to the bowl with the hocks and bacon.

4. Add the onions and garlic to the pan and cook in the bacon fat until tender and starting to brown. Add the juniper berries, peppercorns, cloves, allspice, bay leaves, and sauerkraut. Stir to combine.

Nestle the hocks, bacon, and sausages in the sauerkraut.

5. Pour the wine and enough hock liquid over the top to just cover the meats. Bring to a boil, then cover and place in the oven. Bake for 1½ hours.

6. While the choucroute is baking, prepare the potatoes: Bring a large pot of salted water to a boil. Boil the potatoes until fork-tender, 15 to 20 minutes. Drain the potatoes and toss with the parsley. Keep warm.

7. Serve the choucroute with the boiled potatoes, mustards, horseradish, and cornichons.

SMASHED POTATO SALAD WITH SAUSAGE

What gives this hearty potato salad its distinctive flavor are bits of crisp-fried bacon and some of the drippings from that bacon, as well as some from the added sausage — ideally either Bratwurst (page 142) or All-Beef Summer Sausage (page 137), but also Fresh Chorizo (page 67) or Merguez (page 132). The potatoes are cooked until very tender and, when all the ingredients are combined, the whole thing falls somewhere between a creamy potato salad and mashed potatoes. Mixing the sausages and potatoes with the dressing while they are hot helps the mixture absorb the dressing better.

Serves 6

    Kosher salt
  6 large russet potatoes, peeled and sliced ½ inch thick
  3 tablespoons olive oil or vegetable oil
  1 pound fresh sausage
  1 large onion, thinly sliced
  ¼ cup apple cider vinegar
  ¼ cup grainy mustard

  2 teaspoons minced fresh tarragon
    Freshly ground black pepper

1. Bring a large pot of salted water to a boil. Add the potatoes, reduce the heat, and simmer until the potatoes are fork-tender, about 15 minutes. Drain the potatoes and transfer them to a large bowl.

2. Meanwhile, heat the oil in a large skillet over medium-high heat. Add the sausage and sauté until browned and cooked through, about 20 minutes. Remove the sausage with a slotted spoon, reserving the fat in the pan. Slice the sausage about ½ inch thick and add to the potatoes in the bowl.

3. Add the onion to the skillet, season with salt, and cook over medium heat until softened. Remove from the heat.

4. Add the vinegar and mustard to the hot skillet and stir to scrape up the browned bits from the bottom. Pour the mixture over the potatoes. Add the tarragon and pepper and stir well to combine and smash some of the potatoes.

5. Taste the salad, and add more salt and vinegar as needed. Serve warm or chilled.

BUBBLE AND SQUEAK

A traditional English favorite that should surely get an award for "Best Recipe Name," this beloved combination of potatoes and cabbage gets even tastier with bits of sausage mixed in. The dish evolved as a way to use up leftovers, so feel free to get creative about other additions. Using lard gives the dish the most traditional flavor, but you can substitute olive oil if you prefer.

Serves 4

- 1 tablespoon lard or olive oil
- ½–1 pound fresh sausage, such as English Bangers (page 118), removed from casing
- 1 small onion, thinly sliced
- 2 cups chopped cooked cabbage (see Note)
- 2 cups chopped boiled or mashed potatoes, chilled
- Kosher salt and freshly ground black pepper

1. Heat the lard in a large cast-iron skillet over medium-high heat until melted. Add the sausage and sauté, crumbling it with the back of a spoon, until it is cooked through and browned, about 10 minutes. Remove the sausage with a slotted spoon and set aside.

2. Sauté the onion in the skillet until it is softened, about 5 minutes. Stir in the sausage, cabbage, and potatoes. Add salt and pepper to taste. Press the mixture down into the hot fat.

3. Cook over medium-low heat for about 15 minutes, or until the mixture is browned on the bottom. Turn to brown the other side and cook for about 10 minutes longer. Cut into wedges and serve hot.

NOTE: *If you don't have cooked cabbage on hand, cook some by adding 1 pound sliced cabbage to a large pot of boiling water. Cover the pan and gently boil until tender, about 7 minutes. Drain thoroughly.*

NORTH AFRICAN LAMB TAGINE

Moroccan flavors permeate this cold-weather warmer that's redolent of spices and brightened with ample fresh herbs. Lamb is often paired with dried fruit in North African (and Middle Eastern) cooking, and tagines — named for the earthenware pots they are cooked in — are among the more delicious expressions.

This variation includes lamb in two forms: stew meat and sausage. Merguez (page 132) is your best bet, but you could also use the Lamb, Ginger, and Dried Apricot Sausage (page 124), swapping out the prunes for dried apricots, and then omitting the honey below since that sausage is on the sweeter side. For a change-up from the all-lamb formula, try the Moroccan Goat Sausage (page 128), which shares the same seasonings as the tagine. *See photo, page 125.*

Serves 8 to 10

- 3 large red onions, coarsely chopped
- 4 garlic cloves (peeled)
- 1 tablespoon ground cumin
- 1 tablespoon hot paprika
- 1 teaspoon ground turmeric
- ½ teaspoon cayenne pepper
- 1¼ cups olive oil
- 1¼ cups fresh lemon juice
- 3 tablespoons honey
- 2 cups chopped fresh cilantro (leaves and tender stems)
- ½ cup chopped fresh flat-leaf parsley
- 2 pounds lamb stew meat, cut into 1-inch cubes
- 2 medium carrots, cut into ¾-inch chunks
- 12 prunes, pitted
- Kosher salt
- 2 pounds Merguez (page 132)
- 1 preserved lemon, pulp removed and discarded, rind sliced into thin strips (see next page)

½ cup thinly sliced fresh mint leaves
Harissa, for serving

1. Place 2 of the onions, the garlic, cumin, paprika, turmeric, cayenne, 1 cup of the oil, 1 cup of the lemon juice, 2 tablespoons of the honey, the cilantro, and parsley in a blender. Blend until smooth.

2. Pour into a large bowl, and add the lamb pieces. Stir well to coat, cover, and let marinate in the refrigerator for at least 4 hours, and as long as 12 hours.

3. Preheat the oven to 325°F (165°C).

4. Drain the lamb from the marinade, reserving the marinade. Pat the lamb dry with paper towels.

5. Heat a Dutch oven or heavy pot over high heat. Add 3 tablespoons of the remaining oil to the pan, and carefully add the lamb. Cook, stirring only once or twice, to sear the lamb. Once the lamb has turned dark brown and crusty, add the carrots, the remaining onion, the prunes, and salt to taste to the pan, along with the reserved marinade and 3 cups water.

6. Cover the Dutch oven and transfer it to the oven. Let the tagine braise for 2 to 3 hours, until the lamb is tender. (If there is a lot of liquid remaining, use a skimmer to remove the lamb and vegetables and bring the liquid to a simmer on the stovetop. Let the liquid reduce until you have about 2½ cups.) Stir in the remaining 1 tablespoon honey, the remaining ¼ cup lemon juice, and additional salt to taste. Keep warm while you finish the dish.

7. In a medium skillet, heat the remaining 1 tablespoon oil and cook the merguez until cooked through and the casings are crisp. Add the sausage to the tagine.

8. Transfer the stew to a serving platter or a tagine, if you have one. Garnish with the preserved lemon and mint leaves, and serve harissa on the side.

PRESERVED LEMONS

When a recipe (usually of Moroccan heritage) calls for preserved lemons, you can certainly buy them online or from a local shop, or make them yourself (a better, doable idea) with a bit of planning ahead.

To make them: Soak 4 lemons in lukewarm water for 2 to 3 days to soften the peel, changing the water each day. (You can skip this step if you prefer.) Quarter the lemons from the top, leaving the bottom intact. Sprinkle liberally with kosher salt, then push the wedges back together.

Place the lemons in a container with a ½-inch-layer of salt at the bottom, adding spices as you go (see below for suggestions) and leaving 1 inch of headspace above. Cover tightly with the lid and let ripen in a warm spot for about 30 days, shaking the jar every day.

When ready to use, rinse the lemons and remove the pulp as needed. Preserved lemons will keep at room temperature for months.

Suggested seasonings:
- Cinnamon sticks
- Whole cloves
- Whole peppercorns
- Coriander, mustard, or fennel seeds
- Fresh or dried bay leaves

VIETNAMESE PHO WITH SAUSAGE

Pho, the rightly renowned noodle soup from Vietnam, has a 100-plus-year history (and about as many renditions). Beef and pork are common additions, but sausage is added here. You can't go wrong with any of the fresh sausages with Asian flavor profiles, including Vietnamese Pork and Lemongrass Sausage (page 70), Sai Krok Isan (page 83), or Curry Sausage with Cilantro and Lime (page 82), or try using Lap Cheong (page 102) for a hit of smokiness.

Serves 4

- 2 onions
- 6 garlic cloves (peeled)
- 1 piece (3 inches) fresh ginger, peeled and sliced ⅛ inch thick
- 1 star anise
- 1 teaspoon freshly ground black pepper
- 2 quarts chicken or vegetable stock
- 2 tablespoons brown sugar or coconut sugar
- 1 tablespoon Asian fish sauce, plus more for serving
- 1 pound rice vermicelli
- 1 pound cooked Asian-style sausage, cut into thin coins
- 2 cups bean sprouts
- 1 cup fresh mint leaves
- 1 cup cilantro sprigs
- 1 cup fresh Thai basil or regular basil leaves
- ¼ cup sliced red chiles, such as Thai bird
- 2 limes, cut into wedges

1. Preheat the broiler.

2. Cut 1 onion into ¼-inch-thick slices. Place the cut onion, garlic, and ginger on a baking sheet and broil until the onion is just charred. Remove from the oven.

3. Place the contents of the baking sheet in a 6- to 8-quart saucepan. Add the star anise, pepper, and chicken stock. Bring to a boil, and then reduce the heat so the liquid just simmers. Simmer for 30 minutes.

4. Strain the broth, discarding the solids, and add the sugar and fish sauce.

5. Bring 2 quarts water to a boil in a large pot. Turn off the heat, and add the vermicelli to the pot. Soak the noodles for 5 minutes, then drain well. Portion the noodles into large soup bowls.

6. Cut the remaining onion into paper-thin slices and place on a serving platter that will hold all the garnishes. Add the bean sprouts, mint leaves, cilantro sprigs, Thai basil leaves, chiles, and lime wedges to the platter.

7. Pour the warm broth over the noodles and sausage in each bowl and serve. Diners can customize their pho with the garnishes from the platter.

SCALLOPED POTATOES WITH SAUSAGE

Scalloped potatoes are a homespun favorite. The addition of sausage turns this popular side dish into a satisfying main course. Any spicy fresh sausage works well in this recipe. Think Merguez (page 132), Fresh Chorizo (page 67), or Hot Italian Sausage (page 64).

Serves 4

- 1 pound fresh sausage, in links
- 3 tablespoons unsalted butter
- 2 tablespoons all-purpose flour
- 3 cups milk
  Pinch of cayenne pepper
  Kosher salt and freshly ground black pepper
  Vegetable oil for greasing
- 6 large russet potatoes, peeled and thinly sliced
- 1 onion, thinly sliced

1. In a medium skillet over medium-high heat, cook the sausage in just enough water to cover the bottom of the pan until the links are lightly browned, 15 to 20 minutes. Remove the sausages and set them aside until they are cool enough to handle. Cut them into ½-inch-thick slices.

2. In a saucepan, melt the butter over medium-low heat and whisk in the flour to make a roux. Add the milk, a little at a time, stirring, until the mixture thickens, about 10 minutes. Add the cayenne and salt and black pepper to taste. Remove the sauce from the heat.

3. Preheat the oven to 425°F (220°C). Lightly grease a 2-quart casserole.

4. Layer the sauce, potatoes, onion, and sausage in the casserole, beginning and ending with sauce.

5. Bake, covered, for 40 minutes. Uncover and bake for 20 minutes longer, or until browned. Serve hot.

SAUSAGE STATS

According to the National Hot Dog and Sausage Council website, "Sausages are enjoying unprecedented sales in the United States, as new flavors, convenient products and many great tasting old standards have enjoyed steady category growth." Here's how that trend played out in 2016:

- **IN THE SUPERMARKET:** Americans spent a whopping $3.75 billion on dinner sausages and $581.3 million on breakfast sausages (this does not include sales at butcher shops or other independent outlets).

- **AT THE BALLPARK:** Baseball fans consumed more than 4.3 million sausages, with the top three stadiums being the homes of the San Francisco Giants, St. Louis Cardinals, and Boston Red Sox.

- **ACROSS THE COUNTRY:** Los Angeles topped sausage sales with $110 million, followed by New York City and Chicago, with $102 million and $84 million, respectively.

- **ACCORDING TO REGION:** Italian links were the most popular sausages in the northeastern states, breakfast patties sold the most in the Southeast, and (totally expected) brats reigned in the Midwest.

METRIC EQUIVALENTS AND CONVERSIONS

Unless you have finely calibrated measuring equipment, conversions between US and metric measurements will be somewhat inexact. It's important to convert the measurements for all of the ingredients in a recipe to maintain the same proportions as the original.

General Formula for Metric Conversion	
Ounces to grams	multiply ounces by 28.35
Grams to ounces	multiply grams by 0.035
Pounds to grams	multiply pounds by 453.5
Pounds to kilograms	multiply pounds by 0.45
Cups to liters	multiply cups by 0.24
Fahrenheit to Celsius	subtract 32 from Fahrenheit temperature, multiply by 5, then divide by 9
Celsius to Fahrenheit	multiply Celsius temperature by 9, divide by 5, then add 32

Approximate Metric Equivalents by Volume	
US	**Metric**
1 teaspoon	5 milliliters
1 tablespoon	15 milliliters
¼ cup	60 milliliters
½ cup	120 milliliters
1 cup	230 milliliters
1¼ cups	300 milliliters
1½ cups	360 milliliters
2 cups	460 milliliters
2½ cups	600 milliliters
3 cups	700 milliliters
4 cups (1 quart)	0.95 liter
1.06 quarts	1 liter
4 quarts (1 gallon)	3.8 liters

Approximate Metric Equivalents by Weight

US	Metric
0.035 ounce	1 gram
¼ ounce	7 grams
½ ounce	14 grams
1 ounce	28 grams
1¼ ounces	35 grams
1½ ounces	40 grams
1¾ ounces	50 grams
2½ ounces	70 grams
3½ ounces	100 grams
4 ounces	112 grams
5 ounces	140 grams
8 ounces	228 grams
8¾ ounces	250 grams
10 ounces	280 grams
15 ounces	425 grams
16 ounces (1 pound)	454 grams
1.1 pounds	500 grams
2.2 pounds	1 kilogram

Weight Conversions of Common Ingredients

1 pound kosher salt	=	1½ cups
1 ounce kosher salt	=	2 tablespoons
1 pound granulated sugar	=	2¼ cups
1 ounce curing salts	=	1¾ tablespoons

SPICE WEIGHTS AND MEASURES

This table is for approximate weights and measures of various dried spices and herbs and is intended as a handy compilation in estimating quantities. To use the chart, if a recipe calls for 1 ounce of allspice, then you would use 5 level tablespoons.

	Conversion from Ounces						
	¼	½	¾	1	2	3	4
	Conversion to Tablespoons						
Allspice, ground	1.25	2.5	3.75	5	10	15	20
Basil	1.5	3	4.5	6	12	18	24
Caraway seeds, whole	1.25	1.75	2.66	3.5	7	10.5	14
Cardamom, ground	1	2	3	4	8	12	16
Celery seeds, ground	1	2	3	4	8	12	16
Cinnamon, ground	0.88	1.75	2.63	3.5	7	10.5	14
Cloves, ground	1	2	3	4	8	12	16
Coriander, ground	1	2	3	4	8	12	16
Cumin, ground	1	2	3	4	8	12	16
Dill seeds, whole	1	2	3	3.9	7.8	11.7	15
Fennel seeds, whole	1	2.25	3.33	4.5	9	13.5	18
Garlic powder	0.75	1.5	2.25	3	6	9	12
Ginger, ground	1.25	2.5	3.75	5	10	15	20
Mace, ground	1.33	2.75	4	5.5	11	16.5	22
Marjoram	1.5	3	4.5	6	12	18	24
MSG	0.5	1	1.66	2.2	4.4	6.6	9
Mustard, dry	1	2	3	4	8	12	16
Nutmeg, freshly grated	1	2	3	4	8	12	16
Onion powder	1	2	3	4	8	12	16
Oregano	2	4	6	8	16	24	32
Paprika	1	2	3	4	8	12	16
Parsley flakes	3	6	12	16	32	48	64
Peppercorns, black, whole	1	2	3	4	8	12	16
Pepper, ground	0.93	1.85	2.75	3.7	7.4	11.1	15
Rosemary	1.75	3.5	5.25	7	14	21	28
Sage	2.5	5	7.5	10	20	30	40
Salt	0.5	1	1.5	2	4	6	8
Savory	1.33	2.75	4	5.5	11	16.5	22
Thyme	1.75	3.5	5.25	7	14	21	28
Turmeric, ground	1.17	1.75	2.66	3.5	7	10.5	14

Source: North Dakota State University Extension Service

RESOURCES

MEET THE MAKERS

JAMIE AND AMY AGER
Hickory Nut Gap Farm
Fairview, North Carolina
828-628-1027
www.hickorynutgapfarm.com

LANCE APPELBAUM
Fossil Farms
Boonton Township, New Jersey
973-917-3155
www.fossilfarms.com

CATHY BARROW
Mrs. Wheelbarrow's Kitchen
Washington, D.C.
www.cathybarrow.com

PAUL BERTOLLI
Fra' Mani Handcrafted Foods
Berkeley, California
510-526-7000
www.framani.com

BRYAN BRACEWELL
Southside Market & Barbecue
Elgin, Texas
512-281-4650
http://southsidemarket.com

CRISTIANO CREMINELLI
Creminelli Fine Meats
Salt Lake City, Utah
801-428-1820
www.creminelli.com

HERB AND KATHY ECKHOUSE
La Quercia
Norwalk, Iowa
515-981-1625
http://laquercia.us

DEBRA FRIEDMAN
Old Sturbridge Village
Sturbridge, Massachusetts
800-733-1830
www.osv.org

CHRIS HUGHES
Broken Arrow Ranch
Ingram, Texas
800-962-4263
www.brokenarrowranch.com

JONNY HUNTER AND
CHARLIE DENNO
Underground Meats
Madison, Wisconsin
608-467-2850
www.undergroundmeats.com

KEVIN OUZTS
The Spotted Trotter
Atlanta, Georgia
404-254-4958
thespottedtrotter.com

MIKE PEKARSKI
Pekarski's
South Deerfield, Massachusetts
413-665-4537
http://pekarskis.com

TERRY AND SUSAN RAGASA
Sutter Meats
Northampton, Massachusetts
413-727-3409
www.suttermeats.com

BLAKE ROYER
Toronto, Ontario
www.seriouseats.com/user/profile/
BlakeRoyer

DAVID SAMUELS
**Esposito's Finest Quality
Sausage**
New York, New York
212-868-4142
www.espositosausage.com

JONAH SHAW
Catskill Food Company
Delhi, New York
607-746-8886
http://catskillfoodcompany.com

BEN SIEGEL AND TED PRATER
Banger's Sausage House
& Beer Garden
Austin, Texas
512-386-1656
www.bangersaustin.com

JEREMY STANTON
The Meat Market
Great Barrington, Massachusetts
413-528-2022
www.themeatmarketgb.com

CAROLINA STORY
Straw Stick & Brick Delicatessen
Washington, D.C.
202-726-0102
http://ssbdeli.com

AUBRY AND KALE WALCH
The Herbivorous Butcher
Minneapolis, Minnesota
612-208-0992
www.theherbivorousbutcher.com

BRENT YOUNG AND
BEN TURLEY
The Meat Hook
Brooklyn, New York
718-609-9300
http://the-meathook.com

JARRED ZERINGUE
Wayne Jacob's Smokehouse
La Place, Louisiana
985-652-9990
http://wjsmokehouse.com

SUPPLIES

FRISCO BRAND SPICES
www.friscospices.com
402-339-0550

KALUSTYAN'S
www.kalustyans.com
800-352-3451

LEHMAN'S
www.lehmans.com
800-438-5346

LEM PRODUCTS
www.lemproducts.com
877-336-5895

PS SEASONING & SPICES
www.psseasoning.com
800-328-8313

THE SAUSAGE MAKER
www.sausagemaker.com
888-490-8525

SAUSAGE SOURCE
www.sausagesource.com
800-978-5465

WALTON'S
www.waltonsinc.com
800-835-2832

International Suppliers

CQ BUTCHERS & CATERING
SUPPLIES
Mackay, Queensland, Australia
+61-7-4957-6888
www.cqbutcherssupplies.com.au

DNR SAUSAGE SUPPLIES
Calgary, Alberta, Canada
403-270-9389
http://dnrsausagesupplies.ca

SMOKED AND CURED
Northcote, Victoria, Australia
www.smokedandcured.com.au

STUFFERS SUPPLY COMPANY
Langley, British Columbia, Canada
800-615-4474
www.stuffers.com

TOOWOOMBA BUTCHER'S
SUPPLIES
Toowoomba, Queensland, Australia
+61-7-4659-7399
www.butcherssupplies.com.au

WESCHENFELDER DIRECT
Middlesbrough, United Kingdom
01642-241395
www.weschenfelder.co.uk

ACKNOWLEDGMENTS

Having thumbed through *Home Sausage Making* many times over the many years since it last came out, I set out to update the book with the utmost care. After all, if it ain't broke, don't fix it. It was an honor and a journey that has only increased my appreciation for this old-world craft.

First and foremost, a big thanks to the writers and editors and others who contributed to the prior iterations of this book, and who helped shape it into the respected resource it is today (that includes the devoted sales staff).

I couldn't have asked for a better cohort than Mary Reilly, who lent her considerable time, talents, and tenacity in testing and developing the bulk of the recipes in this edition, and who soulfully styled the food that appears throughout. It was sheer luck that she picked up the phone on my very first try and turned out to be wicked funny and the epitome of organized and just as adept at turning a phrase as cranking her piston stuffer.

Tremendous gratitude to Alethea Morrison for creating the splendid design that graces each and every page, inside and out, and for the others who helped produce the stunning visuals: photographer Joe Keller, prop stylist Christina Lane, and food assistant Deborah Gallagher (who lent the use of her lovely home for the shoot). Thanks to Elena Bulay for providing the charming illustrations.

Heartfelt appreciation for the guidance and support of Deborah Balmuth and the rest of the folks at Storey Publishing: Corey Cusson, Liseann Karandisecky, Kristy MacWilliams, Jennifer Travis, and Regina Velázquez (who plies me with eggs!), as well as Sarah Armour, Alee Moncy, and Megan Posco.

Sincere thanks to the inspiring makers who allowed me to share their stories in these pages. Learn you will from their own experiences — and their indelible contributions to the mosaic (or "face") of sausage today, to borrow a term of the trade. "To boil it down, it all starts with the pig," sayeth Paul Bertolli, who was gracious enough to impart his vast expertise. And to all the others, including Jamie and Amy Ager, Lance Appelbaum, Cathy Barrow, Bryan Bracewell, Cristiano Creminelli, Charlie Denno, Herb and Kathy Eckhouse, Deb Friedman, Chris Hughes, Jonny Hunter, Kevin Ouzts, Mike Pekarski, Ted Prater, Terry and Susan Ragasa, Blake Royer, David Samuels, Jonah Shaw, Ben Siegel, Jeremy Stanton, Carolina Story, Ben Turley, Aubry and Kale Walch, Brent Young, and Jarred Zeringue. Thanks for keeping it real.

Thanks as well to the hardworking farmers who painstakingly tend to their flocks in the most humane and sustainable ways and without whom sausage making wouldn't be the rewarding endeavor it is. Happy pigs, cows, goats, bison, etc., make the most delicious sausage. It's all about honoring the animal. Such a worthy feat.

Finally, thanks to all the home sausage makers who rely on this book in telling their own sausage stories: much obliged.

— **Evelyn Battaglia**

INDEX

Page numbers in *italic* indicate photographs.

EXPAND YOUR CULINARY CREATIVITY

WITH MORE BOOKS FROM STOREY

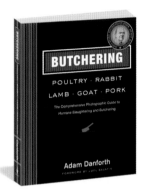

by Adam Danforth

Learn how to slaughter and butcher small livestock humanely with the help of hundreds of detailed step-by-step photos. This award-winning guide includes in-depth coverage of food safety, freezing and packaging, tools and equipment, butchering methods, and preslaughter conditions.

by Paula Marcoux

Rediscover timeless techniques and exciting new ones — from roasting on a spit to baking in a masonry oven — with this comprehensive guide to all the tools, instructions, and recipes you'll need to master wood-fired cooking.

by Kirsten K. Shockey & Christopher Shockey

Get to work making your own kimchi, pickles, sauerkraut, and more with this colorful and delicious guide. Beautiful photography illustrates methods to ferment 64 vegetables and herbs, along with dozens of creative recipes.

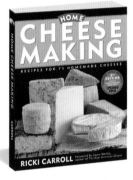

by Ricki Carroll

This classic primer to making artisanal-quality cheeses, from soft to hard styles, also features recipes for sour cream, yogurt, keifer, and buttermilk. Learn everything from purchasing or constructing the right equipment to preparing starter cultures, using rennets, and much more.

Join the conversation. Share your experience with this book, learn more about Storey Publishing's authors, and read original essays and book excerpts at storey.com.

Look for our books wherever quality books are sold or call 800-441-5700.

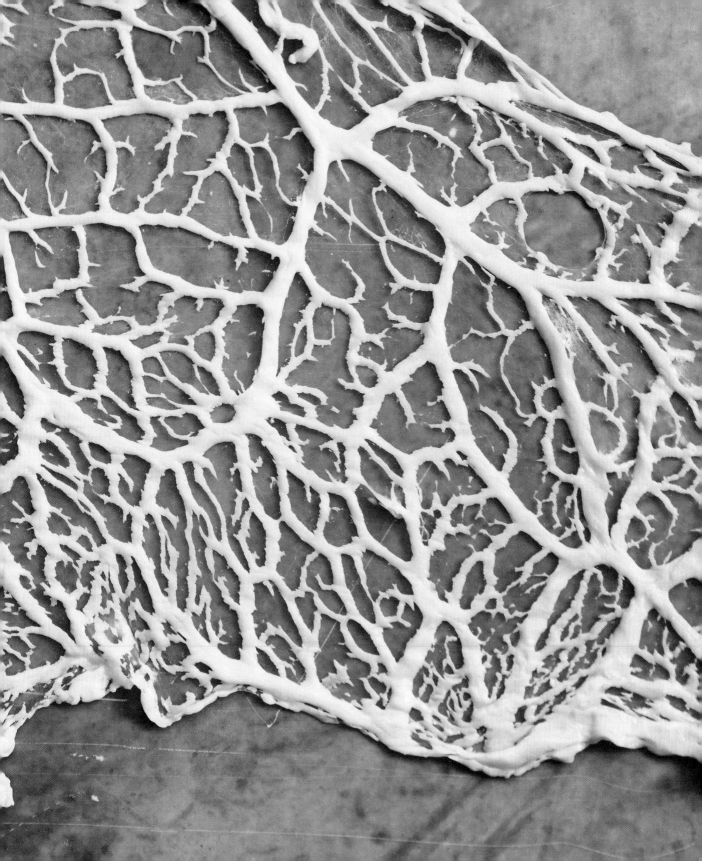